Physical Layer Security in Wireless Communications

WIRELESS NETWORKS AND MOBILE COMMUNICATIONS
Dr. Yan Zhang, Series Editor
Simula Research Laboratory, Norway
E-mail: yanzhang@ieee.org

Physical Layer Security in Wireless Communications

Edited by

Xiangyun Zhou · Lingyang Song · Yan Zhang

CRC Press
Taylor & Francis Group
Boca Raton   London   New York

CRC Press is an imprint of the
Taylor & Francis Group, an **Informa** business

CRC Press
Taylor & Francis Group
6000 Broken Sound Parkway NW, Suite 300
Boca Raton, FL 33487-2742

© 2014 by Taylor & Francis Group, LLC
CRC Press is an imprint of Taylor & Francis Group, an Informa business

No claim to original U.S. Government works

Printed on acid-free paper
Version Date: 20131004

International Standard Book Number-13: 978-1-4665-6700-9 (Hardback)

Visit the Taylor & Francis Web site at
http://www.taylorandfrancis.com

and the CRC Press Web site at
http://www.crcpress.com

Contents

Preface

The ever-increasing demand for private and sensitive data transmission over wireless networks has made security a crucial concern in the current and future large-scale, dynamic, and heterogeneous wireless communication systems. To address this challenge, computer scientists and engineers have tried hard to continuously come up with improved cryptographic algorithms. But typically we do not need to wait too long to find an efficient way to crack these algorithms. With the rapid progress of computational devices, the current cryptographic methods are already becoming more unreliable. In recent years, wireless researchers have sought a new security paradigm termed physical layer security. Unlike the traditional cryptographic approach which ignores the effect of the wireless medium, physical layer security exploits the important characteristics of wireless channel, such as fading, interference, and noise, for improving the communication security against eavesdropping attacks. This new security paradigm is expected to complement and significantly increase the overall communication security of future wireless networks.

Physical Layer Security in Wireless Communications is the book to provide a comprehensive technical guide covering basic concepts, recent advancements, and open issues in providing communication security at the physical layer. The chapters include both high-level discussions and specific examples to provide a systematic overview of the foundations and key developments of physical layer security. References to conference and journal articles are used to allow interested readers to obtain additional reading materials. The book begins with an introductory chapter on the background and fundamentals of physical layer security. The rest of the book is organized into four parts based on the different approaches used for the design and analysis of physical layer security techniques:

1. Information-Theoretic Approaches
2. Signal Processing Approaches
3. Game-Theoretic Approaches
4. Graph-Theoretic Approaches

The first part includes Chapters 2 through 5, focusing on capacity-achieving methods and analysis. Chapters 2 and 3 discuss various coding schemes to achieve near-optimal capacity performance. Chapter 4 summarizes important methods and issues in key generation and agreement over wireless channels. Chapter 5 describes mechanisms to achieve physical layer security with two-way communications.

The second part includes Chapters 6 through 9, presenting the recent progress in applying signal processing techniques to design physical layer security enhancements. Chapters 6 and 7 discuss various multiantenna transmission schemes. Chapter 8 describes efficient resource allocation methods in OFDMA systems. Chapter 9 extends the signal processing techniques from point-to-point systems to relay and cooperative networks.

The third part of the book consists of Chapters 10 through 12, discussing the application of game theory in the design and analysis of secure transmission schemes and protocols. It begins with a summary of both noncooperative and cooperative game settings in a simple two-user network. Then, two specific types of games are described in Chapters 11 and 12 to give further insights into the security design in more complex multiuser networks.

The last part of the book includes Chapters 13 and 14 with the focus on large-scale wireless networks. These chapters present important frameworks developed from graph theory and stochastic geometry to study connectivity and throughput performances.

The book is written for a broad range of audiences with illustrative and easy-to-understand materials. It can serve as a useful reference book for postgraduate students, researchers, and engineers to obtain macro-level understanding of physical layer security and its role in future wireless communication systems. Each chapter is self-contained; hence readers who are interested in a specific topic of physical layer security can read the corresponding chapter(s) without any prior knowledge from other chapters.

We would like to acknowledge the effort and time invested by all contributors for their excellent work. Special thanks go to Taylor & Francis Group for their support, patience, and professionalism since the beginning until the final stage. Last but not least, a special thank you to our families and friends for their constant encouragement, patience, and understanding throughout this project.

Xiangyun (Sean) Zhou, Lingyang Song, Yan Zhang

Authors

Xiangyun (Sean) Zhou is a Lecturer at the Australian National University (ANU), Australia. He received the B.E. (hons.) degree in electronics and telecommunications engineering and the Ph.D. degree in telecommunications engineering from the ANU in 2007 and 2010, respectively. From June 2010 to June 2011, he worked as a postdoctoral fellow at UNIK – University Graduate Center, University of Oslo, Norway. His research interests are in the fields of communication theory and wireless networks, including multiple-input multiple-output (MIMO) systems, relay and cooperative communications, heterogeneous and small cell networks, ad hoc and sensor wireless networks, physical layer security, and wireless power transfer. Dr. Zhou serves on the editorial boards of Security and Communication Networks (Wiley) and Ad Hoc & Sensor Wireless Networks. He was the organizer and chair of the special session on "Stochastic Geometry and Random Networks" in the 2013 Asilomar Conference on Signals, Systems, and Computers. He has also served as the TPC member of major IEEE conferences. He is a recipient of the Best Paper Award at the 2011 IEEE International Conference on Communications.

Lingyang Song is a Professor at Peking University, China. He received his Ph.D. from the University of York, UK, in 2007, where he received the K. M. Stott Prize for excellent research. He worked as a postdoctoral research fellow at the University of Oslo, Norway, and Harvard University, until rejoining Philips Research UK in March 2008. In May 2009, he joined the School of Electronics Engineering and Computer Science, Peking University, China, as a full professor. His main research interests include MIMO, orthogonal frequency division multiplexing (OFDM), cooperative communications, cognitive radio, physical layer security, game theory, and wireless ad hoc/sensor networks. He is coinventor of a number of patents (standard contributions), and author or coauthor of over 100 journal and conference papers. He is the coeditor of two books, "Orthogonal Frequency Division Multiple Access (OFDMA)-Fundamentals and Applications" and "Evolved Network Planning and Optimization for UMTS and LTE," published by Auerbach Publications, CRC Press, USA. Dr. Song received several Best Paper Awards, including one in the IEEE International Conference on Wireless Communications, Networking and Mobile Computing (WiCOM 2007), one in the First IEEE International Conference on Communications in China (ICCC 2012), one in the 7th International Conference on Communications and Networking in China (ChinaCom2012), and one in IEEE Wireless Communication and Networking Conference (WCNC2012). Dr. Song is currently on the Editorial Board of IEEE Transactions on Wireless Communications, Journal of Network and Computer Applications, and IET Communications. He is the recipient of a 2012 IEEE Asia Pacific (AP) Young Researcher Award.

Yan Zhang received a Ph.D. degree from Nanyang Technological University, Singapore. Since August 2006 he has been working with Simula Research Laboratory, Norway. He is currently a senior research scientist at Simula Research Laboratory. He is an associate professor (part-time) at the University of Oslo, Norway. He is a regional editor, associate editor, on the editorial board, or guest editor of a number of international journals. He is currently serving as Book Series Editor for the book series on Wireless Networks and Mobile Communications (Auerbach Publications, CRC Press, Taylor & Francis Group). He has served or is serving as organizing committee chair for many international conferences, including AINA 2011, WICON 2010, IWCMC 2010/2009, BODYNETS 2010, BROADNETS 2009, ACM MobiHoc 2008, IEEE ISM 2007, and CHINACOM 2009/2008. His research interests include resource, mobility, spectrum, energy, and data management in wireless communications and networking.

Contributors

Matthieu Bloch
Georgia Institute of Technology
Lorraine, France

Fuchun Lin
Division of Mathematical Sciences
Nanyang Technological University
Singapore

Frederique Oggier
Division of Mathematical Sciences
Nanyang Technological University
Singapore

Steven McLaughlin
Georgia Institute of Technology
Atlanta, Georgia

Demijan Klinc
Georgia Institute of Technology
Atlanta, Georgia

Jeongseok Ha
Korean Advanced Institute of Science and
    Technology (KAIST)
Daejeon, South Korea

Joao Barros
Faculdade de Engenharia da Universidade
    do Porto
Porto, Portugal

Byung-Jae Kwak
Electronics and Telecommunications
    Research Institute (ETRI)
Daejeon, South Korea

Lifeng Lai
Department of Electrical and Computer
    Engineering
Worcester Polytechnic Institute
Worcester, Massachusetts

Yingbin Liang
Department of Electrical Engineering and
    Computer Science
Syracuse University
Syracuse, New York

H. Vincent Poor
Department of Electrical Engineering
Princeton University
Princeton, New Jersey

Wenliang Du
Department of Electrical Engineering and
    Computer Science
Syracuse University
Syracuse, New York

Xiang He
Electrical Engineering Department
Pennsylvania State University
University Park, Pennsylvania

Aylin Yener
Electrical Engineering Department
Pennsylvania State University
University Park, Pennsylvania

Amitav Mukherjee
Hitachi America Ltd.
San Jose, California

S. Ali A. Fakoorian
University of California Irvine
Irvine, California

Jing Huang
University of California Irvine
Irvine, California

Lee Swindlehurst
University of California Irvine
Irvine, California

Y.-W. Peter Hong
National Tsing Hua University
Hsinchu, Taiwan

Tsung-Hui Chang
National Taiwan University of Science and
    Technology
Taipei, Taiwan

Meixia Tao
Department of Electronic Engineering
Shanghai Jiao Tong University
Shanghai, P. R. China

Jianhua Mo
Department of Electronic Engineering
Shanghai Jiao Tong University
Shanghai, P. R. China

Xiaowei Wang
Department of Electronic Engineering
Shanghai Jiao Tong University
Shanghai, P. R. China

Zhiguo Ding
School of Electrical and Electronic
    Engineering,
Newcastle University
Newcastle, UK

Mai Xu
Department of Electrical Engineering
Beihang University
Beijing, China

Kanapathippillai Cumanan
School of Electrical and Electronic
    Engineering
Newcastle University
Newcastle, UK

Fei Liu
School of Electrical
Engineering, Jiangnan University
Wuxi, China

Eduard Jorswieck
Communications Laboratory,
Communications Theory
Dresden, Germany

Rami Mochaourab
Fraunhofer Heinrich Hertz Institute
Berlin, Germany

Ka Ming (Zuleita) Ho
Communications Laboratory
Communications Theory
Dresden, Germany

Rongqing Zhang
School of Electrical Engineering and
    Computer Science
Peking University
Beijing, China

Lingyang Song
School of Electrical Engineering and
    Computer Science
Peking University
Beijing, China

Zhu Han
Electrical and Computer Engineering
    Department
University of Houston
Houston, Texas

Bingli Jiao
School of Electrical Engineering and
    Computer Science
Peking University
Beijing, China

Xiangyun Zhou
Research School of Engineering
Australian National University
Acton, Australia

Martin Haenggi
University of Notre Dame
South Bend, Indiana

Dennis Goeckel
University of Massachusetts
Amherst, Massachusetts

Cagatay Capar
University of Massachusetts
Amherst, Massachusetts

Don Towsley
University of Massachusetts
Amherst, Massachusetts

Chapter 1

Fundamentals of Physical Layer Security

Matthieu Bloch
Georgia Institute of Technology & GT-CNRS UMI 2958

In virtually all communication systems, the issues of authentication, confidentiality, and privacy are handled in the upper layers of the protocol stack using variations of private-key and public-key cryptosystems. These cryptosystems are usually based upon mathematical operations, such as the factorization into prime factors, believed hard to perform for an attacker with limited computational power; hence, we refer to the security provided by these systems as *computational security*. While computational security has a clear track-record of proven effectiveness, it may be difficult to implement in some emerging network architectures. For example, the advent of wireless networks has fostered the development of mobile adhoc networks comprised of many devices with heterogenous capabilities; the wide range of computing power available in the devices makes it difficult to deploy a public-key infrastructure. In contrast with the established practice of computational security, many results from information theory, signal processing, and cryptography suggest that there is much security to be gained by accounting for the imperfections of the physical layer when designing secure systems. For instance, while noise and fading are usually treated as impairments in wireless communications, information-theoretic results show that they can be harnessed to "hide" messages from a potential eavesdropper or authenticate devices, without requiring a shared secret key. Such results, if they can be implemented in a cost-efficient way without sacrificing much data rate, call for the design of security solutions at the physical layer itself to complement computational security mechanisms.

The study of models, methods, and algorithms that aim at reinforcing the security of communication systems by exploiting the properties of the physical layer has grown into a dynamic research area, colloquially known as *physical layer security*. Physical layer security encompasses many disciplines and topics, from multiple-input multiple-output signaling techniques minimizing probability of interception to error-control codes providing information-theoretic secrecy.

The objective of this introductory chapter is to discuss some of the fundamental aspects of physical layer security and in particular, how to achieve some level of information-theoretic secrecy by exploiting the imperfections of the communication medium. For simplicity, we restrict our discussion to the problem of confidentiality and to the basic models of secret communication [46] and secret-key generation [2, 2], although physical layer security

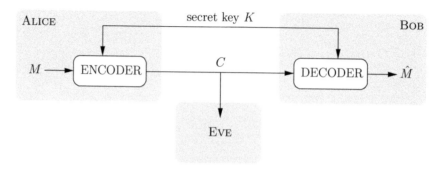

Figure 1.1: Shannon's cipher system.

research has by now evolved much beyond these two problems. Nevertheless, the insights and techniques developed for the basic models find applications in more advanced signal processing and communication techniques, and form building blocks to study other problems, such as biometric authentication [4], privacy in databases [5], bit commitment [6], oblivious transfer [7–9], secure multiparty computation [10], and so forth. All the results discussed in this chapter are provided without formal justifications but with enough high-level intuition to highlight what physical layer security *can* and *cannot* do. The interested reader can refer to the bibliographical references and textbooks [6, 12] for formal derivations.

1.1   Information-theoretic Secrecy

One of the benefits of physical layer security is to be amenable to a precise and quantitative analysis of secrecy in an information-theoretic framework. The objective of this section is to introduce the notion of information-theoretic secrecy and to discuss its meaning from an operational perspective.

1.1.1   Shannon's Cipher System and Perfect Secrecy

The study of confidentiality in information theory was pioneered by Claude Shannon [13] using the model of a cipher system illustrated in Figure 1.1. In this setting, the objective is to communicate a message reliably from a transmitter (Alice) to a legitimate receiver (Bob), while maintaining the message secret from the perspective of an eavesdropper (Eve), who intercepts all transmitted signals. To help guarantee the secrecy of the message, Alice and Bob have access to a shared random key unknown to Eve, which can be used to encode the message into a codeword and decode the codeword back into the message. The message, the key, and the codeword are represented by the random variables M, K, and C, respectively.

The message M is said to be transmitted with *perfect secrecy*, if it is statistically independent of the codeword C intercepted by Eve. Equivalently, perfect secrecy is achieved if the mutual information $I(M; C)$ between the message and the codeword is identically zero. If satisfied, this property guarantees that Eve's best strategy to recover M is to guess its value at random. For instance, if the message is uniformly distributed over k bits, perfect secrecy ensures that the probability of Eve successfully guessing the message is 2^{-k} and decays exponentially fast with k. Although the probability of success is not exactly zero, it is negligible for all practical purposes as soon as k is on the order of a few tens of bits. One should also notice that perfect secrecy is a much stronger requirement than merely preventing Eve from decoding the message correctly; it ensures that the codeword C contains no useful information at all about M.

Perfect secrecy differs from the notions used in computational security on several accounts. It is not only a *quantitative* metric for the information leaked to the eavesdropper, but it is also independent of any assumption regarding Eve's technology or computational power. Unfortunately, these attractive properties are offset by the fact that perfect secrecy for Shannon's cipher system is only achievable provided the entropy $H(K)$ of the shared secret key exceeds the entropy $H(M)$ of the message [13]. In other words, the transmission of a secret message requires the sharing of a secret key as long as the message itself, which significantly limits the usefulness of such a system.

Despite this significant caveat, it is nevertheless insightful to look at how one can achieve perfect secrecy in Shannon's cipher system. Assuming that the messages take binary values, perfect secrecy is obtained with a simple one-time pad encryption [14], which consists in adding uniform key bits K with message bits M modulo two, that is, $C = M \oplus K$. The decoder easily retrieves M from C and K by adding the key K to C since $C \oplus K = M$. That this procedure indeed guarantees perfect secrecy follows from a fairly general result known as the *crypto lemma* [15]. Note that the role of the key K in ensuring perfect secrecy is twofold. First, the key randomizes the encoding of the message. Second, it ensures that, from Eve's perspective, the statistical distribution of C is the same irrespective of the value of M. This idea will reappear again when we discuss information-theoretic secrecy for physical layer security systems in Section 1.2 and Section 1.3.

While Shannon's result sheds a fairly pessimistic light on the usefulness of information-theoretic security, it is important to realize that this conclusion is tied to Shannon's specific model. In particular, one important feature of communication systems that is not captured in the model is the presence of impairments in the communication channel. Shannon's model can be viewed as an "upper-layer" abstraction, which assumes that the correction of transmission errors at the physical layer has already been dealt with. As we discuss in Section 1.2 and Section 1.3, the pivotal idea behind physical layer security models is to revisit this assumption and to explicitly consider the presence of imperfect communication channels.

1.1.2 Information-theoretic Secrecy Metrics

The requirement of *exact* statistical independence between messages and codewords imposed by perfect secrecy is usually too stringent to be tractable. Consequently, as is often done in information theory, it is useful to relax the secrecy requirement and to only ask for statistical independence in an asymptotic sense. Specifically, assume that the message M to be transmitted secretly is encoded in a codeword of n symbols C^n. Instead of requiring that M is statistically independent of C^n, the idea is to demand that M is independent of C^n for infinitely large n.

The shift from exact independence to asymptotic independence introduces a technical difficulty. It is now necessary to specify a metric to measure the statistical dependence between M and C^n. In principle, any metric d measuring the distance between the joint distribution p_{MC^n} and the product of independent distributions $p_M p_{C^n}$ could form a valid measure of statistical dependence, so that the secrecy requirement would be written

$$\lim_{n \to \infty} d(p_{MC^n}, p_M p_{C^n}) = 0.$$

Unfortunately, there are many ways to define such metrics d, all of which do not bear the same significance from a cryptographic perspective. The most commonly used metrics are the following:

- The *outage secrecy metric* $\mathbb{P}\left(\frac{1}{n}\log_2 \frac{p_{M|C^n}(M|C^n)}{p_M(M)} > \epsilon\right)$ for some $\epsilon > 0$, which measures the probability that the mutual information random variable $\frac{1}{n}\log_2 \frac{p_{M|C^n}(M|C^n)}{p_M(M)}$ exceeds some prescribed threshold ϵ.

- The *weak secrecy metric* $\frac{1}{n}I(M;C^n)$, which measures the *rate* of information leaked about M in C^n; the weak secrecy metric can also be written as a normalized Kullback-Leibler divergence between the joint distribution p_{MC^n} and the product of independent distributions $p_M p_{C^n}$.

- The *variational distance secrecy metric* $\mathbb{V}(p_{MC^n}, p_M p_{C^n})$, which is an L_1 distance between the joint distribution p_{MC^n} and the product of independent distributions $p_M p_{C^n}$.

- The *strong secrecy metric* $I(M;C^n)$, which measures the *amount* of information leaked about M in C^n.

All four metrics can be used to express asymptotic statistical independence; however, these metrics are not equivalent, and it can be proved [16] that

$$\lim_{n\to\infty} I(M;C^n) = 0 \Rightarrow \lim_{n\to\infty} \mathbb{V}(p_{MC^n}, p_M p_{C^n})$$

$$\Rightarrow \lim_{n\to\infty} \frac{1}{n}I(M;C^n) = 0$$

$$\Rightarrow \forall \epsilon > 0 \quad \lim_{n\to\infty} \mathbb{P}\left(\frac{1}{n}\log_2 \frac{p_{M|C^n}(M|C^n)}{p_M(M)} > \epsilon\right) = 0.$$

In other words, some metrics are stronger than others from a mathematical perspective. It turns out that this ordering of metrics also has operational significance, and it is possible to construct explicit cryptographic schemes with obvious vulnerabilities that satisfy the weak secrecy condition [6]. As a simple example of such a scheme, consider a message $M = [M_1 \ M_2]$ in which M_1 consists of n bits and M_2 consists of $\sqrt{n}$ bits. The message M is then partially encrypted with a uniformly distributed secret key K of n bits to form $C^n = [M_1 \oplus K \ M_2]$. Such a scheme is clearly flawed since the bits M_2 are transmitted unencrypted. This flaw is captured by the strong secrecy metric, since one can show that $I(M;C^n) = \sqrt{n}$; however, this is not identified by the weak secrecy metric because $\lim_{n\to\infty} \frac{1}{n}I(M;C^n) = 0$.

Consequently, it is now recognized that asymptotic statistical independence should be measured at least in terms of variational distance or in terms of strong secrecy [17]. It is also possible to establish more precise relationships between these metrics in the finite-length regime and, in particular, to show that the strong secrecy metric is related to the notion of semantic security in cryptography [18, 19].

Although the outage secrecy metric is the weakest of all metrics presented, it is sometimes convenient to obtain insight into the security offered by wireless channels, as it allows one to capture the effect of the random fluctuations of fading gains during transmission [20, 21].

1.2 Secret Communication over Noisy Channels

In this section, we introduce the first basic model of physical layer security, called the *wiretap channel*. This model was introduced by Wyner [41] and attempts to generalize Shannon's cipher system by capturing the joint problem of reliable and secret communication over noisy channels. We first introduce the model formally and discuss the fundamental limits of reliable and secure communication in Section 1.2.1. We then discuss in more detail the coding mechanisms that must be implemented to achieve information-theoretic secrecy in Section 1.2.2.

1.2.1 Wiretap Channel Model

Wyner's model of secret communication is illustrated in Figure 1.2. The objective is for the transmitter (Alice) to communicate messages at rate R, represented by the random variable

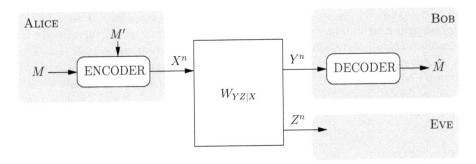

Figure 1.2: Wyner's wiretap channel.

$M \in \{1; 2^{nR}\}$, by encoding them into codewords X^n of length n and transmitting X^n over a noisy memoryless broadcast channel, characterized by a transition probability $W_{YZ|X}$. The receiver observing the signal Y^n is the legitimate receiver (Bob), who should be able to make a correct estimate $\hat{M}$ of M with high probability. The receiver observing Z^n is treated as an eavesdropper (Eve), who should obtain no information about the message M. The encoding of message M is assisted by a local random number generator $M' \in \{1; 2^{nR'}\}$, which we assume is only known by Alice.

A code for this communication problem is called a *wiretap code*. A secret communication R rate is *achievable* if there exists a sequence of wiretap codes with increasing block length n, such that

$$\lim_{n \to \infty} \mathbb{P}(M \neq \hat{M}) = 0 \text{ (reliability)} \quad \text{and} \quad \lim_{n \to \infty} I(M; Z^n) = 0 \text{ (secrecy).}$$

Before discussing the results associated with this model, it is important to fully appreciate the underlying assumptions associated with it.

- The wiretap channel model differs from Shannon's cipher system on two accounts. First, the model includes the presence of noise in the communication channel. Second, the model does not include any shared secret key between Alice and Bob.

- The local random number generator M' in Alice's encoder plays a role similar to the key in a one-time pad, since it randomizes the encoding of the message; however, note that M' is only known by Alice and does not provide any advantage to Alice and Bob over Eve.

- The statistics of the channel and the wiretap code are known to all parties involved, and Eve is a purely passive eavesdropper. The assumption that Alice and Bob can characterize Eve's channel statistics is a weakness of the model; however, recent works have shown that this assumption can be alleviated in part to account for uncertainty about the channel and active attackers [22–24].

- The wiretap channel model only captures the problem of confidentiality and implicitly assumes that authentication is already in place. This assumption is not too restrictive if Alice and Bob initially share a short secret key to authenticate the first transmission. Subsequent message authentication can be implemented by sacrificing a negligible fraction of the previous message rate and using an unconditionally secure authentication scheme [25].

It is not a priori obvious that wiretap codes with nontrivial rates exist. In fact, the reliability requirement for Bob calls for the introduction of redundancy in the encoder to fight the effect of the noise. On the other hand, the secrecy requirement for Eve would intuitively call for limiting such redundancy to avoid leaking information. Perhaps surprisingly, it turns out that simultaneously satisfying both requirements is sometimes possible. In particular, one can characterize the *secrecy capacity*, defined as the supremum of all achievable rates with wiretap codes. The secrecy capacity can be viewed as the counterpart of the traditional channel capacity, when a secrecy requirement is imposed. It can be shown [12] that the secrecy capacity is given by

$$C_s = \max_{V \to X \to YZ} \left(I(V;Y) - I(V;Z) \right). \tag{1.1}$$

The expression is less appealing than channel capacity because of the presence of an auxiliary random variable V, but one can intuitively understand the secrecy capacity as the difference between a rate of reliable communication $I(V;Y)$ between Alice and Bob to which we subtract a rate of information leaked $I(V;Z)$ to the eavesdropper. However, as further discussed in Section 1.2.2, the structure and operation of wiretap codes are quite different from those of standard codes for reliable communications.

The expression of secrecy capacity in Equation (1.1) allows us to make several important observations. The secrecy capacity is positive provided $I(V;Y) - I(V;Z)$ is positive; in particular, if Eve's observations are identical to Bob's observations, $Y = Z$, then the secrecy capacity is zero. Hence, while physical layer security does provide some level of information-theoretic secrecy, it requires Alice and Bob to have an *advantage* over Eve at the physical layer itself. One should therefore not hastily conclude that physical layer security could ever replace computational security. On the one hand, computational security systems cannot provide information-theoretic secrecy, but they rely on few (if any) assumptions regarding the communication channel. On the other hand, physical layer security systems do provide information-theoretic secrecy, but they rely on some a priori knowledge of the communication channels. Realistically, physical layer security systems would therefore *complement* existing cryptographic techniques by adding an additional protection at the physical-layer. For instance, physical layer security could be a means to refresh the secret keys used in upper layers without deploying a traditional key management infrastructure.

For the specific wiretap channel model illustrated in Figure 1.2, the required advantage of Alice and Bob over Eve really means that the communication channel between Alice and Bob should be "less noisy" than the one between Alice and Eve [12, 27]. Nevertheless, this strong assumption can be relaxed by considering more complex communication systems with two-way communication and feedback. We will discuss this further in Section 1.3, and suffice to say here that physical layer security is usually possible as soon as there is an *asymmetry* in communication between Bob and Eve; in particular, physical layer security may be possible even if Eve experiences a better channel than Bob.

1.2.2   Coding Mechanisms for Secret Communication

The objective of this section is to clarify the coding mechanism required to achieve information-theoretic secrecy over the wiretap channel model and hidden in Equation (1.1). For simplicity, we consider a model with a *noiseless* main channel, so that reliable communication is automatically achieved for any injective encoding function. One of the key challenges of code design for physical layer security is to find sufficient operational conditions under which a code will achieve information-theoretic secrecy. In subsequent paragraphs, we describe two different approaches that offer different guidelines for code design and yield different secrecy levels.

Nested structure of wiretap codes and stochastic encoding.   To understand why standard codes for reliability cannot be used as wiretap codes and why randomization in the

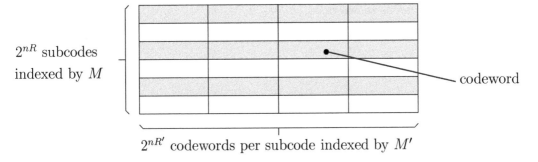

2^{nR} subcodes indexed by M

codeword

$2^{nR'}$ codewords per subcode indexed by M'

Figure 1.3: Nested structure of a wiretap code.

encoding is required, it is useful to consider a simple toy example, in which the eavesdropper observes the output of a binary erasure channel; the ideas developed thereafter readily extend to other channel models. Specifically, Eve observes the codewords through a channel that erases each codeword symbol with probability ϵ, and leaves the symbol unaffected with probability $1 - \epsilon$. We represent an erasure by the symbol "?". According to Equation (1.1), the secrecy capacity of this channel is $C_s = \epsilon$.

Assume that Alice wishes to transmit a binary message $M \in \{0, 1\}$ with a standard code, characterized by the following deterministic encoding:

$$0 \mapsto x_0^n = (x_{0,1}, \dots, x_{0,n}) \in \{0,1\}^n \quad \text{and} \quad 1 \mapsto x_1^n = (x_{1,1}, \dots, x_{1,n}) \in \{0,1\}^n.$$

On average, Eve obtains an observation with $n(1 - \epsilon)$ unerased symbols, which should not allow her to infer which message was sent. By comparing the codewords x_0^n and x_1^n in unerased positions,

$$
\begin{array}{ccccccc}
x_{0,1} & ? & x_{0,3} & \cdots & x_{0,n-1} & ? & \text{(bits of codeword } c_0^n \text{ in unerased positions)} \\
x_{1,1} & ? & x_{1,3} & \cdots & x_{1,n-1} & ? & \text{(bits of codeword } c_1^n \text{ in unerased positions),}
\end{array}
$$

we see that the only way to enforce this is to ensure that symbols in unerased positions are the *same* for x_0^n and x_1^n. Since the positions of unerased symbols are random, this means that the two codewords must be *equal* and Alice cannot transmit two distinct messages. Therefore, the encoding in a wiretap code must randomly choose one of *several* possible codewords for a given message; in other words, the encoding must be *stochastic*.

As illustrated in Figure 1.3, we can therefore understand a wiretap code as having a *nested* structure. For each of the 2^{nR} distinct messages $M \in \{1, 2^{nR}\}$ to be secretly transmitted, there are $2^{nR'}$ possible codewords, chosen at random according to a local random number generator $M' \in \{1, 2^{nR'}\}$. A set of $2^{nR'}$ codewords corresponding to a given message forms a bin, which is a *subcode* of the wiretap code.

To prove that this structure is indeed sufficient to achieve secrecy, one must now understand how to choose the parameters R and R' that control the nested structure. Since we have assumed the main channel to be noiseless, the channel coding theorem guarantees that Bob can retrieve the message as long as the total rate of communication $R + R'$ does not exceed 1 bit. In the remainder of this section, we therefore focus on the determination of the parameter R', which controls the size of each subcode.

Secrecy from channel capacity. To get additional insight into how to choose the subcodes, it is useful to rewrite the information leaked about the message $I(M; Z^n)$ slightly differently. Specifically, with repeated applications of basic information-theoretic equalities and inequalities, it can be shown that

$$I(M; Z^n) = I(X^n; Z^n) - H(M') + H(M'|MX^n). \tag{1.2}$$

The terms on the right hand side of Equation (1.2) have the following interpretation.

- $I(X^n; Z^n)$ represents the information leaked about codewords, which is different from $I(M; Z^n)$ since the encoding is not deterministic.

- $H(M')$ represents the entropy of the local random number generator, which captures the randomness introduced during the encoding process.

- $H(M'|MZ^n)$ measures the uncertainty about the randomization when knowing the message M and the observation Z^n; in other words, this measures Eve's uncertainty *within* the subcode corresponding to message M.

This expression suggests that the information leaked $I(M; Z^n)$ vanishes if two conditions are met: the randomness $H(M')$ *compensates* the information leaked about the codewords X^n and the uncertainty $H(M'|MZ^n)$ is small. This turns out to be possible if each subcode stems from a family of *capacity-achieving codes* for Eve's channel, so that $R' \approx 1 - \epsilon$. In fact, it can be shown [28] that this choice guarantees

$$\frac{1}{n}H(M') \approx \frac{1}{n}I(X^n; Z^n) \quad \text{and} \quad \frac{1}{n}H(M'|MZ^n) \approx 0,$$

so that $\frac{1}{n}I(M; Z^n) \approx 0$. Combining the condition $R + R' < 1$ with $R' \approx 1 - \epsilon$, we see that this approach allows us to construct codes with secrecy rate R as close to $C_s = \epsilon$ as desired.

Since secrecy is achieved by relying on the capacity-achieving properties of codes, we call this approach deriving "secrecy from channel capacity," and we call the corresponding wiretap codes "capacity-based." It should be noted that the result outlined above is only a weak secrecy result. It turns out that this is not merely a shortcoming of the proof, but a fundamental limitation of capacity-based wiretap codes [29–31]. Nevertheless, this approach has the advantage to easily translate into practical designs, as one can rely on good families of error-control codes to construct capacity-based wiretap codes. For example, this has been successfully done with low-density parity-check (LDPC) codes [32, 33] and polar codes [29, 34–36].

Secrecy from channel resolvability. An alternative approach to design the subcodes is to take a perhaps more direct route and to directly tackle the variational distance secrecy metric. Specifically, it can be shown with basic properties of the variational distance that

$$\mathbb{V}(p_{MZ^n}, p_M p_{Z^n}) \leq \sum_m p_M(m)\mathbb{V}(p_{Z^n|M=m}, q_{Z^n}), \tag{1.3}$$

where q_{Z^n} is some arbitrary distribution. Note that the variational distance on the right hand side of Equation (1.3) measures how close the distribution $p_{Z^n|M=m}$ induced by the subcode for message m approximates the distribution q_{Z^n}. Hence, a sufficient condition to ensure secrecy is to guarantee that each subcode induces the same distribution q_{Z^n} at Eve's channel output [31, 37].

The problem of inducing a specific output distribution using a set of codewords is known in information theory as "channel resolvability" [38–40], and it can be shown that $p_{Z^n|M=m}$ can be made arbitrarily close to q_{Z^n} if the subcode rate R' satisfies $R' > 1 - \epsilon$. Combining again the conditions $R + R' < 1$ and $R' > 1 - \epsilon$, we see that the approach allows one to construct codes with secrecy rate R as close to $C_s = \epsilon$ as desired.

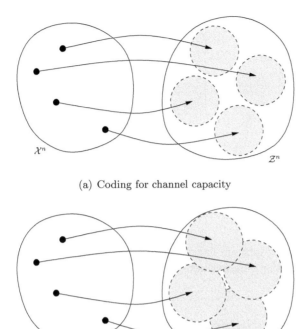

(a) Coding for channel capacity

(b) Coding for channel resolvability

Figure 1.4: Capacity versus resolvability subcodes.

Since secrecy is now achieved by relying on the channel resolvability properties of the subcodes, we call this second approach deriving "secrecy from channel resolvability," and we call the corresponding wiretap codes channel "resolvability-based." Under mild conditions, which are satisfied for memoryless channels, it can be shown that $\mathbb{V}(p_{MZ^n}, p_M p_{Z^n})$ vanishes sufficiently fast so that $I(M; Z^n)$ vanishes as well; hence, deriving secrecy from channel resolvability can provide strong secrecy. This approach has also led to actual code constructions, with duals of LDPC codes [41] and polar codes [29].

One should realize that the similarity of the conditions $R' \approx 1 - \epsilon$ (secrecy from channel capacity) and $R' > 1 - \epsilon$ (secrecy from channel resolvability) hides more profound differences between capacity-based and channel resolvability-based wiretap codes. These differences are shown in Figure 1.4, where we illustrate how the choice of codewords in subcodes affects the eavesdropper's channel output. Upon transmitting a codeword, represented by the symbol "•" in the set of all possible binary sequences of length n, the eavesdropper obtains one of several highly likely sequences in an "uncertainty set," represented by a gray-shaded circle. To obtain a capacity-achieving code, one must carefully choose the codewords so that the corresponding uncertainty sets do not overlap, which is a *sphere packing* problem. In contrast, to obtain a channel resolvability code, one must carefully choose the codewords so that the corresponding uncertainty sets overlap, which is similar to a *sphere covering* problem. This distinction makes the design and analysis of capacity and resolvability codes fundamentally different.

Alternative approaches to secrecy. Although capacity-based and channel resolvability-based constructions are perhaps the simplest conceptually, alternative nested constructions

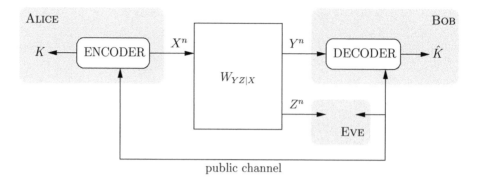

Figure 1.5: Channel model for secret-key generation.

exist that do not readily fit into our definitions. Examples of such constructions include constructions based upon invertible extractors [42, 43], multiplexed coding [44], and matrices with hash properties [45].

1.3   Secret-key Generation from Noisy Channels

In this section, we turn to the second basic model of physical layer security, called the *channel model for secret-key generation*. This model is an extension of the wiretap channel, in which there exists a two-way, noiseless, public, side-channel of unlimited capacity. This model was introduced by Maurer [2] and Ahlswede and Csiszár [2] to analyze the effect of feedback on secret communications. Since the existence of a noiseless side channel eliminates the problem of reliable communication, the focus of this model is on the *generation* of secrecy from the channel in the form of secret keys. As was done for the wiretap channel, we start by formally introducing the model and discussing its implications for physical layer security in Section 1.3.1. We then discuss the coding mechanisms that are required to perform secret-key generation in Section 1.3.2.

1.3.1   Channel Model for Secret-key Generation

The channel model for secret-key generation is illustrated in Figure 1.5. The model consists of a noisy memoryless broadcast channel with transition probability $W_{YZ|X}$, whose input signals X^n are controlled by Alice and whose outputs Y^n and Z^n are observed by Bob and Eve, respectively. In addition, Alice and Bob can communicate over a public, authenticated, two-way side-channel of unlimited capacity. The assumption that the channel is public allows Eve to intercept all messages transmitted over the side-channel, so that the side-channel does not constitute a source of secrecy. However, the assumption that the channel is authenticated prevents Eve from tampering with the messages. The objective is for the legitimate parties Alice and Bob to exchange n symbols over the noisy channel and to transmit messages, collectively denoted by F, over the public channel, so that they eventually agree on the same secret key $K \in \{1, 2^{nR}\}$ unknown to Eve.

A communication scheme for this model is called a *secret-key generation strategy*. We use the term "strategy" to distinguish the coding schemes from the wiretap codes of Section 1.2 that operate exclusively on a noisy channel. A secret-key rate R is achievable if there exists a sequence of secret-key generation strategies with an increasing number of symbols transmitted over the noisy channels n, such that

$$\lim_{n\to\infty} \mathbb{P}(K \neq \hat{K}) = 0 \text{ (reliability)},$$

$$\lim_{n\to\infty} |H(K) - nR| = 0 \text{ (uniformity)},$$

$$\text{and} \quad \lim_{n\to\infty} I(K; Z^n F) = 0 \text{ (secrecy)}.$$

The reliability requirement ensures that, with high probability, Alice and Bob agree on the same key. The uniformity requirement guarantees that the secret key is uniformly distributed in its set, which is a desirable property if the key is to be used for cryptographic applications. Finally, the secrecy requirement ensures that the key is indeed secret with respect to Eve, who observes the noisy signals Z^n and the public messages F.

The remarks made for the wiretap channel model in Section 1.2.1 apply here, as well. In addition, one can observe the following.

- The addition of a public authenticated channel does not trivialize the problem, because it is not a resource for secrecy. The only resource for secrecy remains the noisy communication channel.

- Unlike the wiretap channel model, the channel model for secret-key generation allows for two-way communication and feedback; as we discuss next, feedback turns out to be an essential ingredient for secret-key generation. In addition, the key K is not a message in the traditional sense because its value needs to be fixed at the beginning of a secret-key generation strategy. This allows the key to be generated interactively based on the observations and messages of all legitimate parties, and to be processed with noninvertible functions. This contrasts with the wiretap channel model in which the secret message from the transmitter must be received unaltered.

- Secret-key generation strategies can be extremely sophisticated; for instance, the model allows Alice and Bob to exchange many messages over the public channel between two transmissions over the noisy channel, and the symbols transmitted by Alice over the noisy channel can depend on past received messages.

The supremum of achievable secret-key rates is called the *secret-key capacity*. Because of the presence of the two-way channel, an exact expression of the secret-key capacity remains elusive. Nevertheless, it is possible to establish the following generic bounds.

$$\max_X \max \left(I(X;Y) - I(X;Z), I(X;Y) - I(Y;Z) \right) \leq C_{sk} \leq \max_X I(X;Y|Z). \qquad (1.4)$$

The term $I(X;Y|Z)$ in the upper bound can be rewritten as $I(X;YZ) - I(X;Z)$, so that it can be understood as the secret-key capacity of a model in which Bob has access to Eve's observations. The term $\max\left(I(X;Y) - I(X;Z), I(X;Y) - I(Y;Z)\right)$ in the lower bound reflects the fact that secret-key generation strategies exploit feedback. Specifically, as further discussed in Section 1.3.2, the lower bound follows from Alice and Bob's ability to decide whether Alice's signals X^n or Bob's observations Y^n serve as the reference for secret-key generation.

Most importantly, the expression for secret-key capacity in Equation (1.4) shows that Alice and Bob are not necessarily required to have a "better" channel to generate secret keys. In fact, the term $I(X;Y) - I(Y;Z)$ can be positive even in situations where Eve suffers from less noise than Bob. Intuitively, this is possible because it might be harder for Eve to estimate Bob's observations than it is for Alice. We can therefore conclude that, in most situations, obtaining physical layer security is possible whenever Eve does not obtain the same observations as Bob because feedback, even if public, allows Alice and Bob to gain an advantage.

1.3.2   Coding Mechanisms for Secret-key Generation

The objective of this last section is to discuss the coding mechanisms required to extract provably secret keys, which are hidden in Equation (1.4). We introduce three possible approaches that lead to somewhat different code designs.

Secret-key generation from wiretap codes.   The first approach to construct a secret-key generation strategy consists in using the public channel to construct a *virtual* wiretap channel. Although the principle applies to arbitrary channels, we focus for simplicity on a binary channel, for which $Y = X \oplus E_1$ and $Z = X \oplus E_2$, where E_1 and E_2 are independent Bernoulli random variables with parameters p_1 and p_2. We also assume that $\frac{1}{2} > p_1 > p_2 > 0$, so that Eve's observations suffer from less noise than Bob's and the secrecy capacity of the wiretap channel (without public communication) is $C_s = 0$.

Assume now that Bob wishes to transmit a bit U to Alice over the public channel. Instead of transmitting U directly, Alice first creates a random bit X that she transmits over the noisy channel; Bob and Eve obtain the corresponding observations Y and Z, respectively. Bob then intentionally corrupts his symbol U and transmits the symbol $F = U \oplus Y$ over the public channel. Upon receiving F, Alice attempts to cancel Bob's corruption by forming the estimate $F \oplus X = U \oplus E_1$. It can be shown that Eve's optimal strategy is also to form the estimate $F \oplus Z = U \oplus E_1 \oplus E_2$. Hence, everything happens as if Bob were communicating with Alice over a virtual wiretap channel, in which Alice suffered from noise E_1, while Eve suffered from noise $E_1 \oplus E_2$. In other words, although the true physical channel introduces more noise on Bob's observations than on Eve's, the use of feedback allows Alice and Bob to reverse the situation. If Bob uses a wiretap code designed for the virtual channel, it can be shown that the secrecy rate $I(X;Y) - I(Y;Z)$, which appears in the lower bound in Equation (1.4), is achievable [2].

While the construction of secret-key generation strategies from wiretap codes poses no conceptual problems, it obfuscates the coding mechanism that could be used to directly generate a secret key without constructing a virtual wiretap channel. To obtain additional insight, we therefore investigate two additional coding mechanisms, which can be viewed as the counterparts of those studied in Section 1.2 for wiretap channels. For simplicity, we assume in the following that $X = Y$ and that no public message F is required, so that we can solely focus on the coding mechanism that extracts secret keys from the noise.

Secret-key generation from source codes with side information.   Following the idea behind capacity-based wiretap codes, one can rewrite the secrecy metric $I(K; Z^n)$ in a slightly different way. With repeated applications of basic information-theoretic equalities and inequalities, it can be shown that

$$I(K; Z^n) \leq H(K) + H(X^n | K Z^n) - n H(Z|X). \tag{1.5}$$

The terms on the right side of Equation (1.5) have the following interpretation.

- $H(K)$ represents the entropy of the key $K \in \{1, 2^{nR}\}$ to generate, which is such that $H(K) \leq nR$;

- $H(X^n | K Z^n)$ represents the uncertainty about X^n of an eavesdropper who would observe Z^n and the key K;

- $nH(X|Z)$ represents the uncertainty about X^n of an eavesdropper who would observe Z^n alone.

This expression suggests that $I(K; Z^n)$ vanishes if two conditions are met: $H(K)$ is on the order of $nH(X|Z)$ and the uncertainty $H(X^n|KZ^n)$ vanishes. It turns out that this is possible if the function used to compute K from X^n stems from a family of capacity-achieving source codes for X^n with side information Z^n. In fact, for such codes, the Slepian–Wolf Theorem [46] guarantees that

$$\frac{1}{n}H(K) \approx H(X|Z) \quad \text{and} \quad \frac{1}{n}H(X^n|KZ^n) \approx 0,$$

so that $\frac{1}{n}I(K; Z^n) \approx 0$. Therefore, one can construct secret-key generation strategies from good source codes with side information. As for wiretap channels, the fact that we only claimed weak secrecy results is a fundamental limitation of the approach [47].

Secret-key generation from channel intrinsic randomness. The last construction of secret-key generation strategies that we discuss takes a more direct view at the extraction of secret keys from the observations X^n. If we think of the joint distribution $p_{X^n Z^n} = p_{Z^n} p_{X^n|Z^n}$ as being the joint distribution between an input process Z^n and an output process X^n linked by a channel with transition probability $p_{X^n|Z^n}$, we are effectively trying to extract the randomness that originates from the channel $p_{X^n|Z^n}$ and not from the input Z^n; hence, we call this problem the extraction of the *channel intrinsic randomness*. One can show [48–50] that this operation is possible provided the rate of the key $K \in \{1, 2^{nR}\}$ satisfies $R < H(X|Z)$. Under mild conditions, one can also show that $I(K; Z^n)$ vanishes, so that this approach satisfies a strong secrecy criterion.

Interestingly, explicit powerful coding schemes are known to perform channel intrinsic randomness. In particular, universal hash functions and extractors [51–53] have been used both in theoretical studies and in practical systems to generate secret keys.

1.4   Conclusion

By taking into account the presence of channel imperfections in the design of a secure communication system, physical layer security offers an information-theoretic level of secrecy, which is amenable to precise analysis. The information-theoretic framework also provides quantitative metrics to guide the design and optimization of signaling strategies and network protocols — these various aspects of physical layer security are presented in more detail in the subsequent chapters of this book.

The noise of communication channels turns out to be a *resource* for secrecy. This resource can be harnessed with appropriate coding schemes to either secure communications or generate secret keys. Most importantly, physical layer security coding schemes operate without shared secret keys, except for a small random key required for authentication at initialization. Hence, physical layer security has the potential to considerably simplify key management in communication networks. Nevertheless, one should keep in mind that the ability of physical layer security systems to deliver information-theoretic secrecy heavily depends on the validity of the underlying communication models. Consequently, from a system design perspective, physical layer security should be viewed as a means to complement or simplify existing secure communication architectures.

In this chapter, we have discussed some of the various coding mechanisms that form the backbone of most physical layer security systems. In particular, we have shown how the operation of physical layer security coding schemes is, in general, different from that of traditional codes for reliable communications; for instance, *randomized encoding* plays a crucial role to achieve secrecy. While the theoretical understanding of the coding mechanisms is relatively mature, the design of practical low-complexity codes and efficient signaling schemes for physical layer security remains largely in its infancy. In addition, there is a significant need for information-theoretic models and results that relax some of the restrictive assumptions of the basic models, such as the knowledge of the eavesdropper's channel or the passivity of the attacker.

References

[1] A. D. Wyner, "The wire-tap channel," *Bell System Technical Journal*, vol. 54, no. 8, pp. 1355–1367, October 1975.

[2] U. Maurer, "Secret key agreement by public discussion from common information," *IEEE Trans. Inf. Theory*, vol. 39, no. 3, pp. 733–742, May 1993.

[3] R. Ahlswede and I. Csiszár, "Common randomness in information theory and cryptography. i. Secret sharing," *IEEE Trans. Inf. Theory*, vol. 39, no. 4, pp. 1121–1132, July 1993.

[4] T. Ignatenko and F. M. J. Willems, "Biometric systems: Privacy and secrecy aspects," *IEEE Transactions on Information Forensics and Security*, vol. 4, no. 4, pp. 956–973, 2009.

[5] L. Sankar, S. R. Rajagopalan, and H. V. Poor, "A theory of utility and privacy of data sources," in *Proc. of IEEE International Symposium on Information Theory*, Saint-Petersburg, Russia, August 2011.

[6] A. Winter, A. C. A. Nascimento, and H. Imai, "Commitment capacity of discrete memoryless channels," in *Proc. of 9th IMA International Conference*, Cirencester, UK, 2003, pp. 33–51.

[7] A. Nascimento and A. Winter, "On the oblivious transfer capacity of noisy correlations," in *Proc. of 2006 IEEE International Symposium on Information Theory*, Seattle, WA, July 2006, pp. 1871–1875.

[8] H. Imai, K. Morozov, and A. Nascimento, "On the oblivious transfer capacity of the erasure channel," in *Proc. of 2006 IEEE International Symposium on Information Theory*, Seattle, WA, July 2006, pp. 1428–1431.

[9] R. Ahlswede and I. Csiszár, "On oblivious transfer capacity," in *Proc. 2007 International Symposium on Information Theory*, Nice, France, June 2007, pp. 2061–2065.

[10] H. Tyagi, P. Narayan, and P. Gupta, "When is a function securely computable?" *IEEE Transactions on Information Theory*, vol. 57, pp. 6337–6350, 2011.

[11] M. Bloch and J. Barros, *Physical-Layer Security: From Information Theory to Security Engineering*. Cambridge University Press, October 2011.

[12] Y. Liang, H. V. Poor, and S. Shamai (Shitz), *Information-Theoretic Security*, ser. Foundations and Trends in Communications and Information Theory. Delft, Netherlands: Now Publishers, 2009, vol. 5, no. 1–5.

[13] C. E. Shannon, "Communication theory of secrecy systems," *Bell System Technical Journal*, vol. 28, pp. 656–715, 1948.

[14] G. S. Vernam, "Cipher printing telegraph systems for secret wire and radio telegraphic communications," *Transactions of the American Institute of Electrical Engineers*, vol. 1, pp. 295–301, 1926.

[15] G. D. Forney, Jr., "On the role of MMSE estimation in approaching the information-theoretic limits of linear Gaussian channels: Shannon meets Wiener," in *Proc. of 41st Annual Allerton Conference on Communication, Control, and Computing*, Monticello, IL, October 2003, pp. 430–439.

[16] M. Bloch and J. N. Laneman, "On the secrecy capacity of arbitrary wiretap channels," in *Proceedings of 46th Allerton Conference on Communication, Control, and Computing*, Monticello, IL, September 2008, pp. 818–825.

[17] U. Maurer, *Communications and Cryptography: Two Sides of One Tapestry.* Kluwer Academic Publishers, 1994, ch. The Strong Secret Key Rate of Discrete Random Triples, pp. 271–285.

[18] M. Iwamoto and K. Ohta, "Security notions for information theoretically secure encryptions," in *Proc. of IEEE International Symposium on Information Theory*, St. Petersburg, Russia, August 2011, pp. 1743–1747.

[19] M. Bellare, S. Tessaro, and A. Vardy, "A cryptographic treatment of the wiretap channel," arxiv preprint: 1201.2205, 2012.

[20] Y. Liang and H. V. Poor, "Secure communication over fading channels," in *Proc. of 44th Allerton Conference on Communication, Control and Computing*, Urbana, IL, September 2006.

[21] M. Bloch, J. Barros, M. R. D. Rodrigues, and S. W. McLaughlin, "Wireless information-theoretic security," *IEEE Trans. Inf. Theory*, vol. 54, no. 6, pp. 2515–2534, June 2008.

[22] Y. Liang, G. Kramer, H. V. Poor, and S. Shamai (Shitz), "Compound wiretap channels," *EURASIP Journal on Wireless Communications and Networking*, pp. 1–12, 2009.

[23] X. He and A. Yener, "Providing secrecy when the eavesdropper channel is arbitrarily varying: A case for multiple antennas," in *Proc. 48th Annual Allerton Conf. Communication, Control, and Computing (Allerton)*, 2010, pp. 1228–1235.

[24] E. MolavianJazi, M. Bloch, and J. N. Laneman, "Arbitrary jamming can preclude secure communications," in *Proc. 47th Annual Allerton Conference on Communication, Control, and Computing*, Monticello, IL, September 2009, pp. 1069–1075.

[25] M. N. Wegman and J. Carter, "New hash functions and their use in authentication and set equality," *Journal of Computer Sciences and Systems*, vol. 22, no. 3, pp. 265–279, June 1981.

[26] I. Csiszár and J. Körner, "Broadcast channels with confidential messages," *IEEE Trans. on Inf. Theory*, vol. 24, no. 3, pp. 339–348, May 1978.

[27] S. K. Leung-Yan-Cheong and M. E. Hellman, "The Gaussian wire-tap channel," *IEEE Transactions on Information Theory*, vol. 24, no. 4, pp. 451–456, July 1978.

[28] A. Thangaraj, S. Dihidar, A. R. Calderbank, S. W. McLaughlin, and J.-M. Merolla, "Applications of LDPC codes to the wiretap channels," *IEEE Trans. Inf. Theory*, vol. 53, no. 8, pp. 2933–2945, August 2007.

[29] H. Mahdavifar and A. Vardy, "Achieving the secrecy capacity of wiretap channels using polar codes," *IEEE Transactions on Information Theory*, vol. 57, no. 10, pp. 6428–6443, 2011.

[30] M. R. Bloch, "Achieving secrecy: Capacity vs. resolvability," in *Proc. of IEEE International Symposium on Information Theory*, Saint Petersburg, Russia, August 2011, pp. 632–636.

[31] M. R. Bloch and J. N. Laneman, "Secrecy from resolvability," submitted to *IEEE Transactions on Information Theory*, May 2011. [Online]. Available: arXiv:1105.5419.

[32] V. Rathi, M. Andersson, R. Thobaben, J. Kliewer, and M. Skoglund, "Two edge type LDPC codes for the wiretap channels," in *Proc. of 43rd Asilomar Conference on Signals, Systems and Computers*, Pacific Grove, CA, November 2009, pp. 834–838.

[33] V. Rathi, R. Urbanke, M. Andersson, and M. Skoglund, "Rate-equivocation optimal spatially coupled ldpc codes for the bec wiretap channel," in *Proc. of IEEE International Symposium on Information Theory*, Saint-Petersburg, Russia, August 2011, pp. 2393–2397.

[34] M. Andersson, V. Rathi, R. Thobaben, J. Kliewer, and M. Skoglund, "Nested polar codes for wiretap and relay channels," *IEEE Communications Letters*, vol. 14, no. 4, pp. 752–754, June 2010.

[35] O. O. Koyluoglu and H. E. Gamal, "Polar coding for secure transmission and key agreement," in *Proc. of IEEE International Symposium on Personal Indoor and Mobile Radio Communications*, Istanbul, Turkey, September 2010, pp. 2698–2703.

[36] E. Hof and S. Shamai, "Secrecy-achieving polar-coding," in *Proc. IEEE Information Theory Workshop*, Dublin, Ireland, September 2010, pp. 1–5.

[37] M. Hayashi, "General nonasymptotic and asymptotic formulas in channel resolvability and identification capacity and their application to the wiretap channels," *IEEE Trans. Inf. Theory*, vol. 52, no. 4, pp. 1562–1575, April 2006.

[38] A. Wyner, "The common information of two dependent random variables," *IEEE Trans. Inf. Theory*, vol. 21, no. 2, pp. 163–179, March 1975.

[39] T. Han and S. Verdú, "Approximation theory of output statistics," *IEEE Trans. Inf. Theory*, vol. 39, no. 3, pp. 752–772, May 1993.

[40] T. S. Han, *Information-Spectrum Methods in Information Theory*. Berlin Springer, 2002.

[41] A. Subramanian, A. Thangaraj, M. Bloch, and S. McLaughlin, "Strong secrecy on the binary erasure wiretap channel using large-girth LDPC codes," *IEEE Transactions on Information Forensics and Security*, vol. 6, no. 3, pp. 585–594, September 2011.

[42] M. Cheraghchi, F. Didier, and A. Shokrollahi, "Invertible extractors and wiretap protocols," in *Proc. IEEE International Symposium on Information Theory*, Seoul, Korea, July 2009, pp. 1934–1938.

[43] M. Bellare, S. Tessaro, and A. Vardy, "Semantic security for the wiretap channel," in *Lecture Notes in Computer Science*, vol. 7417, no. 294–311, 2012.

[44] M. Hayashi and R. Matsumoto, "Construction of wiretap codes from ordinary channel codes," in *Proc. IEEE International Symposium on Information Theory*, Austin, TX, June 2010, pp. 2538–2542.

[45] J. Muramatsu and S. Miyake, "Construction of strongly secure wiretap channel code based on hash property," in *Proc. IEEE International Symposium on Information Theory*, Saint-Petersburg, Russia, August 2011, pp. 613–617.

[46] D. Slepian and J. K. Wolf, "Noiseless coding of correlated information sources," *IEEE Trans. Inf. Theory*, vol. 19, no. 4, pp. 471–480, July 1973.

[47] S. Watanabe, R. Matsumoto, and T. Uyematsu, "Strongly secure privacy amplification cannot be obtained by encoder of slepian-wolf code," *IEICE Transactions on Fundamentals of Electronics, Communications and Computer Sciences*, vol. E93-A, no. 9, pp. 1650–1659, September 2010.

[48] I. Csiszár, "Almost independence and secrecy capacity," *Problems of Information Transmission*, vol. 32, no. 1, pp. 40–47, January-March 1996.

[49] J. Muramatsu, H. Koga, and T. Mukouchi, "On the problem of generating mutually independent random sequences," *IEICE Transactions on Fundamentals of Communication*, vol. E-86 A, no. 5, pp. 1275–1284, May 2003.

[50] M. Bloch, "Channel intrinsic randomness," in *Proc. of IEEE International Symposium on Information Theory*, Austin, TX, June 2010, pp. 2607–2611.

[51] C. H. Bennett, G. Brassard, C. Crépeau, and U. Maurer, "Generalized privacy amplification," *IEEE Trans. Inf. Theory*, vol. 41, no. 6, pp. 1915–1923, November 1995.

[52] U. M. Maurer and S. Wolf, "Information-theoretic key agreement: From weak to strong secrecy for free," in *Advances in Cryptology—Eurocrypt 2000*, Lecture Notes in Computer Science, B. Preneel, 2000, p. 351.

[53] R. Renner, "Security of Quantum Key Distribution," Ph.D. dissertation, Swiss Federal Institute of Technology, Zurich, 2005.

Chapter 2

Coding for Wiretap Channels

Fuchun Lin
Nanyang Technological University

Frederique Oggier
Nanyang Technological University

This chapter illustrates constructions of wiretap codes with examples from different types of physical channels. Coding criteria with respect to different types of channels are discussed, followed by explicit code constructions. Section 2.1 deals with the best understood example of explicit wiretap codes, referred to as *wiretap II codes*. They are built from classical error-correcting codes, and are used on a *bounded erasure channel*. Recent progress in the wiretap code design has been made thanks to the introduction of polar codes, which will be discussed in Section 2.2. Wiretap polar codes are best suited for symmetric channels. The approach presented in Section 2.3 is different from many points of view: first, it deals with additive white Gaussian noise channels (all the other cases considered are Discrete Memoryless Channels), and second, it approaches the notion of secrecy from a coding point of view, through an error probability computation. Wiretap lattice codes will be built.

2.1 Coding for the Wiretap Channel II

Figure 2.1 illustrates the original wiretap channel model as introduced by A. D. Wyner [1]. The original wiretap channel assumes that both the *main channel* and the *wiretapper channel* are Discrete Memoryless Channels (DMC), with respective channel capacity C_M and C_W, and the wiretapper channel is assumed to be a degraded version of the main channel. Alice encodes a message $\mathbf{s} = \{s_1, s_2, \cdots, s_K\}$, where every s_i comes from a discrete memoryless source S with entropy H_S,[1] into a codeword $\mathbf{x} = \{x_1, x_2, \cdots, x_N\}$, which is sent through both channels. Bob and Eve's outputs are $\mathbf{y} = \{y_1, y_2, \cdots, y_N\}$ and $\mathbf{z} = \{z_1, z_2, \cdots, z_N\}$, respectively.

Definition 2.1.1. The *equivocation* of the message experienced by Eve is measured by the conditional entropy

$$\Delta = H(\mathbf{S}|\mathbf{Z}), \tag{2.1}$$

where $\mathbf{S}$ and $\mathbf{Z}$ are random vectors corresponding to $\mathbf{s}$ and $\mathbf{z}$, respectively. One immediately has that $0 \leq \Delta \leq H(\mathbf{S}) = K H_S$. When $\Delta = 0$, Eve completely knows the secret message,

---

[1]The logarithm in the entropy function is as all logarithms in this chapter taken to the base 2.

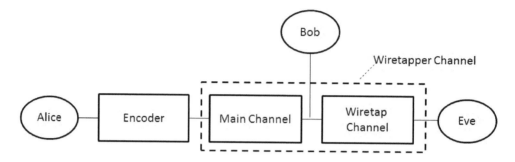

Figure 2.1: The Wiretap Channel Model: in Wyner's original terminology, the channel to Eve was described as the concatenation of the main channel and another channel, which he called "the wiretap channel," though today a wiretap channel refers to the whole communication channel.

while Eve might as well guess $\mathbf{S}$ when $\Delta = H(\mathbf{S})$, corresponding to the best possible confidentiality.

It took a decade after the publication of the original wiretap channel for an explicit wiretap coding scheme to be proposed [2] for a wiretap channel where the main channel is a binary noiseless channel and the wiretapper channel is a binary *bounded erasure channel*, in which $\eta = N\varepsilon$ signals are erased in every N transmissions, or alternatively, as put in [2], the eavesdropper Eve observes her choice of $\mu = N(1 - \varepsilon)$ transmissions in every N transmissions, a model suited for eavesdropping over a network (see, e.g.,) [3]. This binary bounded erasure channel is different from the usual BEC of erasure rate ε, where $\eta = N\varepsilon$ erasures occur with high probability but not with certainty. Finally, the source is assumed to output independent, identically distributed binary random variables, for simplicity. This scenario is traditionally called the "wiretap channel II."

It is a common practice in coding to fix K and N (hence R) and compare the performance of a code against the achievable bound. Figure 2.2 shows the trade-off between the achievable normalized equivocation $\delta = \frac{\Delta}{K}$ and the wiretapping rate $\alpha = \frac{\mu}{N}$ for fixed R. A wiretap II code achieving the upper and upper right boundaries is considered optimal.

2.1.1 Basics of Error-Correcting Codes

Coding for the wiretap channel II relies on binary linear codes. Recall that a binary $[n, k]$ *code* C is a k-dimensional subspace of the vector space $\{0, 1\}^n$. Vectors in C are called *codewords* of C and are written as row vectors.

Definition 2.1.2. The *(Hamming) weight* of a codeword counts the number of its nonzero coefficients. The *(Hamming) distance* between two codewords $\mathbf{x}$ and $\mathbf{y}$ is the weight of $\mathbf{x} - \mathbf{y}$, and the *minimum distance* of a code C is the minimum of the distance between any two different codewords in C.

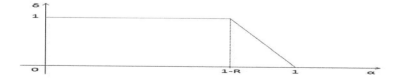

Figure 2.2: Achievable (δ, α) region for fixed R.

We sometimes write an $[n, k, d]$ code C to emphasize the minimum distance d of C.

Definition 2.1.3. Two codes are called *equivalent* if one code can be obtained from the other by permutation of coordinates.

A binary $[n, k, d]$ code can be defined either by a full-rank $k \times n$ binary matrix G, called a *generator matrix*:

$$C = \{\mathbf{m}G \mid \mathbf{m} \in \{0, 1\}^k\}, \tag{2.2}$$

or by a full-rank $(n - k) \times n$ binary matrix, called a *parity check matrix* H:

$$C = \{\mathbf{x} \mid H\mathbf{x}^T = \mathbf{0}\}, \tag{2.3}$$

where $\mathbf{x}^T$ is the transpose of $\mathbf{x}$.

Remark 2.1.1. *Note that the row vectors of G form a basis of C, hence one can replace G with any k linearly independent codewords. Moreover, by definition of equivalent codes, one can exchange columns of G to get an equivalent code. To summarize, one can perform row/column exchanges, and add one row to another row of G safely without getting a different code.*

Definition 2.1.4. The *dual* code $C^\perp$ of an $[n, k, d]$ code C is

$$C^\perp = \{\mathbf{y} \mid \mathbf{y}\mathbf{x}^T = 0, \text{ for every } \mathbf{x} \in C\}.$$

Remark 2.1.2. *A generator matrix for C is a parity check matrix for its dual $C^\perp$ and a parity check matrix for C is a generator matrix for its dual $C^\perp$.*

There are various bounds on the minimum distance of a code [4].

Proposition 2.1.1. *The Singleton bound states that*

$$d \le n - k + 1. \tag{2.4}$$

Proof. According to (2.3), any $d - 1$ columns of H are linearly independent (otherwise, there exists a codeword $\mathbf{x}$ of C with weight less than d). But this is only possible when $d - 1 \le n - k$ because H is an $(n - k) \times n$ matrix and any $n - k + 1$ columns of H are linearly dependent. □

An $[n, k, d]$ code achieving equality in (2.4) is called a *Maximum Distance Separable* code, or an MDS code.

Proposition 2.1.2. *The dual of an MDS code is still an MDS code [4].*

Proof. Let H be a parity check matrix of an $[n, k, n - k + 1]$ code C, which is at the same time a generator matrix of the $[n, n - k]$ code $C^\perp$. We can choose H in such a way that it contains a row of the minimum weight $d^\perp$. In order to show that $C^\perp$ is MDS, we need to show that $d^\perp = n - (n - k) + 1 = k + 1$. Now assume by contradiction that $d^\perp \le k$, namely, there are at least $n - k$ zeroes in that specific row. Take $n - k$ columns of H that correspond to a zero in that row. They are obviously linearly dependent since there is a zero row in the $(n - k) \times (n - k)$ submatrix. Evoking the fact that H is a parity check matrix of the MDS code C, hence any $n - k$ columns of H are linearly independent, we have a contradiction. □

Here are some examples of MDS codes, called trivial MDS codes. They are in fact the only binary MDS codes [4].

Example 2.1.1. The $[n, n, 1]$ universe code, which contains every vector of the vector space $\{0, 1\}^n$, is MDS.

Example 2.1.2. The even weight code $[n, n-1, 2]$, consisting of all even weight vectors in $\{0, 1\}^n$, and its dual, the repetition code $[n, 1, n]$, consisting of the all zero vector $\mathbf{0}$ and the all one vector $\mathbf{1}$, are both MDS.

Here are more examples of codes.

Example 2.1.3. A $[2^m - 1, 2^m - m - 1, 3]$ Hamming code can be defined by its $m \times (2^m - 1)$ parity check matrix, whose columns consist of all the nonzero binary m-tuples. For example, if $m = 3$, H is given by

$$H = \begin{pmatrix} 0 & 0 & 0 & 1 & 1 & 1 & 1 \\ 0 & 1 & 1 & 0 & 0 & 1 & 1 \\ 1 & 0 & 1 & 0 & 1 & 0 & 1 \end{pmatrix}. \tag{2.5}$$

That $d = 3$ follows from the facts that any two columns are different and that the sum of two different columns is again a column.

Example 2.1.4. Consider the 2×2 matrix

$$F_2 = \begin{pmatrix} 0 & 1 \\ 1 & 1 \end{pmatrix}. \tag{2.6}$$

Let $\otimes$ denote the Kronecker product of matrices. Let $F_{2^m} = F_2^{\otimes m}$. For example,

$$F_{2^3} = \begin{pmatrix} 0_{4 \times 4} & F_{2^2} \\ F_{2^2} & F_{2^2} \end{pmatrix} = \begin{pmatrix} 0 & 0 & 0 & 0 & 0 & 0 & 0 & 1 \\ 0 & 0 & 0 & 0 & 0 & 0 & 1 & 1 \\ 0 & 0 & 0 & 0 & 0 & 1 & 0 & 1 \\ 0 & 0 & 0 & 0 & 1 & 1 & 1 & 1 \\ 0 & 0 & 0 & 1 & 0 & 0 & 0 & 1 \\ 0 & 0 & 1 & 1 & 0 & 0 & 1 & 1 \\ 0 & 1 & 0 & 1 & 0 & 1 & 0 & 1 \\ 1 & 1 & 1 & 1 & 1 & 1 & 1 & 1 \end{pmatrix}. \tag{2.7}$$

A Reed-Müller code with length $n = 2^m$, $m \geq 0$, and dimension $k, 0 \leq k \leq n$, denoted by $RM(n, k)$, can be defined by a generator matrix obtained from F_{2^m} by deleting $2^m - k$ rows such that none of the deleted rows has a larger weight than any of the remaining k rows. For example, a generator matrix for $RM(8, 4)$ is

$$G = \begin{pmatrix} 0 & 0 & 0 & 0 & 1 & 1 & 1 & 1 \\ 0 & 0 & 1 & 1 & 0 & 0 & 1 & 1 \\ 0 & 1 & 0 & 1 & 0 & 1 & 0 & 1 \\ 1 & 1 & 1 & 1 & 1 & 1 & 1 & 1 \end{pmatrix}. \tag{2.8}$$

2.1.2 Wiretap II Codes

Encoding for wiretap II codes relies on coset coding. An $[N, N - K, d]$ code C is used to partition $\{0, 1\}^N$ into 2^K disjoint subsets, namely the code C and all its cosets, each of size 2^{N-K}. More precisely, let $\mathbf{s}$ be the confidential message and H be a parity check matrix of C. The codeword $\mathbf{x}$ is chosen uniformly at random from the 2^{N-K} solutions of

$$H\mathbf{x}^T = \mathbf{s}^T. \tag{2.9}$$

Let Δ_η denote Eve's minimum equivocation (see Definition 2.1) when her channel experiences η erasures. The rest of this section characterizes Δ_η.

Proposition 2.1.3. *The equivocation* $\Delta = H(\mathbf{S}|\mathbf{Z})$ *is alternatively expressed as*

$$H(\mathbf{S}|\mathbf{Z}) = \eta - H(\mathbf{X}|\mathbf{S}, \mathbf{Z}). \tag{2.10}$$

Proof. By repeated usages of the chain rule of the entropy function,

$$
\begin{aligned}
H(\mathbf{S}|\mathbf{Z}) &= H(\mathbf{S}, \mathbf{Z}) - H(\mathbf{Z}) \\
&= H(\mathbf{X}, \mathbf{S}, \mathbf{Z}) - H(\mathbf{X}|\mathbf{S}, \mathbf{Z}) - H(\mathbf{Z}) \\
&= H(\mathbf{S}|\mathbf{X}, \mathbf{Z}) + H(\mathbf{X}, \mathbf{Z}) - H(\mathbf{X}|\mathbf{S}, \mathbf{Z}) - H(\mathbf{Z}) \\
&= H(\mathbf{S}|\mathbf{X}, \mathbf{Z}) + H(\mathbf{X}|\mathbf{Z}) - H(\mathbf{X}|\mathbf{S}, \mathbf{Z}).
\end{aligned}
$$

By (2.9), $\mathbf{S} = \mathbf{s}$ can be recovered once $\mathbf{X} = \mathbf{x}$ is known, hence $H(\mathbf{S}|\mathbf{X}, \mathbf{Z}) = 0$. Now $\mathbf{Z}$ is a corrupted version of $\mathbf{X}$ with exactly η bits missing and moreover, according to the coding scheme, $\mathbf{X}$ is uniformly distributed on $\{0, 1\}^N$ and its coordinates are independent identically distributed uniform binary variables. Thus $H(\mathbf{X}|\mathbf{Z}) = \eta$. (2.10) then follows. $\square$

Write $H = (H_1, H_2, \cdots, H_N)$ and let $H|_{\mathcal{I}} = (H_i | i \in \mathcal{I})$ be a submatrix of H consisting of the columns specified by $\mathcal{I} \subset \{1, 2, \cdots, N\}$.

Proposition 2.1.4. *Let $\mathcal{I}$ with $|\mathcal{I}| = \eta$ be the indices of the erased positions. Then*

$$H(\mathbf{X}|\mathbf{S}, \mathbf{Z}) = \eta - rank(H|_{\mathcal{I}}). \tag{2.11}$$

Proof. To compute $H(\mathbf{X}|\mathbf{S}, \mathbf{Z})$, suppose that $\mathbf{S} = \mathbf{s}$ and $\mathbf{Z} = \mathbf{z}$ are known, and count the number of $\mathbf{x}$'s for which (2.9) holds, i.e., the number of solutions of

$$\sum_{i \in \mathcal{I}} x_i H_i = \mathbf{s} + \sum_{i \notin \mathcal{I}} x_i H_i. \tag{2.12}$$

This is a linear system with η unknowns and coefficient matrix $H|_{\mathcal{I}}$. The number of solutions is exactly $2^{\eta - \text{rank}(H|_{\mathcal{I}})}$. We observe that given any $\mathbf{S} = \mathbf{s}$ and $\mathbf{Z} = \mathbf{z}$, all these solutions are equally likely and (2.11) follows. $\square$

Substituting (2.11) into (2.10) yields $H(\mathbf{S}|\mathbf{Z}) = \text{rank}(H|_{\mathcal{I}})$ and

$$\Delta_\eta = \min_{|\mathcal{I}| = \eta} \text{rank}(H|_{\mathcal{I}}), \tag{2.13}$$

and the values of $\min_{|\mathcal{I}| = \eta} \text{rank}(H|_{\mathcal{I}})$ for all η characterize the performance of such wiretap II codes.

Proposition 2.1.5. *The minimum equivocation Δ_η is further given by*

$$\min_{|\mathcal{I}| = \eta} rank(H|_{\mathcal{I}}) = \begin{cases} \eta, & \text{if } 0 < \eta \le d - 1 \\ K, & \text{if } N - d^\perp + 1 \le \eta \le N. \end{cases}$$

Proof. The first assertion follows from H being a parity check matrix of C. To show the second assertion, we view H as a generator matrix of $C^\perp$. Assume by contradiction that $\min_{|\mathcal{I}| = \eta} \text{rank}(H|_{\mathcal{I}}) < K$ for $\eta \ge N - d^\perp + 1$. There exists a series of row manipulations that produces a zero row in a $K \times \eta$ matrix $H|_{\mathcal{I}}$. Then the same row manipulations will produce a codeword of $C^\perp$ that has weight at most $N - (N - d^\perp + 1) = d^\perp - 1$, which is a contradiction. $\square$

Remark 2.1.3. This theorem only characterizes Δ_η for small η and large η. There are possibly Δ_η not known for η in the gap between $d-1$ and $N - d^\perp + 1$ except when $d - 1 = N - d^\perp + 1$, namely $d + d^\perp = N + 2$. Now applying the Singleton bound to both C and $C^\perp$ we have

$$\begin{cases} d & \leq N - K + 1 \\ d^\perp & \leq K + 1 \end{cases} \implies d + d^\perp \leq N + 2$$

and equality is achieved if and only if C is MDS. By comparing with the achievable region, a wiretap II code is optimal if and only if C is MDS.

We next study how well some known binary linear codes perform as wiretap II codes, though these wiretap II codes will not be optimal, since there are no nontrivial binary MDS codes [4]. The notion of *generalized Hamming weights* of a linear code, defined next, relates the drops in Eve's equivocation with the parameters of the linear code. The *support* $\mathcal{I}(\mathbf{c})$ of a codeword $\mathbf{c}$ in an $[n, k, d]$ code C is a set collecting the positions where a one occurs in $\mathbf{c}$, and the size $|\mathcal{I}(\mathbf{c})|$ is then its Hamming weight (see Definition 2.1.2). The concept of support is extended to a subcode D of C, which by definition is a subspace of C, by taking the union of the supports of every codeword in D, namely, $\mathcal{I}(D) = \bigcup_{\mathbf{c} \in D} \mathcal{I}(\mathbf{c})$.

Example 2.1.5. Let C be an $[n, k, d]$ code. The support of any one-dimensional subcode generated by a codeword $\mathbf{c}$ coincides with the support of $\mathbf{c}$. On the other hand, the support of C is $\{1, 2, \cdots, n\}$.

Now the size of the support of an r-dimensional subcode D is used to define an rth weight of C, as follows.

Definition 2.1.5. The minimum rth weight of C, denoted by $d_r(C)$, is defined as

$$d_r(C) = \min_{D \subset C;\ \dim(D) = r} |\mathcal{I}(D)|.$$

The weight hierarchy of C refers to $d_r(C)$, $1 \leq r \leq k$. From Example 2.1.5, we immediately have $d_1(C) = d$ and $d_k(C) = n$.

The role of generalized weights for wiretap coding was studied by Wei [5].

Proposition 2.1.6.

$$\min_{|\mathcal{I}| = \eta} rank(H|_{\mathcal{I}}) = K - r, \quad N - d_{r+1}(C^\perp) < \eta \leq N - d_r(C^\perp).$$

Proof. We give a different proof than that of [5]. For $r = 0$, we need to show that $\min_{|\mathcal{I}| = \eta} rank(H|_{\mathcal{I}}) = K$ for any η satisfying $N - d_1(C^\perp) < \eta \leq N$. This was shown in the previous theorem. From the argument in that proof, if we perform a series of row manipulations and column exchanges to transform H into H' with a maximal length 0-string in the first row, then the 0-string is of length $N - d_1(C^\perp)$. For $r = 1$, we need to show that $\min_{|\mathcal{I}| = \eta} rank(H|_{\mathcal{I}}) = K - 1$ for any η satisfying $N - d_2(C^\perp) < \eta \leq N - d_1(C^\perp)$. Now we leave alone the first row of H' and perform row manipulations to the other rows. We do column exchanges, too. But we only exchange columns corresponding to the string of zeroes in the first row. We should be able to transform H' into H'' with the first row unaltered and the second row with a maximum length 0-string. We claim that this 0-string is of length at most $N - d_2(C^\perp)$. Otherwise, we would be able to generate a two-dimensional subcode of $C^\perp$ with the first and second row of H'' and the weight of this subcode is less than $d_2(C^\perp)$. Now by just looking at H'' we know that $\min_{|\mathcal{I}| = \eta} rank(H|_{\mathcal{I}}) \leq K - 1$ for any η satisfying $N - d_2(C^\perp) < \eta \leq N - d_1(C^\perp)$. The other side of inequality, that $\min_{|\mathcal{I}| = \eta} rank(H|_{\mathcal{I}}) \geq K - 1$, follows from the fact that the length of the 0-strings are maximum. Finally, by transforming H'' in a similar way till we reach the Kth row and repeating the argument over and again, we have proven the proposition. $\qquad\square$

Table 2.1: Weight Hierarchy of Binary Linear Codes

Code C	$d_r(C)$
Dual Hamming	$2^m - 2^{m-r}$, $r = 0, 1, \cdots, m$
$RM(2^m, m+1)$	$2^m - 2^{m-r}$, $r = 0, 1, \cdots, m$ and $d_{m+1} = 2^m$
Hamming	$d_r(C) \in \{1, 2, \cdots, n\} \setminus \{2^i \mid 0 \leq i < m\}$
Dual $RM(2^m, m+1)$	$d_r(C) - 1 \in \{1, 2, \cdots, n\} \setminus \{2^i \mid 0 \leq i < m\}$

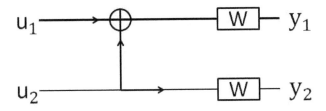

Figure 2.3: Generic combining unit.

The weight hierarchy of several well-known linear codes was computed in [5]. They are summarized in Table 2.1 for the Hamming and Reed-Müller codes (see Examples 2.1.3 and 2.1.4) and their duals.

Although we do not obtain implementable codes for real life channels, wiretap II codes provide a good understanding on how coding for secrecy is realized.

2.2 Wiretap Coding with Polar Codes

This section is devoted to polar wiretap codes, which are secrecy capacity-achieving wiretap codes. It was shown by Leung-Yan-Cheong [6] that the secrecy capacity of a wiretap channel is $C_S = C_M - C_W$ if both the main channel and the wiretapper channel are symmetric[2] and binary-input, and the wiretapper channel is a degraded version of the main channel. This result was rediscovered by Van Dijk [7] and shown to hold for a slightly more general setting, namely when the main channel is *less noisy* than the wiretapper's channel in the sense of Csiszár and Körner [8], where the *Broadcast Channel with Confidential messages (BCC)* was proposed generalizing Wyner's wiretap channel model by relaxing the condition of one channel being a degraded version of the other.

We start by presenting polar codes, which when implemented in a binary symmetric wiretap channel through coset coding give secrecy capacity-achieving wiretap codes. Polar codes have been recently introduced by Arikan [9]. Our exposition follows his work.

2.2.1 Polar Codes

Let us first look at a toy example. Let W be a Binary Erasure Channel (BEC) with erasure rate $\frac{1}{2}$. Suppose we have two copies of W lined up vertically (see the right hand side of Figure 2.3). To send two bits u_1 and u_2 through these two channels, we are allowed to first "combine" them, to obtain the input x_1 and x_2 of each respective copy of W. The idea behind polar coding is to understand how a combining process affects the transmission of u_1 and u_2. Figure 2.3 suggests $x_1 = u_1 \oplus u_2$ and $x_2 = u_2$ as a way of combining u_1 and u_2.

---

[2]A channel is symmetric if its channel transition matrix has rows which are permutations of each other, and columns which are permutations of each other.

Table 2.2: Truth Table of Combining Two Channels

(y_1, y_2)	$(u_1 \oplus u_2, u_2)$	$(u_1 \oplus u_2, \ )$	$(\ , u_2)$	$(\ , \ )$
u_1	$\checkmark$	$\times$	$\times$	$\times$
$u_2\|u_1 = 0$	$\checkmark$	$\checkmark$	$\checkmark$	$\times$

If one observes the outputs y_1 and y_2 to estimate u_1, u_1 can only be recovered when both transmissions are successful, $y_1 = x_1$ and $y_2 = x_2$, which happens with probability $\frac{1}{4}$, as shown in Table 2.2. Whenever either x_1 or x_2 is erased, u_1 cannot be recovered. This combining process thus decreases u_1's probability of successful transmission from $\frac{1}{2}$ to $\frac{1}{4}$. Now assume that the transmitter, knowing that u_1 is unlikely to be received, decides not to send any message using u_1 by agreeing with the receiver that u_1 is always zero. The receiver, with the knowledge of y_1 and y_2 which he observes, and of $u_1 = 0$, which he knows even before the transmission, can recover u_2 with probability $\frac{3}{4}$, as shown in Table 2.2. The combining process increases u_2's probability of successful transmission, though at the cost of freezing u_1. How u_2's probability of successful transmission is increased in this example is not difficult to understand. We can see that the actual inputs of the two copies of W are $x_1 = 0 \oplus u_2 = u_2$ and $x_2 = u_2$. This is just repetition coding (see Example 2.1.2) in classical coding theory.

This idea was generalized by Arikan [9] into a recursive combining process denoted by G_N, where N is a power of two (see Figure 2.4), which systematically decreases the probability of successful transmission of a portion of the u_i's (these will be frozen) while increasing the probability of successful transmission of the rest of them, used to transmit the actual message. Let $W : \{0,1\} \to \mathcal{Y}$, $x \mapsto y$ be a generic Binary-input Discrete Memoryless Channel (B-DMC) with transition probabilities $W(y|x)$ and denote by a_1^k the row vector $(a_1, a_2, \cdots, a_k)$. Let $W^N : \{0,1\}^N \to \mathcal{Y}^N$, $x_1^N \mapsto y_1^N$ denote N uses of W where $W^N(y_1^N|x_1^N) = \Pi_{i=1}^N W(y_i|x_i)$. Now define a sequence of N new channels $W_N^{(i)}$ corresponding to each input bit u_i, assuming that when estimating u_i, u_1^{i-1} are successfully obtained, some being frozen bits whose values are agreed on before the transmission, the others being successively deduced with high probability from observing the outputs y_1^N and the u_j's obtained before them. In this sense, y_1^N together with u_1^{i-1} are considered as the output of a new channel $W_N^{(i)}$, from which a decision about the input bit u_i is derived, where $W_N^{(i)} : \{0,1\} \longrightarrow \mathcal{Y}^N \times \{0,1\}^{i-1}$, $1 \leq i \leq N$ is defined by the transition probabilities

$$W_N^{(i)}(y_1^N, u_1^{i-1}|u_i) = \sum_{u_{i+1}^N \in \{0,1\}^{N-i}} \frac{1}{2^{N-1}} W^N(y_1^N|u_1^N G_N).$$

Arikan's combining strategy plays a key role here and hence is worth a careful description. The repeating generic combining unit on the left hand side of Figure 2.4 is exactly the one we have seen in Figure 2.3. We then have the starting stage (when $N = 2$) of this recursive process, which is described by the matrix equation $x_1^2 = u_1^2 G_2$, where

$$G_2 = \begin{bmatrix} 1 & 0 \\ 1 & 1 \end{bmatrix}.$$

A typical N-stage ($N \geq 4$) combining process as shown in Figure 2.4 consists of two steps. Step one, pair u_1^N into $(u_1^2, u_3^4, \cdots, u_{N-1}^N)$ and apply G_2 to each pair. The transformation in this step can be described by the matrix

$$I_{\frac{N}{2}} \otimes G_2 = \begin{bmatrix} G_2 & & \\ & \ddots & \\ & & G_2 \end{bmatrix}.$$

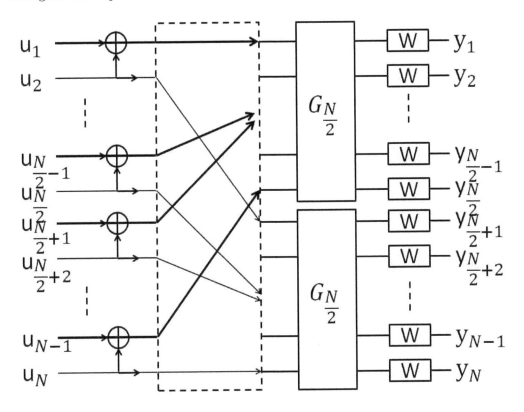

Figure 2.4: Arikan's combining strategy.

Step two, feed the $\frac{N}{2}$ sum bits $(u_1 \oplus u_2, u_3 \oplus u_4, \cdots, u_{N-1} \oplus u_N)$ and the original $\frac{N}{2}$ even bits $(u_2, u_4, \cdots, u_N)$ separately into two copies of the $\frac{N}{2}$-stage of this combining process. The transformation in this step can be described by the matrix $R_N(I_2 \otimes G_{\frac{N}{2}})$, where R_N denotes the permutation matrix that separates the sum bits and the original even bits.

Arikan [9] was able to prove the following result.

Theorem 2.2.1. For any B-DMC W, the channels $\{W_N^{(i)}\}$ polarize in the sense that, for any fixed $\delta \in (0, 1)$, as N goes to infinity through powers of two, the fraction of indices $i \in \{1, \cdots, N\}$ such that the channel capacity $I(W_N^{(i)})$ falls in $(1 - \delta, 1]$ goes to $I(W)$ and the fraction of i such that $I(W_N^{(i)})$ falls in $[0, \delta)$ goes to $1 - I(W)$.

Now we want to digress a little and present a connection between polar codes and Reed-Müller codes. The matrix F_2 (see (2.6)) we used to construct Reed-Müller codes in Section 2.1.1 is the transpose of the matrix G_2, and both constructions are recursive. More precisely, by evoking two properties of the Kronecker product $A \otimes B = P(B \otimes A)P^T$ (for square matrices A and B and for some permutation matrix P) and $(A \otimes B)(C \otimes D) = AC \otimes BD$, the matrix G_N can be analyzed as follows:

$$\begin{aligned} G_N &= (I_{\frac{N}{2}} \otimes G_2)R_N(I_2 \otimes G_{\frac{N}{2}}) \\ &= R_N(G_2 \otimes I_{\frac{N}{2}})R_N^T R_N(I_2 \otimes G_{\frac{N}{2}}) \\ &= R_N(G_2 \otimes I_{\frac{N}{2}})(I_2 \otimes G_{\frac{N}{2}}) \\ &= R_N(G_2 \otimes G_{\frac{N}{2}}). \end{aligned}$$

We thus obtain the recursive formula, and making use of it we further deduce, evoking

$(A \otimes B)(C \otimes D) = AC \otimes BD$ again, that

$$
\begin{aligned}
G_N &= R_N \left(G_2 \otimes R_{\frac{N}{2}} (G_2 \otimes G_{\frac{N}{4}}) \right) \\
&= R_N (I_2 \otimes R_{\frac{N}{2}}) \left(G_2 \otimes (G_2 \otimes G_{\frac{N}{4}}) \right) \\
&\quad \vdots \\
&= R_N (I_2 \otimes R_{\frac{N}{2}}) \cdots (I_{\frac{N}{2}} \otimes R_2) G_2^{\otimes n}.
\end{aligned}
$$

It can be shown that the matrix $R_N (I_2 \otimes R_{\frac{N}{2}}) \cdots (I_{\frac{N}{2}} \otimes R_2)$ is a permutation matrix, revealing that G_N is $G_2^{\otimes n}$ with rows permuted. Now both polar codes and Reed-Müller codes are constructed by choosing rows of $G_2^{\otimes n}$ (or equivalently, $F_2^{\otimes n}$). The only difference is that the choice in polar coding is polarization oriented while that in Reed-Müller coding is weight oriented.

2.2.2   Polar Wiretap Codes

Polar codes have been proposed to build wiretap codes for symmetric binary-input channels by Mahdavifar and Vardy [10].

Precoding: Compute the channel capacity $I(W_N^{(i)})$ for all $1 \leq i \leq N$ with respect to the main channel W_M to locate the good and bad polarized channels for Bob; compute the channel capacity $I(W_N^{(i)})$ for all $1 \leq i \leq N$ with respect to the wiretapper's channel W_W to locate the good and bad polarized channels for Eve.

Coding: Use the u_i's corresponding to the polarized channels that are good for Bob but bad for Eve to transmit message; flood the u_i's corresponding to the polarized channels that are good for both Bob and Eve with random bits; send zeroes through u_i's corresponding to the polarized channels that are bad for both Bob and Eve.

That this scheme works depends much on the fact that when W_W is a degraded version of W_M and they are both symmetric binary-input channels, the good polarized channels for Eve form a subset of the good polarized channels for Bob. This fact is critical for the constructed wiretap codes to be secrecy capacity achieving. Since polar codes are channel capacity-achieving codes, the fraction of the polarized channels that are good for Bob but bad for Eve is exactly $C_M - C_W$, which is the secrecy capacity of the wiretap channel for the channels considered. That Bob can correctly decode is also a direct consequence of polar coding. Bob receives all the information bits sent through his good polarized channels and the message bits are among them. What is not obvious is that the polarized channels that are good for both Bob and Eve have to be flooded with random bits to achieve perfect secrecy. Mahdavifar and Vardy [10] showed that if one sends zeroes through the polarized channels that are good for both Bob and Eve, instead of random bits, in the coding scheme, at least KC_W bits of information will be exposed to Eve, where K is the size of the message.

Remark 2.2.1. This coding scheme can be interpreted as an instance of a coset coding scheme. Denote the polar codes induced by the main channel W_M and the wiretapper's channel W_W by C_m and C_w, respectively. A message vector is now corresponding to a coset of C_w in C_m and is encoded to a randomly chosen codeword of C_m in the corresponding coset.

Remark 2.2.2. This coding scheme only achieves weak secrecy. Mahdavifar and Vardy [10] then proposed a modified scheme that achieves strong secrecy while still operating at a rate that approaches the secrecy capacity. But Bob's ability to correctly decode the message is not guaranteed in this case, unless the main channel is noiseless.

2.3 Coding for Gaussian Wiretap Channels

This last section, dedicated to Gaussian channels, is different from the rest of the chapter in several ways. First, we encounter the first example of continuous channels (the information-theoretical aspect of this wiretap channel scenario was done by Leung-Yan-Cheong in [11]). A more significant difference is that we will present an error probability point of view. Finally, codes are not defined over finite alphabets anymore. We will consider lattice codes over the reals.

2.3.1 Error Probability and Secrecy Gain

Let Λ be an n-dimensional real lattice, that is a discrete set of points in $\mathbb{R}^n$, which can be described in terms of its generator matrix M [12] by

$$\Lambda = \{\mathbf{x} = \mathbf{u}M \mid \mathbf{u} \in \mathbb{Z}^n\},$$

where the rows $\mathbf{v}_1, \cdots, \mathbf{v}_n$ of M are a linearly independent set of vectors in $\mathbb{R}^n$ which form a basis of the lattice. Let us denote by Λ_b the lattice that Alice will use to communicate with Bob. Alice encodes her K-bit message $\mathbf{s} = \{s_1, \ldots, s_K\}$ into a point $\mathbf{x} \in \Lambda_b$. In order to get confusion at the eavesdropper, coset coding is used as always, except that here it is realized with two nested lattices $\Lambda_e \subset \Lambda_b$, where Λ_e is a lattice contained in Λ_b: the lattice Λ_b is partitioned into Λ_e and a union of disjoint cosets of the form $\Lambda_e + \mathbf{c}$ with $\mathbf{c}$ an n-dimensional vector not in Λ_e. We need 2^K disjoint subsets to be labeled by the information vector $\mathbf{s} \in \{0, 1\}^K$:

$$\Lambda_b = \cup_{j=1}^{2^K}(\Lambda_e + \mathbf{c}_j)$$

which means that Λ_e is chosen so that $|\Lambda_b/\Lambda_e| = 2^K$. Once the mapping $\mathbf{s} \mapsto \Lambda_e + \mathbf{c}_{j(\mathbf{s})}$ is done, Alice randomly chooses a point $\mathbf{x} \in \Lambda_e + \mathbf{c}_{j(\mathbf{s})}$ and sends it over the wiretap channel. This is equivalent to choosing a random vector $\mathbf{r} \in \Lambda_e$, and the transmitted lattice point $\mathbf{x} \in \Lambda_b$ is finally of the form

$$\mathbf{x} = \mathbf{r} + \mathbf{c}_{j(\mathbf{s})} \in \Lambda_e + \mathbf{c}_{j(\mathbf{s})}. \tag{2.14}$$

In classical lattice coding over an Additive White Gaussian Noise (AWGN) channel, a lattice point $\mathbf{x} \in \Lambda$ is correctly decoded if and only if the received point falls in the Voronoi region $\mathcal{V}_\Lambda(\mathbf{x})$ of Λ, which is by definition all the real vectors which are closer to a given lattice point $\mathbf{x}$ than to any other. The probability is computed by integrating, over $\mathcal{V}_\Lambda(\mathbf{x})$, the Gaussian distribution function $\frac{e^{-\|\mathbf{u}\|^2/2\sigma^2}}{(\sqrt{2\pi}\sigma)^n}$ of a Gaussian channel with noise variance σ^2:

$$P_c = \frac{1}{(\sqrt{2\pi}\sigma)^n} \int_{\mathcal{V}_\Lambda(\mathbf{x})} e^{-\|\mathbf{u}\|^2/2\sigma^2} d\mathbf{u}.$$

Now in wiretap lattice coding, coset coding is implemented and one message is corresponding to a coset of lattice points instead of one lattice point. Assume that Eve is wiretapping through an AWGN channel with noise variance σ_e^2. As long as her received point falls in the Voronoi region of any of the lattice points in the coset of the sent lattice point, she can correctly decode. So the probability $P_{c,e}$ that Eve correctly decodes her received message is

$$P_{c,e} \leq \frac{1}{(\sqrt{2\pi}\sigma_e)^n} \sum_{\mathbf{t} \in \Lambda_e} \int_{\mathcal{V}_{\Lambda_b}(\mathbf{0})} e^{-\|\mathbf{u}+\mathbf{t}\|^2/2\sigma_e^2} d\mathbf{u}.$$

It was shown in [13, 15] that $P_{c,e}$ is bounded by

$$P_{c,e} \leq \frac{vol(\Lambda_b)}{(\sqrt{2\pi}\sigma_e)^n} \sum_{\mathbf{t} \in \Lambda_e} e^{-\|\mathbf{t}\|^2/2\sigma_e^2} = \frac{vol(\Lambda_b)}{(\sqrt{2\pi}\sigma_e)^n} \Theta_{\Lambda_e}\left(\frac{1}{2\pi\sigma_e^2}\right)$$

where the *volume vol*(Λ) of a lattice Λ with generator matrix M is by definition $\sqrt{\det(MM^T)}$, and Θ_Λ is the *theta series* of Λ [12] defined by

$$\Theta_\Lambda(z) = \sum_{\mathbf{x} \in \Lambda} q^{\|\mathbf{x}\|^2}, \quad q = e^{i\pi z}, \mathrm{Im}(z) > 0, \tag{2.15}$$

where we set $y = -iz$, and thus consider $\Theta_\Lambda(y)$, for $y > 0$. In what follows, we will write $\Theta_\Lambda(q)$ whenever it does not matter whether we consider z or y.

Example 2.3.1. Let us compute the theta series of the lattice $\mathbb{Z}^n$:

$$\Theta_{\mathbb{Z}^n}(q) = \sum_{\mathbf{x} \in \mathbb{Z}^n} q^{\|\mathbf{x}\|^2} = \sum_{x_1 \in \mathbb{Z}} q^{x_1^2} \cdots \sum_{x_n \in \mathbb{Z}} q^{x_n^2} = \left(\sum_{m \in \mathbb{Z}} q^{m^2} \right)^n = \Theta_{\mathbb{Z}}(q)^n.$$

To evaluate the benefit of coding (using a specifically designed lattice Λ_e) with respect to using $\Lambda_e = \lambda\mathbb{Z}^n$ ($\mathbb{Z}^n$ scaled to the same volume), we compare the behavior of the theta series of $\lambda\mathbb{Z}^n$ with that of Λ_e, and consequently define the secrecy function of a given lattice Λ as the ratio of the theta series of $\lambda\mathbb{Z}^n$ and its theta series.

Definition 2.3.1. *[13] Let Λ be an n-dimensional lattice of volume λ^n. The secrecy function of Λ is given by*

$$\Xi_\Lambda(y) = \frac{\Theta_{\lambda\mathbb{Z}^n}(y)}{\Theta_\Lambda(y)}$$

defined for $y > 0$.

We are interested in the secrecy function at the chosen point $y = \frac{1}{2\pi\sigma_e^2}$. However, by considering σ_e^2 as a variable, and since we want to minimize Eve's probability of correct decision, we further maximize the secrecy function over $y > 0$. This leads to the notion of secrecy gain.

Definition 2.3.2. *The (strong) secrecy gain $\chi_{\Lambda,strong}$ of an n-dimensional lattice Λ is defined by*

$$\chi_{\Lambda,strong} = \sup_{y>0} \Xi_\Lambda(y).$$

The role of the theta series Θ_{Λ_e} at the point $y = \frac{1}{2\pi\sigma_e^2}$ has been independently confirmed in [14], where it was shown for the mod-Λ Gaussian channel that the mutual information $I(\mathbf{S}; \mathbf{Z})$ is bounded by a function that depends on the channel parameters and on $\Theta_{\Lambda_e}\left(\frac{1}{2\pi\sigma_e^2}\right)$.

To summarize, a good Gaussian wiretap lattice code is a pair of nested lattices $\Lambda_e \subset \Lambda_b$ such that Λ_b enables good error correction and Λ_e has a big secrecy gain. The design criterion for Λ_b has been well studied while the secrecy gain of Λ_e is new.

2.3.2 Unimodular Lattice Codes

The secrecy gain is so far best understood for a class of lattices called *unimodular lattices*.

Definition 2.3.3. *A lattice is said to be* unimodular *if the matrix MM^T has integer coefficients, and if the determinant of this matrix is one.*

Figure 2.5 illustrates (in log scale) the typical shape of the secrecy function of a unimodular lattice. The problem of determining the (strong) secrecy gain is hard in general. For

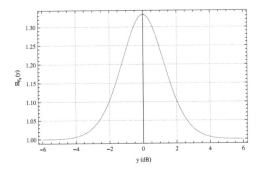

Figure 2.5: The secrecy function of the eight-dimensional unimodular lattice E_8 where the x-axis is in decibels $(10\log_{10}(y))$.

unimodular lattices, it has been shown in [15] that they have a (multiplicative) symmetry point at $y_0 = 1$, by which we mean a point y_0 such that

$$\Xi_\Lambda(y_0 \cdot y) = \Xi_\Lambda(y_0/y)$$

for all $y > 0$, or, in $\log y$ and $\log y_0$,

$$\Xi_\Lambda(\log y_0 + \log y) = \Xi_\Lambda(\log y_0 - \log y).$$

The existence of this symmetry point motivated the introduction of a *weak secrecy gain*:

Definition 2.3.4. *Suppose that Λ is an n-dimensional lattice, whose secrecy function has a symmetry point y_0. Then the weak secrecy gain χ_Λ of Λ is given by*

$$\chi_\Lambda = \Xi_\Lambda(y_0).$$

Weak secrecy gain was computed and analyzed in [15] for a subclass of unimodular lattices called *even* unimodular lattices, those with the extra property that every lattice point has a squared norm which is a multiple of two. An achievable lower bound on the secrecy gain of such unimodular lattices is thus obtained from the fact that the strong secrecy gain is always no less than the weak secrecy gain. Let Ω_n be the set of all inequivalent such unimodular n-dimensional lattices. In that case, we have that

$$\max_{\Lambda \in \Omega_n} \chi_\Lambda \gtrsim \frac{1.086^n}{2}.$$

On the other hand, it is conjectured that both the weak and strong secrecy gains coincide for the case of unimodular lattices, though this is proven only for some particular cases including some special lattices called *extremal* even unimodular lattices and all the unimodular lattices up to dimension 23, thanks to a method given by Ernvall-Hytönen [16]. The behavior of unimodular lattices in small dimensions (up to $n \le 23$) with respect to the secrecy gain have been studied in [17, 18], relying on the fact that until dimension 23, unimodular lattices are classified [12]. It was shown that the secrecy function of all these lattices reach their maximum at $y = 1$, and lattices with the best secrecy gain in each dimension have been determined.

Let us illustrate how wiretap unimodular lattice codes are constructed and analyzed with the eight-dimensional lattice E_8. The Gosset lattice E_8 is a famous eight-dimensional lattice which can be described by vectors of the form $(x_1, \ldots, x_8)$, $x_i \in \mathbb{Z}$, or $x_i \in \mathbb{Z} + 1/2$

(mixing integers and half-integers is not allowed), such that $\sum x_i \equiv 0 \mod 2$ [12]. Its theta series is

$$\Theta_{E_8}(q) = \frac{1}{2}\left(\vartheta_2(q)^8 + \vartheta_3(q)^8 + \vartheta_4(q)^8\right),$$

where

$$\vartheta_2(q) = \sum_{n=-\infty}^{+\infty} q^{\left(n+\frac{1}{2}\right)^2}, \vartheta_3(q) = \sum_{n=-\infty}^{+\infty} q^{n^2}, \vartheta_4(q) = \sum_{n=-\infty}^{+\infty} (-1)^n q^{n^2}.$$

Its secrecy function is illustrated in Figure 2.5, which shows that the symmetry point is $y_0 = 1$. We next evaluate the value of the secrecy function Ξ_{E_8} at the point $y = 1$. We have that

$$\Xi_{E_8}(y) = \frac{\vartheta_3(e^{-y\pi})^8}{\frac{1}{2}[\vartheta_2(e^{-y\pi})^8 + \vartheta_3(e^{-y\pi})^8 + \vartheta_4(e^{-y\pi})^8]}.$$

It is easier to look at $(\Xi_{E_8}(y))^{-1}$, which we evaluate in $y = 1$:

$$\begin{aligned}
\frac{1}{\Xi_{E_8}(1)} &= \frac{\frac{1}{2}\left(\vartheta_2(e^{-\pi})^8 + \vartheta_3(e^{-\pi})^8 + \vartheta_4(e^{-\pi})^8\right)}{\vartheta_3(e^{-\pi})^8} \\
&= \frac{1}{2}\left(1 + \frac{2\vartheta_4(e^{-\pi})^8}{4\vartheta_4(e^{-\pi})^8}\right) = \frac{3}{4}
\end{aligned}$$

using

$$\vartheta_2\left(e^{-\pi}\right) = \vartheta_4\left(e^{-\pi}\right), \vartheta_3\left(e^{-\pi}\right) = \sqrt[4]{2}\vartheta_4\left(e^{-\pi}\right).$$

We thus deduce that the secrecy gain of E_8 is

$$\boxed{\chi_{E_8} = \Xi_{E_8}(1) = \frac{4}{3} = 1.33333}.$$

Now, we need to make sure that Alice is able to actually use $\Lambda_e = E_8$ in her coset coding scheme. This is done using the so-called Construction A [12], which is defined via the modulo 2 map ρ from $\mathbb{Z}^n$ to $\mathbb{F}_2^n$. The preimage $\rho^{-1}(C)$ of a binary linear code C is obviously a lattice. Moreover, $\frac{1}{\sqrt{2}}\rho^{-1}(C)$ is a unimodular lattice if and only if $C = C^{\perp}$. The Reed-Müller code $RM(8,4)$ of Example 2.1.4 turns out to be one satisfying $C = C^{\perp}$. Actually one can show that

$$E_8 = \sqrt{2}\mathbb{Z}^8 + \frac{1}{\sqrt{2}}RM(8,4). \tag{2.16}$$

On the other hand, since

$$\mathbb{Z}^8 = 2\mathbb{Z}^8 + [8,8,1],$$

where $[8,8,1]$ is the universe code of Example 2.1.1, we have that

$$\frac{1}{\sqrt{2}}\mathbb{Z}^8 = \sqrt{2}\mathbb{Z}^8 + \frac{1}{\sqrt{2}}[8,8,1],$$

which combined with (2.16) yields

$$\begin{aligned}
E_8 &= \frac{1}{\sqrt{2}}\mathbb{Z}^8 + \frac{1}{\sqrt{2}}[8,8,1] + \frac{1}{\sqrt{2}}RM(8,4) \\
&= \frac{1}{\sqrt{2}}\mathbb{Z}^8 + \frac{1}{\sqrt{2}}RM(8,4)^{\dagger},
\end{aligned}$$

where by definition $RM(8,4) + RM(8,4)^{\dagger} = [8,8,1]$. Scaling this last equation, we further obtain

$$2E_8 = \sqrt{2}\mathbb{Z}^8 + \sqrt{2}RM(8,4)^{\dagger}$$

which together with (2.16) gives

$$E_8 = 2E_8 + \sqrt{2}RM(8,4)^\dagger + \frac{1}{\sqrt{2}}RM(8,4).$$

We can choose $\Lambda_b = E_8$ and $\Lambda_e = 2E_8$.

2.4 Conclusion

Wiretap codes are coding techniques that promise not only reliability between a legitimate sender and a legitimate receiver, but also confidentiality of the transmission in the presence of an eavesdropper. Despite many information-theoretic studies, explicit constructions of wiretap codes remain sparse in the literature.

In this chapter, we summarize several known coding techniques. (1) Wiretap II codes are arguably the most well-understood wiretap codes, they provide a good intuition on what coding for wiretap channel is, and are related to classical error-correcting codes. (2) Polar codes are recently introduced capacity-achieving codes. We reviewed their constructions and their application to wiretap coding. (3) We discussed a completely different approach, based on error probability, and presented some new results on lattice wiretap codes for Gaussian channels. There are of course other types of code constructions not presented here, especially those using low-density parity check (LDPC) codes, which will be discussed in detail in the next chapter.

Acknowledgment

The work of Fuchun Lin and Frédérique Oggier is supported by the Singapore National Research Foundation under the Research Grant NRF-RF2009-07.

References

[1] A. D. Wyner, "The wire-tap channel," *Bell Syst. Tech. Journal*, vol. 54, pp. 1355–1387, 1975.

[2] L. H. Ozarow and A. D. Wyner, "Wire-tap channel II," *Bell Syst. Tech. Journal*, vol. 63, no. 10, pp. 2135–2157, December 1984.

[3] S. El Rouayheb, E. Soljanin, A. Sprintson, "Secure Network Coding for Wiretap Networks of Type II," *IEEE Transactions on Information Theory*, vol. 58, no. 3, March 2012.

[4] F. J. MacWilliams and N. J. A. Sloane, "The Theory of Error-Correcting Codes," Amsterdam, The Netherlands: North-Holland, 1977.

[5] V. K. Wei, "Generalized hamming weights for linear codes," *IEEE Trans. Inform. Theory*, vol. 37, no. 5, pp. 1412–1418, September 1991.

[6] S. K. Leung-Yan-Cheong, "On a special class of wire-tap channels," *IEEE Trans. Inform. Theory*, vol. IT-23, no. 5, pp. 625–627, September 1977.

[7] M. van Dijk, "On a special class of broadcast channels with confidential messages," *IEEE Trans. Inform. Theory*, vol. 43, no. 2, pp. 712–714, March 1997.

[8] I. Csiszár and J. Körner, "Broadcast channels with confidential messages," *IEEE Trans. Inform. Theory*, vol. IT-24, no. 3, pp. 339–348, May 1978.

[9] E. Arikan, "Channel polarization: A method for constructing capacity-achieving codes for symmetric binary-input memoryless channels," *IEEE Trans. Inf. Theory*, vol. 55, pp. 3051–3073, July 2009.

[10] H. Mahdavifar and A. Vardy, "Achieving the Secrecy Capacity of Wiretap Channels Using Polar Codes," *IEEE Trans. Inf. Theory*, vol. 57, no. 10, pp. 6428–6443, October 2011.

[11] S. K. Leung-Yan-Cheong, "The Gaussian wire-tap channel," *IEEE Trans. Inform. Theory*, vol. IT-24, no. 4, pp. 451–456, July 1978.

[12] J. H. Conway, N. J. A. Sloane, "*Sphere Packings, Lattices and Groups*," Third edition, Springer-Verlag, New York, 1998.

[13] J.-C. Belfiore, F. Oggier, "Secrecy Gain: A Wiretap Lattice Code Design," *International Symposium on Information Theory and Its Applications (ISITA)*, pp. 174–178, 2010.

[14] C. Ling, L. Luzzi, J.-C. Belfiore, "Lattice codes with strong secrecy over the mod-Λ Gaussian channel," *International Symposium on Information Theory (ISIT)*, pp. 2306–2310, 2012.

[15] F. Oggier, P. Solé, J.-C. Belfiore, "Lattice Codes for the Wiretap Gaussian Channel: Construction and Analysis," arXiv:1103.4086 [cs.IT].

[16] A.-M. Ernvall-Hytönen, "On a Conjecture by Belfiore and Solé on some Lattices," *IEEE Trans. Inf. Theory*, vol. 58, no. 9, pp. 5950-5955, September 2012.

[17] F. Lin, F. Oggier, "Secrecy Gain of Gaussian Wiretap Codes from Unimodular Lattices," *Information Theory Workshop (ITW)*, pp. 718–722, 2011.

[18] F. Lin, F. Oggier, "A Classification of Unimodular Lattice Wiretap Codes in Small Dimensions," vol 55, issue 6, June 2013, *IEEE Transactions on Information Theory*.

Chapter 3

LDPC Codes for the Gaussian Wiretap Channel

Demijan Klinc
Georgia Institute of Technology

Jeongseok Ha
Korean Advanced Institute of Science and Technology

Steven McLaughlin
Georgia Institute of Technology

Joao Barros
Faculdade de Engenharia da Universidade do Porto

Byung-Jae Kwak
Electronics and Telecommunications Research Institute

This chapter discusses one approach to the design of error-correction codes for the Gaussian wiretap channel. More specifically, we will consider one type of powerful error-correction codes called low-density parity-check (LDPC) codes and show how they should be designed to provide a high level of data security at the physical layer, which is left unsecured in conventional communication systems. After a short review of basic principles in physical layer security we proceed to show that well-performing codes for physical layer security can be designed by a technique called puncturing.

3.1 Channel Model and Basic Notions

Consider the model depicted in Figure 3.1. There are three persons involved: Alice tries to send a message M to Bob, while Eve is eavesdropping on their communication. In order to secure the message from Eve, Alice encodes M into the ciphertext X using the secret key K. Bob has access to the secret key K and uses it to decipher X. Eve, on the other hand, is assumed to have full knowledge about the encoding and the decoding process, but she does not have access to the secret key K and thus cannot obtain M directly. Shannon proposed [19] that the secrecy of ciphertext X be evaluated in information-theoretic terms. More specifically, he proposed that a ciphertext X be considered perfectly secure if the

33

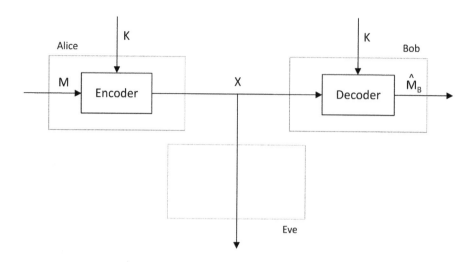

Figure 3.1: The classic Shannon model for secure communication.

mutual information $I(M; X)$ between M and X equals zero. Then, we have

$$H(M|X) = H(M), \tag{3.1}$$

where $H(M|X)$ is the entropy of M given X, also referred to as equivocation, and $H(M)$ is the entropy of M. Equation (3.1) means that Eve does not obtain any information about M by observing ciphertext X.

Shannon proved that communication in perfect secrecy in this setting is possible only if $H(K) \geq H(M)$ [19]. An example of a perfectly secure scheme is the one-time pad, for which Eve who does not have access to the secret key is provably unable to extract any information about the message. Unfortunately, the one-time pad scheme only translates the problem of sharing a message to sharing a secret key. To circumvent this difficulty, a variety of cryptographic algorithms were invented that employ shorter secret keys, but rely on unproved mathematical assumptions and limited computational resources at Eve for secrecy.

Observe that Shannon's model in Figure 3.1 assumes that Bob's and Eve's observations of the transmitted ciphertexts are identical. In many cases that assumption is not realistic due to the stochastic nature of many communication channels. A few decades after Shannon's work it was shown in [3,14,22] that information theoretically secure communication is possible exclusively by means of coding at the physical layer if Eve has a worse channel than Bob. These works assumed a slightly weaker measure of secrecy from Shannon's: instead of the absolute value of mutual information, they require that the mutual information rate

$$\frac{1}{n} I(M; X) \tag{3.2}$$

goes to zero, as n, the number of bits in X, goes to infinity.

Consider the wiretap channel model in Figure 3.2. Alice wants to transmit an s-bit message M^s to Bob. She uses an error-correction code to encode M^s to an n-bit codeword X^n and transmits it over Channel 1 to Bob. Eve listens to the transmission over Channel 2, which is noisier than Channel 1, and tries to reconstruct the message M^s. The error-correction code should be designed such that Bob can reconstruct M^s reliably, while Eve is unable to extract any information about M^s. A more exact definition of these two conditions will be specified later.

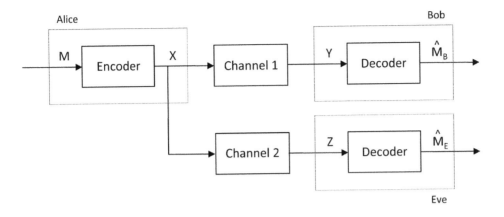

Figure 3.2: The wiretap channel.

Designing codes for this channel has proven to be a difficult problem. One of the major obstacles is that it is difficult to evaluate the abilities of an eavesdropper with boundless power by means of simulation. As a result, the proposed coding schemes for the wiretap channel have been restricted to codes that possess sufficient structure that their secrecy properties can be proven mathematically. One such type of code that is amenable to analysis is called the block code, and the specific type of block code that will be of most interest to us in this chapter is called the low-density parity-check (LDPC) code [4]. A common approach to the wiretap channel problem has been to use nested codes, where the codewords of an error-correction code are partitioned into multiple bins and each bin is assigned an index. Suppose there is one such bin for each message M^s. Then Alice could encode M^s by choosing a random codeword from the bin corresponding to M^s. The nested code must be designed such that Bob can recover the transmitted codeword and once the transmitted codeword is known, Bob can look up the bin index that the codeword belongs to to recover the message. On the other hand the nested code must be such that given Eve's observation, which is noisier than Bob's, the transmitted codeword could belong to any bin with equal probability. The first LDPC-based coding scheme for the wiretap channel was presented in [21], where the authors show how LDPC codes can be used effectively when Channel 1 and Channel 2 are binary erasure channels (BECs) and when Channel 1 is noiseless and Channel 2 is a binary symmetric channel (BSC). An interested reader can find more details and references about these ideas in [2, Chapter 6]. In this chapter we will focus on the Gaussian wiretap channel [11], where Channel 1 and Channel 2 are additive white Gaussian noise (AWGN) channels.

Equivocation and equivocation rate at Eve are established metrics for information-theoretic security [2], but they are difficult to analyze and measure on noisy coded sequences, especially at finite block lengths. That may be one of the main reasons that no practical code constructions at finite block lengths for secure communication exist at this point. The existing code constructions [12,21] based on the insight provided from information-theoretic proofs do not directly apply to continuous channels, like the Gaussian wiretap channel. To get around this problem, the bit-error-rate (BER) over message bits, which is much easier to analyze and measure, is used as a measure for security. For example, if Eve observes data through a channel with BER close to 0.5 (errors are independent and identically distributed [i.i.d.]), then she would not be able to extract much information about the message. It should be noted at the outset that BER is a different metric than the equivocation; therefore, this work does not address information-theoretic security, but rather physical layer security.

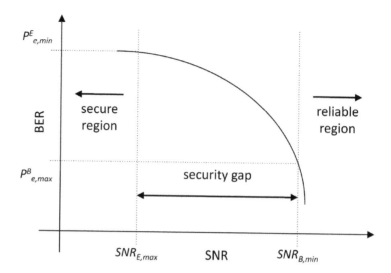

Figure 3.3: The security gap. A typical BER vs. SNR performance curve of an error-correction code is shown. $\mathrm{SNR}_{B,\min}$ is the threshold for reliability (between Alice and Bob) and $\mathrm{SNR}_{E,\max}$ is the point below which Eve has a very high error rate, typically close to 0.5. The security gap (in dB) is the SNR difference between $\mathrm{SNR}_{B,\min} - \mathrm{SNR}_{E,\max}$ that must be maintained between Bob and Eve in order to achieve both reliability and security constraints in conditions (a) and (b).

Nevertheless, it is argued that a high BER at Eve is useful and can, possibly in conjunction with standard cryptographic techniques, deliver improved resilience against eavesdropping. Existing works that have considered this approach include [1, 10, 13].

Consider now the Gaussian wiretap channel in Figure 3.2, where Channel 1 and Channel 2 are AWGN channels. Let an average BER over the Bob's estimate $\hat{M}_B^s$ be P_e^B and let an average BER over the Eve's estimate $\hat{M}_E^s$ be P_e^E. It is desired that P_e^B be sufficiently low to ensure reliability and that P_e^E be high. If P_e^E is close to 0.5 and the errors are i.i.d., then Eve will not be able to extract much information from the received sequence Z^n. Thus, for fixed $P_{e,\max}^B(\approx 0)$ and $P_{e,\min}^E(\approx 0.5)$, it must hold that

(a) $P_e^B \le P_{e,\max}^B$ (reliability),

(b) $P_e^E \ge P_{e,\min}^E$ (security).

Let $\mathrm{SNR}_{B,\min}$, or the reliability threshold, be the lowest signal-to-noise ratio for which (a) holds. This parameter has been subject to considerable scrutiny, because it constitutes one of the main performance metrics associated with error-correction codes. In this chapter we will be interested in another threshold, the security threshold $\mathrm{SNR}_{E,\max}$, which we define as the highest SNR for which (b) holds. In other words, the security threshold tells us when an error-correction code fails miserably. Rather than investigating the absolute values of these two thresholds, we will focus on their ratio $\mathrm{SNR}_{B,\min}/\mathrm{SNR}_{E,\max}$, which we call the security gap and which can alternatively be expressed in dB by taking the logarithm of the ratio. Thus, the size of the security gap in dB (see Figure 3.3) tells us the minimum required difference between Bob and Eve's SNRs for which secure communication in our context is possible. Note that conventional error-correction codes require large (> 20 dB) security gaps when $P_{e,\min}^E > 0.4$. Our objective is to design a coding scheme that exhibits a small security gap.

The main idea is to hide data from Eve by means of puncturing. Instead of transmitting message bits, they are punctured in the encoder and must be deduced from the channel observations of the transmitted bits at the decoder. If the receiver (Eve) has a low SNR, the channel observations are expected to be very noisy; therefore, the reconstruction of punctured message bits is expected to be hard.

Binary low-density parity-check (LDPC) codes are chosen as the coding scheme for two reasons: (i) their excellent error-correction performance and (ii) availability of powerful tools for asymptotical analysis of bit-wise MAP decoders both below and above the reliability threshold. Bob and Eve are assumed to use the belief propagation decoder, which is asymptotically equal to the bit-wise MAP decoder and hence very powerful. It will be shown that transmitting messages over punctured bits can significantly reduce security gaps and can thus be efficiently used for increased security of data. Security gaps as low as a few dB are sufficient to force Eve to operate at BER above 0.49. The suggested coding scheme can be employed as a standalone solution or in conjunction with existing cryptographic schemes that operate on higher layers of the protocol stack.

Choosing a mother code is an important part of the design process. It will be seen that for some choices the excellent security gap performance comes with a significant increase in the reliability threshold $\text{SNR}_{B,\min}$ (compared to an unpunctured code), which in effect results in higher power consumption. Even though the main focus of this chapter is to design codes with small security gaps, we will show measures that can be taken to keep the increase in the reliability threshold low.

LDPC codes were introduced in [4] and shown to reach or perform close to capacity over many channels [18]. An LDPC code can be specified by means of a bipartite graph, composed of variable nodes representing codeword bits and check nodes representing the constraints imposed on the codeword bits. An important parameter that describes a bipartite graph of an LDPC code is the degree distribution, which is given by two polynomials $\lambda(x) = \sum_{i=2}^{d_v} \lambda_i x^{i-1}$ and $\rho(x) = \sum_{i=2}^{d_c} \rho_i x^{i-1}$. The values d_v and d_c represent the maximum variable and check node degrees, while λ_i and ρ_i denote the fractions of edges connected to variable and check nodes of degree i, respectively. From the node perspective, the fraction of variable nodes of degree i is denoted by Λ_i, where $\Lambda_i = (\lambda_i/i)/(\sum_{i=2}^{d_v} \lambda_i/i)$.

If an LDPC code is punctured, some of its variable nodes are not transmitted. One way of describing how a binary LDPC is punctured is by means of a puncturing distribution $\pi(x) = \sum_{i=2}^{d_v} \pi_i x^{i-1}$, where π_i denotes the fraction of variable nodes of degree i that are punctured [5]. The fraction of all punctured bits is denoted by $p^{(0)}$, so that we have $p^{(0)} = \sum_{i=2}^{d_v} \Lambda_i \pi_i$.

Let s be the number of message bits, let k be the dimension of an LDPC code, and let n be the number of bits transmitted over the channel. Define the secrecy rate as $R_s = s/n$ and the design rate as $R_d = k/n$. Usually, in error-correction coding, the number of message bits s is equal to the dimension k of an error-correction code and thus $R_s = R_d$. However, in this chapter all messages are transmitted exclusively over the punctured bits; therefore, it may occur that $s < k$ and, in effect, $R_s < R_d$. In such cases, the unpunctured independent bit locations of a codeword are set randomly by dummy bits as depicted in Figure 3.4.

In the following we focus on the design of secure LDPC codes in the asymptotic case and show how puncturing distributions can be optimized to significantly reduce security gaps. Some results for secure LDPC codes at finite block lengths are reported as well.

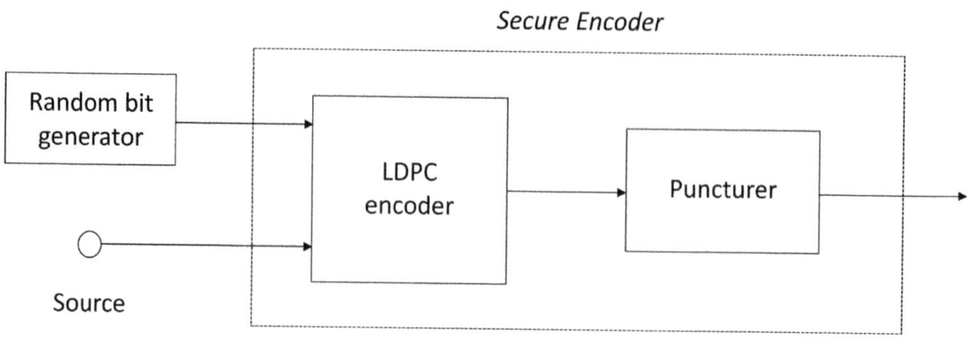

Secure Encoder

Figure 3.4: Block diagram of the proposed encoder.

3.2   Coding for Security

The use of puncturing for improved data security is investigated in this section. The main objectives are to provide a better sense of the level of security that the proposed coding scheme delivers by analyzing the BER over (punctured) message bits and to show how codes with small security gaps can be designed. It is assumed that the decoding algorithm is belief propagation and the analysis is asymptotic, where belief propagation decoding is equivalent to bit-wise MAP decoding [18].

3.2.1   Asymptotic Analysis

The analysis can be performed with density evolution (DE) [16,17], which is known for its accurate asymptotic analysis of the belief propagation decoder. It enables accurate evaluation of the bit-error probability both above and below the reliability threshold. Density evolution from [16,17] can be extended to account for puncturing [5] and further to calculate bit-error probability over punctured bits [10]. One may be tempted to perform the analysis using the Gaussian approximation (GA) [9], a computationally less demanding alternative to DE; however, while GA was shown to perform well around and above the reliability threshold, approximation errors can be considerable when the decoder is operating below the reliability threshold. For simplicity, we refer the interested reader to the above-cited references for technical details and focus on the presentation and discussion of results.

Consider a mother code of rate 1/2 with the degree distribution pair:

$$\lambda_1(x) = 0.25105x + 0.30938x^2 + 0.00104x^3 + 0.43853x^9, \tag{3.3}$$
$$\rho_1(x) = 0.63676x^6 + 0.36324x^7 \tag{3.4}$$

that is punctured randomly according to

$$\pi(x) = 0.4x + 0.4x^2 + 0.4x^3 + 0.4x^9. \tag{3.5}$$

The overall fraction of punctured bits $p^{(0)}$ is 0.4 and all punctured bits are assumed to carry messages; therefore, $R_s = p^{(0)}/(1 - p^{(0)}) = 2/3$. The unpunctured mother code has $R_d = 1/2$, while the punctured code has $R_d = k/(2k(1 - p^{(0)})) = 5/6$. Since R_s and R_d are not equal for the punctured code, we must randomly set $k - s = n/6$ variable nodes in the encoder.

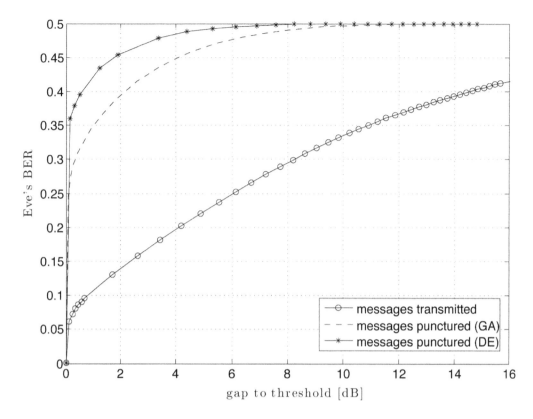

Figure 3.5: Eve's BER performance when operating below the reliability threshold $\text{SNR}_{B,\min}$ and message bits are (i) transmitted and (ii) punctured.

For comparison, consider an unpunctured LDPC code with $R_s = R_d = 2/3$ and degree distribution pair[1]

$$\lambda_2(x) = 0.17599x + 0.40223x^2 + 0.42178x^9 \tag{3.6}$$
$$\rho_2(x) = 0.61540x^{10} + 0.38460x^{11}. \tag{3.7}$$

Eve's BER over message bits for these two codes are shown in Figure 3.5. Observe Eve's BER as her SNR decreases from the reliability threshold. If the message bits are punctured, Eve's BER increases much faster with the growing gap to the reliability threshold than it does when the messages are transmitted. For instance, for $P^{\text{E}}_{e,\min}$ set to 0.40, 0.45, and 0.49, the security gaps amount to 0.6 dB, 1.8 dB, and 4.1 dB, respectively. In contrast, if the message bits are transmitted over the channel, the security gaps are considerably larger at 14 dB, 20 dB, and 34 dB, respectively. These results manifest the benefits of protecting the message bits by puncturing. They show that small security gaps are sufficient to force Eve to operate at relatively high BERs even if she has the capability of using a bit-wise MAP decoder.

While not apparent from Figure 3.5, it must be noted that the increased security is leveraged at the expense of an increased reliability threshold $\text{SNR}_{B,\min}$. As we mentioned earlier, some of the independent variable nodes need to be set randomly in Bob's encoder when $R_s < R_d$; thus, we use an LDPC code with design rate R_d, but messages are transmitted at a lower rate R_s. Consequently, an unpunctured code with all independent bits used

---

[1]Most degree distribution pairs in this Chapter were obtained at http://lthcwww.epfl.ch/research/ldpcopt/

Table 3.1: Puncturing Distributions Optimized for Security; $P^E_{e,\min}$ was set to 0.49

$p^{(0)}$	0.10	0.25	0.33	0.40
Secrecy rate R_s	0.11	0.33	0.50	0.67
π_2	0.1073	0.3105	0.4930	0.5519
π_3	0.1310	0.0010	0.0004	0.1378
π_4	0.8703	0.0121	0.1170	0.4765
π_{10}	0.0015	0.6639	0.6400	0.5814
Security gap [dB]	4.346	4.086	4.034	4.386

for messages is expected to have a lower reliability threshold. In our example, the reliability threshold $\mathrm{SNR}_{B,\min}$ for the unpunctured code is -0.48 dB, whereas for the punctured code it is 2.28 dB; an increase of 2.76 dB. In the following, we refer to the difference in $\mathrm{SNR}_{B,\min}$ between a punctured code and an unpunctured code with the same R_s as SNR loss.

3.2.2 Optimized Puncturing Distributions

A natural question at this point is whether lower security gaps are achievable by optimizing the puncturing distribution for security instead of using a random one. To get an answer, a mother code with the degree distribution pair in Equations (3.3) and (3.4) is punctured in two different manners: (i) randomly, and (ii) according to a puncturing distribution optimized to minimize the security gap and obtained with differential evolution [20]. The optimized puncturing distributions are given in Table 3.1.

The performance comparison between random and optimized puncturing is shown in Figure 3.6, where four different puncturing fractions are considered: 0.10, 0.25, 0.33, and 0.40, which correspond to secrecy rates 0.11, 0.33, 0.50, and 0.67, respectively. SNR loss at each considered rate is depicted as well.

The benefit of optimized puncturing distributions for security is most pronounced at high secrecy rates. The gains over random puncturing of up to 0.4 dB were achieved, a considerable improvement at asymptotic block lengths. The security gap can be reduced at the expense of lower secrecy rate; however, reducing the secrecy rate below 0.43 in this case may not be reasonable due to negative effects on the security gap and SNR loss.

3.2.3 Reducing SNR Loss

The SNR loss translates into higher power consumption at the transmitter, and here we describe how the SNR loss can be reduced for systems subject to stringent power consumption constraints.

Two main factors that cause the SNR loss are puncturing loss and rate loss. Puncturing loss occurs due to inferior performance of punctured codes as compared to unpunctured codes, as observed in [5, 8, 15], but can generally be kept relatively low (< 1 dB). On the other hand, rate loss occurs when $R_s < R_d$, that is when the number of message bits transmitted per codeword is smaller than the dimension of a code.

While empirical evidence in [5, 15] suggests that puncturing loss cannot be prevented, rate loss can be eliminated completely by codes with $R_s = R_d$. For this purpose we design secure LDPC codes by using mother codes with design rates 0.10, 0.25, 0.33, and 0.40, where all independent variable nodes are punctured in order to achieve equality $R_s = R_d$. Since neither R_s nor R_d should be higher than one, the rate of the mother code must not exceed 0.5.

The degree distribution pairs for these mother codes (listed in the same order as above) are

$$\lambda_3(x) = 0.551251x + 0.203119x^2 + 0.0917565x^4 + 0.00428x^6 + 0.01705x^7 +$$
$$0.09970x^8 + 0.03284x^9, \tag{3.8}$$
$$\rho_3(x) = x^2 \tag{3.9}$$

$$\lambda_4(x) = 0.38961x + 0.199689x^2 + 0.110605x^3 + 0.00971174x^4 + 0.290384x^9, \tag{3.10}$$
$$\rho_4(x) = 0.8x^3 + 0.2x^4 \tag{3.11}$$

$$\lambda_5(x) = 0.334539x + 0.242082x^2 + 0.054702x^3 + 0.0000052x^4 + 0.368671x^9, \tag{3.12}$$
$$\rho_5(x) = x^4 \tag{3.13}$$

$$\lambda_6(x) = 0.29445x + 0.257133x^2 + 0.448417x^9, \tag{3.14}$$
$$\rho_6(x) = x^5 \tag{3.15}$$

We optimize puncturing distributions to minimize the security gap, while making sure that R_s equals R_d. The puncturing fractions $p^{(0)}$ for the above listed degree distribution pairs are fixed at 0.10, 0.25, 0.33, and 0.40, respectively, and the corresponding secrecy rates R_s are 0.11, 0.33, 0.50, and 0.67. The new optimized puncturing distributions are given in Table 3.2.

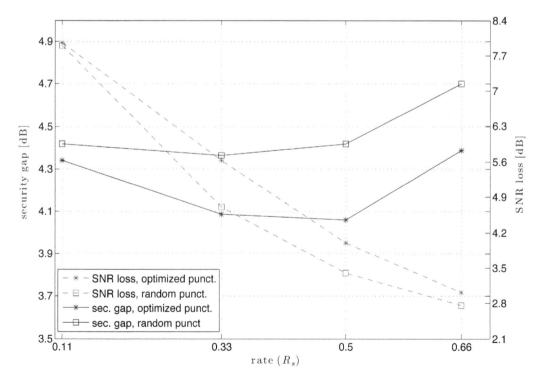

Figure 3.6: The performance comparison between random and optimized puncturing. The probability $P^E_{e,\min} = 0.49$.

Table 3.2: Optimized Puncturing Distributions for $R_s = R_d$ Codes and $P^E_{e,\min} = 0.49$

$p^{(0)}$	0.10	0.25	0.33	0.40
Secrecy rate R_s	0.11	0.33	0.50	0.67
π_2	0.1370	0.2619	0.3589	0.5105
π_3	0.0015	0.0014	0.0063	0.0013
π_4	—	0.0002	0.0818	—
π_5	0.0003	0.0424	0.3594	—
π_7	0.0093	—	—	—
π_8	0.0021	—	—	—
π_9	0.0003	—	—	—
π_{10}	0.0013	0.9916	0.9993	0.7992
Security gap [dB]	9.992	6.644	4.906	4.2193

The new codes are compared in Figure 3.7 with the punctured codes from Section 3.2.2, which were derived from a mother code with rate 0.5. Note that for those codes $R_s < R_d$ and rate loss is nonzero. It is seen that the SNR loss is significantly reduced when the code does not suffer rate loss. At most rates the SNR loss is kept below 1 dB, while with $R_s < R_d$ codes it ranged from 4 dB to 8 dB. For codes with no rate loss, the SNR loss increases with the increasing secrecy rate. These losses are incurred by puncturing and increase with the increasing fraction of punctured bits $p^{(0)}$, much like it was observed in [5, 15].

The $R_s = R_d$ codes are superior in terms of power consumption, while the $R_s < R_d$ codes have smaller security gaps for most considered rates, as is shown in Figure 3.8, where all security gaps are measured for $P^E_{e,\min} = 0.49$. These results indicate there is a trade-off between the SNR loss and the security gap.

3.2.4 Finite Block Lengths

The performance of codes from Figure 3.5 is evaluated for random puncturing at finite block lengths and presented in Figure 3.9, where the number of message bits is 1576, the number of transmitted bits in each block is 2364, and $P^B_{e,\max}$ is set to 10^{-5}. With random puncturing, security gaps as low as a few dB are attainable for $P^E_{e,\min} = 0.4$. Since at finite block lengths Tanner graphs of LDPC codes usually have cycles, belief propagation decoding does not yield exact bit-wise MAP probabilities of codeword bits. Nevertheless, the belief propagation was shown to exhibit excellent performance.

3.3 System Aspects

The proposed coding scheme operates on the physical layer of the protocol stack and can be viewed as a first step toward a somewhat unconventional bottom-up architecture for secure communications. In traditional cryptography the physical layer merely provides the higher layers with a virtually error-free channel abstraction, and it is argued that stronger levels of secrecy are achievable by redesigning the channel coding modules according to the aforementioned security metrics. Intuitively it is to be expected that a noisy ciphertext is more difficult to break than its error-free counterpart. At BER close to 0.5 there is little correlation left between the signals observed by the eavesdropper and the original message. To break the cipher she would have to guess both the key and the random error sequence introduced by the channel, which leads to a significant increase in the search space while performing cryptanalysis.

Ultimately, a wireless link should be at least as difficult to eavesdrop as an Ethernet cable, so that the attacker would have to gain physical access to the channel at very close proximity to be able to acquire information-bearing signals. Even if this threshold is exceeded, the attacker would still have to break hard cryptographic primitives. In other words, the proposed coding scheme does not necessarily replace cryptography, yet it adds one more layer of protection that is targeted at the lower and possibly most vulnerable stage of wireless devices.

While wiretap codes, whose practical construction is still elusive for most cases of interest, would use part of the rate to confuse the eavesdropper and to achieve information-theoretic security (at least asymptotically), the proposed LDPC codes can be readily implemented to induce higher BER at the eavesdropper with a controlled reduction of the rate of communication.

One could argue that cryptographic primitives such as the Advanced Encryption Standard (AES) are designed for the worst case in which the eavesdropper acquires an error-free ciphertext. In applications, such as RFID systems and wireless sensor networks, where strong ciphers like AES are too costly computationally, the proposed codes for security can be combined with lightweight ciphers, while still ensuring sufficient levels of confidentiality. Thus, joint design of channel codes and cryptographic primitives emerges as a worthwhile line of research [6, 7].

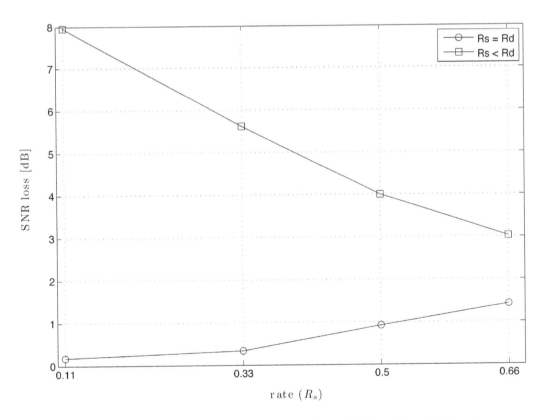

Figure 3.7: SNR loss comparison between codes with $R_s = R_d$ and $R_s < R_d$.

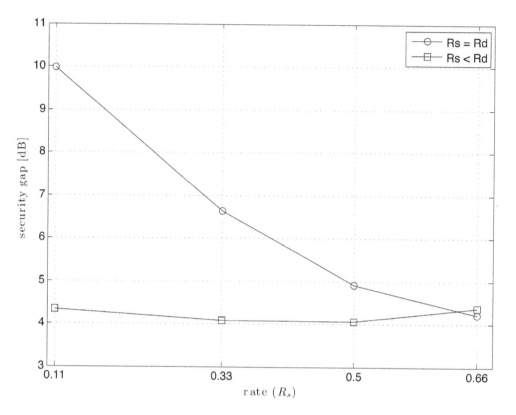

Figure 3.8: Security gap comparison between $R_s = R_d$ and $R_s < R_d$ codes.

3.4   Concluding Remarks

In this chapter an alternate approach to the design of error-correction codes for the Gaussian wiretap channel is explored. Instead of equivocation, which is believed to be challenging to analyze and measure, we choose BER over message bits at the eavesdropper as the security metric given that a bit-wise MAP decoder is available. We define the security gap as a measure of separation between the reliability and security thresholds of an error-correction code and propose codes that minimize the security gap. These codes exhibit a very sharp increase in BER toward 0.5 as the signal deteriorates, which prevents eavesdroppers from extracting the message even if the wiretapped signal is only slightly weaker than that of the intended receiver.

In proposed punctured LDPC codes all message bits are communicated exclusively over punctured bits. The asymptotic analysis shows this to be effective as it yields security gaps as small as a few dB, which is a significant improvement over conventional error-correction codes or uncoded transmission. It is worth noting that puncturing the messages creates a nonsystematic code. We conjecture that most nonsystematic error-correction codes would exhibit good security gap performance, but asymptotic analysis of bit-wise MAP decoders is hard for many types of error-correction codes. The benefit of the proposed LDPC coding scheme is in that it can be effectively analyzed using existing tools and therefore reliable assessment can be made about the eavesdropper's performance above and below the reliability threshold.

The proposed codes can be effectively applied to practical, finite-block length systems.

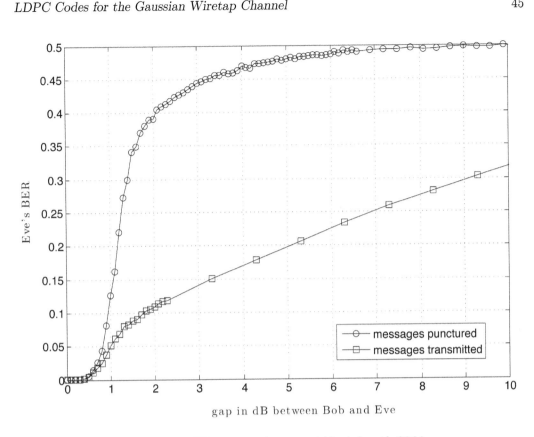

Figure 3.9: BER vs. security gap at block length 2364.

While in some cases such codes can be used as a stand-alone security solution, it is perhaps more prudent to view them as an addition to the existing layered approach, which adds security to the traditionally unsecured physical layer.

References

[1] J-C Belfiore and P Sole. Unimodular lattices for the gaussian wiretap channel. In *Proc. Information Theory Workshop (ITW)*, 2010.

[2] M Bloch and J Barros. *Physical Layer Security: From Information Theory to Security Engineering*. Cambridge University Press, 2011.

[3] I Csiszar and J Korner. Broadcast channels with confidential messages. *IEEE Transactions on Information Theory*, 24(3):339–348, May 1978.

[4] R G Gallager. *Low-density parity-check codes*. MIT Press, 1963.

[5] J Ha, J Kim, and S McLaughlin. Rate-compatible puncturing of low-density parity-check codes. *IEEE Transactions on Information Theory*, 50(11):2824–2836, Nov 2004.

[6] W K Harrison and S W McLaughlin. Physical-layer security: Combining error control coding and cryptography. In *Proc. IEEE International Conference on Communications*, Dresden, Germany, 2009.

[7] W K Harrison and S W McLaughlin. Tandem coding and cryptography on wiretap channels: EXIT chart analysis. In *Proc. International Symposium on Information Theory (ISIT)*, Seoul, South Korea, 2009.

[8] D Klinc, J Ha, and S W McLaughlin. Optimized puncturing and shortening distributions for nonbinary LDPC codes over the binary erasure channel. In *Proc. Allerton Conference on Communication, Control, Computers*, Monticello, IL, Oct 2008.

[9] D Klinc, J Ha, S W McLaughlin, Joao Barros, and Byung-Jae Kwak. LDPC codes for physical layer security. In *Proc. Global Telecommunications Conference (GLOBECOM)*, Honolulu, HI, Dec 2009.

[10] D Klinc, J Ha, S W McLaughlin, Joao Barros, and Byung-Jae Kwak. LDPC codes for the gaussian wiretap channel. *IEEE Transactions on Information Forensics and Security*, 6(3):532–540, Sep 2011.

[11] S K Leung-Yan-Cheong and M E Hellman. The Gaussian wire-tap channel. *IEEE Transactions on Information Theory*, 24(4):451–456, July 1978.

[12] H Mahdavifar and A Vardy. Achieving the secrecy capacity of wiretap channels using polar codes. In *Proc. International Symposium on Information Theory (ISIT)*, Austin, TX, 2010.

[13] F Oggier and J-C Belfiore. Secrecy gain: A wiretap lattice code design. In *Proc. International Symposium on Information Theory and its Applications (ISITA)*, 2010.

[14] L H Ozarow and A D Wyner. Wire-tap channel II. *AT&T Bell Laboratories Technical Journal*, 63(10):2135–2157, Dec 1984.

[15] H Pishro-Nik and F Fekri. Results on punctured low-density parity-check codes and improved iterative decoding techniques. *IEEE Transactions on Information Theory*, 53(2):599–614, Feb 2007.

[16] T Richardson and R Urbanke. The capacity of low-density parity-check codes under message-passing decoding. *IEEE Transactions on Information Theory*, 47(2):599–618, Feb 2001.

[17] T Richardson and R Urbanke. Design of capacity-approaching irregular low-density parity-check codes. *IEEE Transactions on Information Theory*, 47(2):619–637, Feb 2001.

[18] T Richardson and R Urbanke. *Modern Coding Theory*. Cambridge University Press, 2008.

[19] C Shannon. Communication theory of secrecy systems. *Bell Systems Technical Journal*, 29:656–715, Jan 1949.

[20] R Storn and K Price. Differential evolution—A simple and efficient heuristic for global optimization over continuous spaces. *Journal of Global Optimization*, 11:341–359, Dec 1997.

[21] A Thangaraj, S Dihidar, A R Calderbank, S W McLaughlin, and J M Merolla. Applications of ldpc codes to the wiretap channels. *IEEE Transactions on Information Theory*, 53(8):2933–2945, Aug 2007.

[22] A D Wyner. The wire-tap channel. *The Bell System Technical Journal*, 54(8):1355–1387, Oct 1975.

Chapter 4

Key Generation from Wireless Channels

Lifeng Lai
Worcester Polytechnic Institute

Yingbin Liang
Syracuse University

H. Vincent Poor
Princeton University

Wenliang Du
Syracuse University

This chapter introduces basic ideas of information-theoretic models for generating secret keys via public discussion, and reviews recent research on applying these ideas to wireless networks. In these applications, wireless channels between transceivers are exploited as random sources for key generation, and the keys generated in such a way can be shown to be provably secure with information-theoretic guarantees. Classic information-theoretic key generation models are first introduced and applications of these fundamental results to key generation over wireless channels are then discussed. Some new directions that generalize and improve the basic scenarios for key generation are described. These new directions include a joint source-channel approach for key generation, a relay-assisted key generation approach, and key generation under active attacks.

4.1 Introduction

In recent years, a physical layer (PHY) approach that exploits reciprocity of wireless channels for generating symmetric secret keys has attracted considerable attention [1–11]. Here, channel reciprocity refers to the case in which the channel response of the forward channel (from the transmitter to the receiver) is the same as the channel response of the backward channel (from the receiver to the transmitter) in time-division duplex (TDD) systems. Such a random channel serves as a source of common randomness from which the parties can generate secret keys. Eavesdroppers in such situations experience physical channels independent of the legitimate users' channel as long as they are a few wavelengths away from these legitimate nodes, which is generally the case in wireless networks [12]. Thus, the keys are provably secure with an information-theoretic guarantee due to the nature of the key generation scheme, as opposed to crypto keys whose security depends on the assumption of intractability of certain mathematical problems. The key generated using such a PHY approach can be made to be uniformly distributed [13], and can hence be used for encryption using the one-time pad scheme. Furthermore, two major properties of such keys greatly facilitate one-time pad encryption using these keys: (1) these keys are already shared by two legitimate receivers via the generation process, which overcomes the typical challenge of key distribution in using one-time pad encryption; and (2) these keys can be replenished dynamically as wireless channels vary over time, and the key rate can be improved via certain schemes such as relay-assisted schemes (see Section 4.5), which greatly improves the transmission rate via one-time pad encryption. Thus, the PHY approach not only produces information-theoretically secure keys, but also facilitates information-theoretically secure encryption. Here, we note that if the source is not uniformly distributed, a one-time pad may no longer be optimal [14].

The PHY approach for key generation is developed based on the information-theoretic source model studied in [13] and [15], in which two nodes receive correlated random source sequences and wish to distill a secret key, and on its generalization to a network setting studied in [16] and [17]. Such a model naturally captures scenarios in wireless networks, in which two remote nodes may obtain correlated random sequences via estimation of the random wireless channel that they share. Hence, information-theoretic approaches for key generation for source models can be applied for developing key generation schemes over wireless networks as studied in [1, 2, 4–6] and [7]. Sections 4.2 and 4.3 below introduce the basic information-theoretic source models and key generation approaches, and their applications to wireless networks.

However, key generation approaches developed based on the basic model have limitations for many wireless scenarios. Since the random sources for key generation arise from wireless channel estimation, each coherence time interval (during which the channel is constant) provides essentially only one random symbol. Hence when wireless channels vary slowly, the generated key rate can be very small. In fact, it is shown in [1] that the key rate goes to zero when the coherence time of the channel becomes large. In these cases, secret keys cannot be updated frequently. Two approaches have recently been proposed to address this issue in [3] and [18].

The first approach is based on a joint source-channel scheme for key generation [3]. In addition to exploiting the wireless channel as a random source (source-type), such an approach also uses opportunities for secure transmission of keys using wireless channels for establishing a common secret key between nodes (channel-type). Such secure channel transmission can happen when the channel gain between legitimate users is better than that of the channel between the legitimate transmitter and eavesdroppers. Schemes achieving secure communication are based on the wiretap channel model [19]. It is shown in [3] that the channel-type approach is asymptotically optimal as the coherence time of the channel becomes long, whereas the source-type approach is asymptotically optimal in the high power regime. We will introduce this approach in more detail in Section 4.4.

The second approach to address the lack of sufficiently rich random sources is a relay-assisted key generation scheme proposed in [18]. As discussed above, one of the reasons for the limited key rate is that the wireless channel provides essentially only one random observation for each fading block. The main idea of the relay-assisted key generation algorithms is to exploit the presence of multiple relay nodes in wireless networks and use the random channels associated with these relay nodes as additional random sources for key generation. It is shown in [18] that the relay-assisted key generation approach enjoys several advantages. First, the key rate scales linearly with the number of relay nodes. Such a scalability property is very important for consistently replenishing large-size keys for securing wireless transmissions of large packets. Second, such a scheme can be adapted to enjoy the help from relays while keeping the generated key secure from relays, thus improving the robustness to node compromise. Third, such an approach does not require precise synchronization among nodes, which is often required by cooperative schemes for information transmission [20]. We will introduce this approach in more detail in Section 4.5.

In standard information-theoretic source and channel models for secret key generation [1, 2, 4–6], it is assumed that the attacker is *passive*, i.e., it only overhears (but does not transmit over) the channel and tries to infer information about the generated key. This assumption implies that the messages exchanged between Alice and Bob are authenticated and will not be modified by the attacker. In reality, an *active* attacker might modify the messages exchanged between Alice and Bob. For example, when Alice and Bob try to learn the channel gain, an eavesdropper Eve can send attack signals to make the channel estimation imprecise. Similarly, when Alice and Bob exchange information over the channel, Eve can modify the message exchanged over the wireless channel.

The last section of this chapter introduces a key generation approach [3] for scenarios with an active attacker, whose goal is to minimize the key rate that can be generated using the PHY key agreement protocol. The attacker can design the signal it transmits based on the signal overheard over the channel. It is assumed that the attacker uses an independent and identically distributed (i.i.d.) attack strategy. The optimal attack strategy under given key generation schemes is derived, and the key rate that can be generated from the fading wireless channel in the presence of an attacker that employs the optimal attack strategy is characterized. Such an approach demonstrates that legitimate nodes can establish a key over the wireless fading channels even in the presence of an active attacker under certain circumstances. We note that the active attacker considered in [3] is more benign than those considered in arbitrary varying channels [21, 22].

4.2 Information-theoretic Models for Key Generation

In this section, we review information-theoretic models for key generation via public discussion [13, 15–17]. These studies fall mainly into two categories: those based on source models and those based on channel models. We here mainly focus on the source model setting. These results are extensively used for designing key generation schemes using wireless fading channels. We discuss basic ideas and results for three increasingly sophisticated scenarios: (1) key generation with unlimited public discussion; (2) key generation with limited public discussion; and (3) key generation with side-information at Eve. Interested readers can refer to [13, 15, 16] and [17] for details.

4.2.1 Key Generation via Unlimited Public Discussion

In the basic setting considered in key generation via public discussion [13, 15], the two terminals Alice and Bob observe sequences X^n and Y^n, which are correlated discrete i.i.d. source sequences with generic component pair (X, Y) having joint distribution P_{XY}. That is,

$$P_{X^n Y^n}(x^n, y^n) = \prod_{i=1}^{n} P_{XY}(x_i, y_i). \tag{4.1}$$

Alice and Bob can communicate over a noiseless public channel with unlimited capacity. The eavesdropper Eve has perfect access to the public communication channel. Alice and Bob need to agree on a common random key such that Eve obtains a negligible amount of information about the key.

We denote the message transmitted by Alice to Bob over k uses of the public channel by Φ_i for $i = 1, \ldots, k$, and denote the message transmitted by Bob to Alice over k uses of the public channel by Ψ_i for $i = 1, \ldots, k$. There is no limit on the number of messages (or on the communication rate) that may be transmitted over the public channel. The terminals are allowed to generate two independent random variables U_1 and U_2 for initial randomization. All following communication messages are deterministic functions which may depend on these two random variables. The terminals generate their communication messages causally based on all information available to them up until transmission time. In the end, Alice and Bob determine the random key to be K and L, respectively, based on their received information, where

$$K = K(U_1, X^n, \Psi^k) \quad \text{and} \quad L = L(U_2, Y^n, \Phi^k).$$

These two terminals need to agree on this random key, i.e., K should equal L with high probability. Since Eve has full access to the public communication channel, in order to keep the key secret from Eve, the public communication messages should be independent of the key.

Definition 4.2.1. *A secret key (SK) rate R_s is achievable if for every $\epsilon > 0$ and sufficiently large n, there exists a public communication strategy such that*

$$Pr\{K \neq L\} \;\; < \;\; \epsilon, \tag{4.2}$$

$$\frac{1}{n} I(\Phi^k, \Psi^k; K) \;\; < \;\; \epsilon, \tag{4.3}$$

$$\frac{1}{n} H(K) \;\; > \;\; R_s - \epsilon \quad and \tag{4.4}$$

$$\frac{1}{n} \log |\mathcal{K}| \;\; < \;\; \frac{1}{n} H(K) + \epsilon. \tag{4.5}$$

The largest achievable key rate is the SK *capacity.*

In the above definition, (4.4) measures the rate of the key, and other conditions specify the properties that the generated key should satisfy including that the keys generated by the two terminals should agree with high probability, the generated key is independent of the public communication, and the generated key is approximately uniformly distributed.

Theorem 4.2.1. [13,15] *The secret key capacity with unlimited public discussion is*

$$C_s = I(X;Y). \tag{4.6}$$

Furthermore, the capacity is achieved by only one transmission from Alice.

To generate a uniformly distributed key with the rate in (4.6), one needs to employ Slepian–Wolf coding to send helper information from Alice to Bob through the public channel. More specifically, Alice randomly divides the typical X^n sequences into nonoverlapping bins, with each bin having $2^{nI(X;Y)}$ typical sequences. Hence, each sequence has two indices: the bin number and the index within the bin. Now, after observing the vector X^n, Alice sets the key to be the index of this sequence within her bin. Alice then sends the bin number as the helper data to Bob through the public channel. That is, Alice needs to send $nH(X|Y)$ bits of information through the public channel, where $H(X|Y)$ denotes the conditional entropy of X given Y. After combining the information observed from the public channel with Y^n, it can be shown that Bob can recover the value of X^n with probability arbitrarily close to 1. Then Bob can recover the key. It can also be shown that the bin number and index within each bin are independent of each other. Hence, even though Eve can observe the bin number transmitted over the public channel, she learns no information about the generated key.

4.2.2 Key Generation with Rate Constraint in Public Discussion

In certain scenarios, there might be rate constraints on the capacity of the public channel. The problem of key generation from correlated sources through a public channel with limited capacity has been studied in [16].

Theorem 4.2.2. [16] *If the public channel has a rate constraint R, then the secret key capacity is*

$$C_s = \max_U I(U;Y) \tag{4.7}$$

$$s.t. \quad U \to X \to Y, \tag{4.8}$$

$$and \quad I(U;X) - I(U;Y) \le R, \tag{4.9}$$

where U is an auxiliary random variable subject to the Markov chain relationship given to it in (4.8). Furthermore, this rate can be achieved by sending from Alice only.

To achieve this, for any given U, Alice generates $2^{nI(U;X)}$ typical U sequences. It then divides these typical sequences into bins, each bin containing $2^{nI(U;Y)}$ sequences. Hence, each U^n sequence can be specified by two indices: the bin number (ranging from 1 to $2^{n(I(U;X)-I(U;Y))}$), and the index of the sequence within each bin. Now, after observing X^n, Alice finds a U^n sequence that is jointly typical with X^n. (This step will be successful with very high probability.) Alice sets the key value as the index of the sequence in the bin and sends the bin number to Bob, which requires a rate of $I(U;X) - I(U;Y)$. This rate can be accommodated by the public channel since the capacity of the public channel is larger than this rate requirement. After receiving the bin number, Bob obtains an estimate $\hat{U}^n$ by looking for a unique sequence in the bin specified by the bin number that is jointly typical with his observation Y^n. $\hat{U}^n$ will be equal to U^n with probability 1, and thus Bob can then recover the key value.

4.2.3 Key Generation with Side-information at Eve

In certain applications, Eve might also observe a sequence Z^n that is correlated with (X^n, Y^n). This problem was studied in [13, 15, 16]. In this case, the security condition in (4.3) should be replaced with

$$\frac{1}{n} I(\Phi^k, \Psi^k, Z^n; K) \quad < \quad \epsilon. \tag{4.10}$$

The secret key capacity in the general case has not been characterized. Here, we cite an achievable rate that will be used in the later part of this chapter.

Theorem 4.2.3. [16] *For key generation from a public channel with rate constraint R and side-information Z^n at Eve, the following secret key rate is achievable:*

$$R_s = [I(U;Y) - I(U;Z)]^+ \tag{4.11}$$

$$s.t. \quad U \to X \to Y, \tag{4.12}$$

$$and \quad I(U;X) - I(U;Y) \le R, \tag{4.13}$$

in which $[x]^+ = \max\{x, 0\}$ and U is an auxiliary random variable subject to the Markov chain relationship given to it in (4.12). Furthermore, this rate can be achieved by sending from Alice only.

To achieve this rate, Alice generates $2^{nI(U;X)}$ typical U sequences. She then divides these typical sequences into bins, each bin containing $2^{nI(U;Y)}$ sequences. Hence, each U^n sequence can be specified by two indices (i,j) with i being the bin number (ranging from 1 to $2^{n(I(U;X)-I(U;Y))}$), and j being the index of the sequence within each bin. Now, after observing X^n, Alice finds a U^n sequence that is jointly typical with X^n. (This step will be successful with probability very close to one.) Alice sets the key value as $j \mod 2^{nI(U;Z)}$, and sends the value of i to Bob, which requires a rate of $I(U;X) - I(U;Y)$. After receiving the bin number i, Bob obtains an estimate $\hat{U}^n$ by looking for a unique sequence in the bin specified by the bin number that is jointly typical with his observation Y^n. $\hat{U}^n$ will be equal to U^n with probability 1, and thus Bob can then recover the key value by setting it as $\hat{j} \mod 2^{nI(U;Z)}$.

4.3 Basic Approaches for Key Generation via Wireless Networks

In this section, we introduce recent research on key generation for wireless networks, which applies the information theoretic approaches discussed in Section 4.2 by exploiting channel reciprocity [1,2,4–6]. Most of this work applies the results discussed in Section 4.2.1, namely key generation via public discussion with unlimited capacity.

As before, the two terminals Alice (A) and Bob (B) wish to agree on a key via a wireless fading channel between Alice and Bob in the presence of the eavesdropper Eve (E). Both Alice and Bob can transmit over the wireless channel. They can also send information using the public channel. It is assumed that Eve listens to transmissions of Alice and Bob over wireless channels, and also has full access to the information transmitted over the public channel. It is also assumed that Eve is passive, and does not send signals to interfere with legitimate transmissions. The key generation process is not covert, i.e., Eve knows that there is a key agreement process going on, although she does not learn any information about the generated key. The basic idea is to obtain the correlated observations from the common fading channel gain between them via channel training, then apply the approach for key generation via public discussion.

If Alice transmits a signal X_A over the wireless channel, Bob and Eve receive

$$
\begin{aligned}
Y_B &= h_1 X_A + N_B \\
\text{and } Y_E &= h_{AE} X_A + N_E
\end{aligned}
\tag{4.14}
$$

respectively, in which h_1 and h_{AE} are the respective random (fading) channel gains from Alice to Bob and Eve, and N_B and N_E are zero mean additive Gaussian noises with variances σ^2. Alternatively, if Bob transmits a signal X_B, Alice and Eve receive

$$
\begin{aligned}
Y_A &= h_1^* X_B + N_A \\
\text{and } Y_E &= h_{BE} X_B + N_E
\end{aligned}
\tag{4.15}
$$

respectively, in which h_1^* and h_{BE} are the respective channel gains from Bob to Alice and Eve, and N_A and N_E are zero mean additive Gaussian noises with variances σ^2. It is reasonable to assume that all additive Gaussian noises are independent of each other. It is further assumed that the channel is reciprocal, i.e., $h_1 = h_1^*$. The protocols are still applicable even if this reciprocity does not hold perfectly, as long as the forward channel and backward channel are correlated. Since the signals arriving at different wireless receivers experience different transmission paths, and hence different random phases, the channel gain h_1 is independent of h_{AE} and h_{BE}. We consider an ergodic block fading model, in which

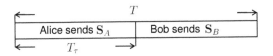

Figure 4.1: Training-based scheme.

the channel gains are fixed for a block of T symbols and change randomly to other values at the beginning of the next block. For simplicity, in this paper, we assume that h_1 is a Gaussian random variable with zero mean and variance σ_1^2, i.e., $h_1 \sim \mathcal{N}(0, \sigma_1^2)$. The results can be easily extended to other fading models. *It is assumed that none of the terminals knows the value of the fading gains initially.*

There are two steps in key generation via wireless fading channel reciprocity: (1) channel estimation, in which Alice and Bob estimate the common channel gain h_1 via training symbols; and (2) key agreement, in which Alice and Bob agree on a common random key based on their correlated but imperfect estimates of the channel gain by employing Slepian–Wolf source coding.

Let T_τ denote the amount of time spent by Alice on training, and let $T - T_\tau$ denote the amount of time used by Bob on training, as depicted in Figure 4.1. Suppose Alice sends a known sequence $\mathbf{S}_A$ with T_τ components; then Bob receives

$$\mathbf{Y}_B = h_1 \mathbf{S}_A + \mathbf{N}_B. \tag{4.16}$$

After that, Bob sends a known sequence $\mathbf{S}_B$ with $T - T_\tau$ components; then Alice receives

$$\mathbf{Y}_A = h_1 \mathbf{S}_B + \mathbf{N}_A. \tag{4.17}$$

The observations at Eve during this process are independent of the observations at both Alice and Bob due to her independent channel gains from the transmitters.

To generate a key from these observations, Alice first generates an estimate of the channel gain h_1

$$\tilde{h}_{1,A} = \frac{\mathbf{S}_B^T}{||\mathbf{S}_B||^2} \mathbf{Y}_A = h_1 + \frac{\mathbf{S}_B^T}{||\mathbf{S}_B||^2} \mathbf{N}_A, \tag{4.18}$$

in which $|| \cdot ||$ denotes the norm of its argument. Similarly, Bob computes an estimate $\tilde{h}_{1,B}$ of the channel gain h_1

$$\tilde{h}_{1,B} = \frac{\mathbf{S}_A^T}{||\mathbf{S}_A||^2} \mathbf{Y}_B = h_1 + \frac{\mathbf{S}_A^T}{||\mathbf{S}_A||^2} \mathbf{N}_B. \tag{4.19}$$

Note that $\tilde{h}_{1,A}$ is a zero mean Gaussian random variable with variance $\sigma_1^2 + \frac{\sigma^2}{||S_B||^2}$, and similarly $\tilde{h}_{1,B}$ is a zero mean Gaussian random variable with variance $\sigma_1^2 + \frac{\sigma^2}{||S_A||^2}$. Assuming that Alice and Bob transmit with power P during the training period, we have $||\mathbf{S}_B||^2 = (T - T_\tau)P$ and $||\mathbf{S}_A||^2 = T_\tau P$.

From $(\tilde{h}_{1,A}, \tilde{h}_{1,B})$, applying the result in Theorem 4.2.1, one can generate a key with rate

$$R_s = \frac{1}{T} I(\tilde{h}_{1,A}; \tilde{h}_{1,B}) = \frac{1}{2T} \log\left(\frac{(\sigma^2 + \sigma_1^2 P T_\tau)(\sigma^2 + \sigma_1^2 (T - T_\tau)P)}{\sigma^4 + \sigma^2 \sigma_1^2 PT}\right), \tag{4.20}$$

in which the normalization factor $1/T$ is due to the fact that the channel gain is fixed for T symbols, and hence the communicating parties can observe *only one* value of the channel

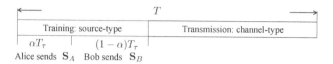

Figure 4.2: Training-based scheme.

statistics for every T symbols. It is obvious that the value of T_τ that maximizes (4.20) is $T/2$, with which (4.20) is simplified to

$$R_s = \frac{1}{T}I(\tilde{h}_{1,A}; \tilde{h}_{1,B}) = \frac{1}{2T}\log\left(1 + \frac{\sigma_1^4 P^2 T^2}{4(\sigma^4 + \sigma^2\sigma_1^2 PT)}\right). \qquad (4.21)$$

It is easy to see that the key rate depends on the power P and the coherence time T. As the available power P increases, the key rate increases at an order of $\frac{1}{2T}\log P$. However, as the coherence time T increases (meaning that the channel changes slowly), the key rate decreases at an order of $\frac{1}{2T}\log T$, which approaches zero.

4.4 A Joint Source-Channel Key Agreement Protocol

In this section, we present a joint source-channel key agreement protocol developed in [3] that enables us to achieve a better key rate compared to approaches based on either the source or channel model. The key idea is to use the wireless channel in two ways: (1) to obtain correlated channel estimates for key generation via public discussion (source-type), as in the existing approach; and (2) to send another independently generated key through the wireless channel (channel-type). The second way of using the wireless channel allows us to exploit the benefit brought by time instants when the channel gain between legitimate users is large.

4.4.1 Key Agreement with a Public Channel

We first consider a scenario in which, in addition to the wireless channel, there is a public channel with infinite capacity.

As shown in Fig 4.2 the developed scheme has two stages: (1) channel training, which is the same as the existing protocol; and (2) transmission using the channel, in which Alice also sends another randomly generated key using the noisy wireless channel. Let T_τ denote the amount of time spent on training, and let $T - T_\tau$ denote the amount of time that is used in the second stage.

The channel training has two purposes: (1) to generate a key from these two correlated observations using the source model through the public channel; and (2) to generate an estimate of the channel gain h_1 in the given coherence block, which will be used for key generation using the channel model.

Key generation from the training phase

We first look at the source-type key generation. This step is the same as the existing approach, in which Alice and Bob send training sequences $\mathbf{S}_A$ and $\mathbf{S}_B$. The relationships between the signals $\mathbf{Y}_B$ and $\mathbf{Y}_A$ at the corresponding receivers and the training sequences are the same as (4.16) and (4.17), except that in this case the length of the training sequence is shorter than the existing approach. At the end of training, Alice and Bob obtain estimates

$\tilde{h}_{1,A}$ and $\tilde{h}_{1,B}$ respectively using (4.18) and (4.19). Note that $\tilde{h}_{1,A}$ is a zero mean Gaussian random variable with variance $\sigma_h^2 + \frac{\sigma^2}{||\mathbf{S}_B||^2}$, and similarly $\tilde{h}_{1,B}$ is a zero mean Gaussian random variable with variance $\sigma_h^2 + \frac{\sigma^2}{||\mathbf{S}_A||^2}$. Assuming that Alice and Bob transmit with power P_τ during the training period, we have $||\mathbf{S}_B||^2 = (1-\alpha)T_\tau P_\tau$ and $||\mathbf{S}_A||^2 = \alpha T_\tau P_\tau$.

Applying the results in Theorem 4.2.1, we know that from $(\tilde{h}_{1,A}, \tilde{h}_{1,B})$, Alice and Bob can generate a key with rate

$$R_s = \frac{1}{T}I(\tilde{h}_{1,A}; \tilde{h}_{1,B}) = \frac{1}{2T}\log\left(\frac{(\sigma^2 + \sigma_h^2\alpha P_\tau T_\tau)(\sigma^2 + \sigma_h^2(1-\alpha)P_\tau T_\tau)}{\sigma^4 + \sigma^2\sigma_h^2 P_\tau T_\tau}\right). \quad (4.22)$$

Key generation after the training phase

After the training period of T_τ symbols, Alice can send another randomly generated key to Bob using the scheme developed for the fading eavesdropper channel [23]. More specifically, Bob obtains a Minimum Mean Square Error (MMSE) estimate $\hat{h}_{1,B}$ of the channel gain h_1 in the given coherence block,

$$\hat{h}_{1,B} = \frac{\sigma_h^2}{\sigma^2 + \alpha P_\tau T_\tau \sigma_h^2}\mathbf{S}_A^T\mathbf{Y}_B, \quad (4.23)$$

and treats this as the true value of the channel gain. We can write

$$h_1 = \hat{h}_{1,B} + \bar{h}_1,$$

in which $\bar{h}_1$ is the estimation error. It is easy to verify that $\bar{h}_1$ is a zero mean Gaussian random variable with variance $\sigma_h^2/(\sigma_h^2\alpha P_\tau T_\tau + \sigma^2)$.

We consider a simple scheme in which Alice does not perform power control or rate control. Clearly, one can improve this rate by allowing Alice to adapt her transmission scheme based on her estimate of the channel. But this simple strategy allows us to decouple the key generation problem into these two stages. If Alice adapts her transmission scheme based on her estimated channel gain, the eavesdropper might be able to learn some information about the channel gain h_1 during the second stage, which complicates the key generation from the source model. Alice sends a key to Bob, using a constant power P_d. Then the following secrecy rate is achievable [23]:

$$R_{ch} = \frac{T - T_\tau}{T}[I(X_A; Y_B|\hat{h}_{1,B}) - I(X_A; Y_E|h_{AE})]^+ \quad (4.24)$$

$$= \frac{T - T_\tau}{2T}\left[\mathbb{E}\left\{\log\left(1 + \frac{\hat{h}_{1,B}^2 P_d}{\sigma^2 + \frac{\sigma_h^2 P_d}{\sigma_h^2\alpha P_\tau T_\tau + \sigma^2}}\right) - \log\left(1 + \frac{h_{AE}^2 P_d}{\sigma^2}\right)\right\}\right]^+. \quad (4.25)$$

Here, the first term is the rate that Bob can decode using a mismatched decoder [24, 25]. The second term is an upper bound on the mutual information that Eve can accumulate. We obtain this upper bound by assuming that Eve has perfect knowledge of h_{AE}. We note here that Alice and Bob do not need to know the instantaneous value of h_{AE}.

In summary, we have the following result.

Theorem 4.4.1. [3] *In a wireless fading channel with a public channel, the following secret key rate is achievable using the training-based scheme:*

$$R_{key} = \max_{\alpha, P_\tau, T_\tau} \{R_s + R_{ch}\} \quad (4.26)$$

$$s.t. \quad T_\tau P_\tau + (T - T_\tau)P_d \leq TP, \quad (4.27)$$

in which R_s and R_{ch} are given by (4.22) and (4.24), respectively.

One can optimize the key rate by choosing appropriate values of α, P_τ, and T_τ. If T_τ is small, one has more time left for transmitting a key using the channel-type approach. But the channel gain estimates at Alice and Bob will be coarse, which will affect both key generation processes using the source-type and channel-type approaches. On the other hand, if T_τ is large, one can generate a larger key rate using the source-type approach, since the estimates of the channel at Alice and Bob are more precise. But, in this case the time left for sending a key from Alice to Bob is reduced. For general values of the available power P and the coherence length T, it is difficult to obtain closed-form expressions for the optimal values of α, P_τ, and T_τ. In the following, we consider two asymptotic regimes to gather insight into the behavior of these quantities.

1) *Long coherence time regime*, in which $T \to \infty$.

We have the following inequalities, which can be verified easily:

$$
\begin{aligned}
R_s &\leq \max_{\alpha, P_\tau, T_\tau} \frac{1}{2T} \log\left(\frac{(\sigma^2 + \sigma_h^2 \alpha P_\tau T_\tau)(\sigma^2 + \sigma_h^2 (1-\alpha) P_\tau T_\tau)}{\sigma^4 + \sigma^2 \sigma_h^2 P_\tau T_\tau} \right) \\
&\leq \frac{1}{2T} \log\left(\frac{(\sigma^2 + \frac{1}{2}\sigma_h^2 PT)^2}{\sigma^4 + \sigma^2 \sigma_h^2 PT} \right).
\end{aligned}
\tag{4.28}
$$

Thus, as $T \to \infty$, $R_s \to 0$. That is, in this regime, the channel-type approach is asymptotically optimal. As a result, to maximize R_{key}, we can choose α, P_τ, and T_τ to maximize R_{ch}. It is easy to see that we should set $\alpha = 1$; that is, only Alice sends a training sequence, since even if Bob sends a training sequence, the key rate that can be generated from the correlated observations will be zero.

2) *High power regime*, in which $P \to \infty$.

It can be shown that R_{ch} is upper bounded by a constant term

$$
R_{ch} \leq \mathbb{E}\{h_1^2\} + C_1 C_2,
\tag{4.29}
$$

in which $C_1 = \sup f(h_1^2)$ and $C_2 = \sup f(h_{AE}^2)$.

Hence, the R_{ch} term is bounded by a constant when P increases. On the other hand, it is easy to see that R_s increases with P. Thus, in the high power regime, the source-type approach is asymptotically optimal. As a result, in order to maximize the key rate, we choose the parameters to maximize R_s. Simple calculation shows that the optimal parameter values are $\alpha = 1/2$, $P_\tau = P$, and $T_\tau = T$. As a result,

$$
R_{key} \sim \frac{1}{2T} \log P.
$$

Hence, in the high power regime, if the coherence time is fixed, the secret-key rate increases logarithmically with P.

4.4.2   Key Agreement without a Public Channel

A more realistic scenario is also studied in [3], in which there is no public channel available. Similarly to the development in Section 4.4.1, we consider a training-based scheme, in which both Alice and Bob send training sequences over the wireless channel during the training period. Then, Alice and Bob generate a key from the correlated observations using the source model. Alice also sends another randomly generated key to Bob after the training period using the channel model. Hence, the total key rate that can be generated from the wireless channel is the sum of the two key rates.

If there is no public channel, the key generation problem during the channel-type stage is the same as that of Section 4.4.1, since no public resources were used. On the other hand, due to the absence of the public channel, the key generation process from the correlated observations should be revised. As discussed in Section 4.4.1, to generate a key with a rate of $I(\tilde{h}_{1,A}; \tilde{h}_{1,B})/T$ from the correlated estimates of the channel gain, Alice needs to send $H(\tilde{h}_{1,A}^{\triangle}|\tilde{h}_{1,B}^{\triangle})$ bits of information (more precisely, the bin number of her observations) to Bob, in which $\tilde{h}_{1,A}^{\triangle}$ and $\tilde{h}_{1,B}^{\triangle}$ are quantized channel gains of $h_{1,A}$ and $h_{1,B}$ with quantization interval $\triangle$. To achieve the above mentioned key rate, $\triangle \downarrow 0$, in which case, $H(\tilde{h}_{1,A}^{\triangle}|\tilde{h}_{1,B}^{\triangle})$ is infinite. If there is a public channel with infinite capacity, this is not an issue. If there is no public channel, one has to send the bin number over the wireless channel. Since the wireless channel has limited capacity, the key rate that one can generate from these correlated observations is less than $I(\tilde{h}_{1,A}; \tilde{h}_{1,B})/T$.

Now, if we do not have a public channel at our disposal, we can use the wireless channel after the training stage to send the bin number needed for the key generation from the correlated observations. In Section 4.4.1, we use the wireless channel after the training stage to send another randomly generated key from Alice to Bob using the wiretap channel model. An important observation is that in coding for the wiretap channel, one needs to use randomization. Roughly speaking, the randomization rate is the same as the mutual information between Alice and Eve. In the coding scheme used in Section 4.4.1, this randomization rate does not convey any information, although Bob is able to decode these randomization bits. Hence, the basic idea here is that instead of randomly generating randomization bits, we use the bin number to specify the random bits. In this way, we can use the wireless channel after the training phase to send a new key and the bin number simultaneously.

One can use the results in Theorem 4.2.2 to characterize the key rate one can generate from the correlated observations. In the proposed scheme, we set $U = \tilde{h}_{1,A} + Z$, in which Z is a zero mean Gaussian random variable with variance σ_z^2 and is independent of other random variables considered in this analysis. The variance is chosen to satisfy the condition that the wireless channel is able to support the rate of the helper data necessary for the key generation from the correlated noisy observations. It is easy to check that $U \to \tilde{h}_{1,A} \to \tilde{h}_{1,B}$. In this case, the key rate one can generate from the correlated observations is

$$2TR_s = 2I(U; \tilde{h}_{1,B}) = \log\left(\frac{(\sigma_h^2 + \frac{\sigma^2}{(1-\alpha)P_\tau T_\tau} + \sigma_z^2)(\sigma_h^2 + \frac{\sigma^2}{\alpha P_\tau T_\tau})}{(\sigma_h^2 + \frac{\sigma^2}{(1-\alpha)P_\tau T_\tau} + \sigma_z^2)(\sigma_h^2 + \frac{\sigma^2}{\alpha P_\tau T_\tau}) - \sigma_h^4}\right). \quad (4.30)$$

To achieve this rate, one needs to transmit at rate

$$\frac{1}{T}(I(U; \tilde{h}_{1,A}) - I(U; \tilde{h}_{1,B})) = \frac{1}{2T}\log\left(1 + \frac{\sigma_h^2\sigma^2}{\sigma_z^2(\sigma^2 + \sigma_h^2\alpha P_\tau T_\tau)} + \frac{\sigma^2}{\sigma_z^2(1-\alpha)P_\tau T_\tau}\right) (4.31)$$

over the wireless channel. Hence, the value of σ_z^2 should be chosen carefully.

Theorem 4.4.2. [3] *Using a fading wireless channel without a public channel, a key rate of*

$$R_{key} = \max_{\alpha, P_\tau, T_\tau} \{R_s + R_{ch}\} \quad (4.32)$$

is achievable. Here, we require that

$$P_\tau T_\tau + (T - T_\tau)P_d \leq PT. \quad (4.33)$$

At the same time, R_{ch} and R_s are given in (4.24) *and* (4.30), *respectively, and σ_z^2 should* *be chosen to satisfy the following condition:*

$$\frac{I(U;\tilde{h}_{1,A}) - I(U;\tilde{h}_{1,B})}{T - T_\tau}$$

$$\leq \min\left\{\mathbb{E}\left\{\log\left(1 + \frac{\hat{h}_{1,B}^2 P_d}{\sigma^2 + \frac{\sigma_h^2 P_d}{\sigma_h^2 \alpha P_\tau T_\tau + \sigma^2}}\right)\right\}, \mathbb{E}\left\{\log\left(1 + \frac{h_{AE}^2 P_d}{\sigma^2}\right)\right\}\right\}. \quad (4.34)$$

Similarly to the situation in Section 4.4.1, for general values of the available power P and the coherence length T, it is difficult to obtain closed-form expressions for the optimal values of these parameters. In the following, we again consider two asymptotic regimes to gather insight.

1) *Long coherence time regime*, in which $T \to \infty$.

We first look at the R_s term. For any values of P_τ, T_τ, and α, a simple calculation shows that $\frac{dR_s}{d\sigma_z^2} < 0$. Hence

$$2TR_s \leq \max_{\alpha,T_\tau,P_\tau} \log\left(\frac{(\sigma_h^2 + \frac{\sigma^2}{(1-\alpha)P_\tau T_\tau} + \sigma_z^2)(\sigma_h^2 + \frac{\sigma^2}{\alpha P_\tau T_\tau})}{(\sigma_h^2 + \frac{\sigma^2}{(1-\alpha)P_\tau T_\tau} + \sigma_z^2)(\sigma_h^2 + \frac{\sigma^2}{\alpha P_\tau T_\tau}) - \sigma_h^4}\right)$$

$$\leq \max_{\alpha,T_\tau,P_\tau} \log\left(\frac{(\sigma_h^2 + \frac{\sigma^2}{(1-\alpha)P_\tau T_\tau})(\sigma_h^2 + \frac{\sigma^2}{\alpha P_\tau T_\tau})}{(\sigma_h^2 + \frac{\sigma^2}{(1-\alpha)P_\tau T_\tau})(\sigma_h^2 + \frac{\sigma^2}{\alpha P_\tau T_\tau}) - \sigma_h^4}\right)$$

$$\leq \log\left(\frac{\sigma^2}{\sigma_h^2 PT} + 1 + \frac{1}{4\sigma^2}\sigma_h^2 PT\right). \quad (4.35)$$

Thus, as $T \to \infty$, $R_s \to 0$. As a result, in this regime, the channel-type approach is asymptotically optimal. The R_{ch} term is the same as that of the scenario with a public channel. Hence in the long coherence time regime, the key rate is the same as that of the scenario with a public channel.

2) *High power regime*, in which $P \to \infty$.

We can bound the R_{ch} term in the same manner as that of Section 4.4.1. Hence, in the high power regime, the source-type approach is asymptotically optimal. In the following, we study how R_s scales as P increases. From Section 4.4.1, we know that if there is a public channel with infinite capacity, R_s scales logarithmically with P. Hence, in the absence of the public channel, R_s scales at most logarithmically with P. In the following, we show that R_s indeed scales logarithmically with P. We set $P_\tau = P$, $P_d = P$, $T_\tau = T/2$, and $\alpha = 1/2$. Note that these parameters are not necessarily optimal.

Note that in the high power regime,

$$\mathbb{E}\left\{\log\left(1 + \frac{h_{AE}^2 P_d}{\sigma^2}\right)\right\} \sim \log P, \quad (4.36)$$

$$\mathbb{E}\left\{\log\left(1 + \frac{\hat{h}_{1,B}^2 P_d}{\sigma^2 + \frac{\sigma_h^2 P_d}{\sigma_h^2 \alpha P_\tau T_\tau + \sigma^2}}\right)\right\} \sim \log P. \quad (4.37)$$

Hence, if we choose $\sigma_z^2 = P^{-2}$, (4.34) will be satisfied. Now, we substitute these choices of parameters into (4.30) and obtain

$$R_s = \frac{1}{2T}\log\left(\frac{(\sigma_h^2 + \frac{4\sigma^2}{PT} + P^{-2})(\sigma_h^2 + \frac{4\sigma^2}{PT})}{(\sigma_h^2 + \frac{4\sigma^2}{PT} + P^{-2})(\sigma_h^2 + \frac{4\sigma^2}{PT}) - \sigma_h^4}\right) \sim \frac{1}{2T}\log P. \quad (4.38)$$

Hence, $R_{key} \sim \frac{1}{2T}\log P$ in the high power regime, which is the same as that in the case with a public channel.

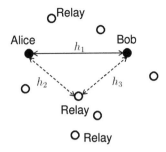

Figure 4.3: A network with multiple terminals.

4.5 Relay-assisted Key Generation with a Public Channel

As discussed in Section 5.1, in the key generation approach based on the source model, the key rate is limited when the coherence time T is large. In this section, we present a relay-assisted scheme for key generation proposed in [18], which improves the key rate by exploiting relay nodes in networks. The main idea is to exploit the presence of multiple relay nodes in wireless networks and use the random channels associated with these relay nodes as additional random sources for key generation. As illustrated in Figure 4.3, besides the direct fading link connecting the legitimate users Alice and Bob, the fading channels between Alice and relays and between Bob and relays (depicted as dotted lines in Figure 4.3) can also serve as additional random sources for key generation.

4.5.1 Relay-assisted Key Generation with One Relay

We first consider a simple network with one relay (see Figure 4.4) to illustrate the basic idea. We then discuss more general networks with multiple relays. The relay is assumed to follow the designed transmission protocols, and not to leak information to Eve (i.e., not to cooperate with Eve). The generated key is required to be secure only from Eve, not necessarily from the relay. We denote the channel gain between Alice and the relay by h_2 and the channel gain between the relay and Bob by h_3. We assume $h_2 \sim \mathcal{N}(0, \sigma_2^2)$ and $h_3 \sim \mathcal{N}(0, \sigma_3^2)$, i.e., they both are Gaussian with zero means and variances σ_2^2 and σ_3^2, respectively, although the schemes developed here are also applicable to models with fading coefficients taking other distributions. We note that throughout this section, we assume for the sake of simplicity that there is a public channel via which legitimate nodes can communicate for generating keys. Interested readers can refer to [18] for studies of scenarios without public discussion channels.

The key agreement protocol has two steps: (1) channel estimation and (2) key agreement. In the channel estimation step, these three nodes take turns sending training sequences, as

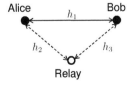

Figure 4.4: Key generation with the assistance of one relay.

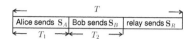

Figure 4.5: Time frame for the training-based scheme with a relay.

shown in the time frame in Figure 4.5. In particular, Alice sends a known sequence $\mathbf{S}_A$, from which Bob and the relay obtain estimates $\tilde{h}_{1,B}$ and $\tilde{h}_{2,R}$, respectively. Then Bob sends a known sequence $\mathbf{S}_B$, from which Alice and the relay obtain estimates $\tilde{h}_{1,A}$ and $\tilde{h}_{3,R}$, respectively. Finally, the relay sends a known sequence $\mathbf{S}_R$, from which Alice and Bob obtain estimates $\tilde{h}_{2,A}$ and $\tilde{h}_{3,B}$, respectively. In the key agreement step, Alice and Bob first agree on a key K_1 with a rate $I(\tilde{h}_{1,A}; \tilde{h}_{1,B})/T$ using the correlated information $(\tilde{h}_{1,A}, \tilde{h}_{1,B})$. Similarly, Alice and the relay generate a key K_2 with a rate $I(\tilde{h}_{2,A}; \tilde{h}_{2,R})/T$ using the correlated information $(\tilde{h}_{2,A}, \tilde{h}_{2,R})$, and Bob and the relay generate a key K_3 with a rate $I(\tilde{h}_{3,B}; \tilde{h}_{3,R})/T$ using the correlated information $(\tilde{h}_{3,B}, \tilde{h}_{3,R})$. Finally, the relay sends $K_2 \oplus K_3$ over the public channel. The steps are shown in Figure 4.6. After these steps, both Alice and Bob know (K_1, K_2, K_3). If the size of K_2 is smaller than the size of K_3, Alice and Bob set (K_1, K_2) as the key, otherwise they set (K_1, K_3) as the key. The resulting key rate is given by

$$R_{co} = \frac{1}{T} \left\{ \min\{I(\tilde{h}_{2,R}; \tilde{h}_{2,A}), I(\tilde{h}_{3,R}; \tilde{h}_{3,B})\} + I(\tilde{h}_{1,A}; \tilde{h}_{1,B}) \right\}. \qquad (4.39)$$

We note that K_2 and K_3 cannot simultaneously serve as the final key since the eavesdroppers learn $K_2 \oplus K_3$ from the public channel.

The following theorem states that the key generated in the above algorithm is secure and the protocol is optimal.

Theorem 4.5.1. [18] *The generated key is provably secure from any eavesdropper that experiences an independent wireless channel from the legitimate nodes. Furthermore, among the training-based approaches for key generation, the above algorithm generates a key with the largest possible key rate, and is hence optimal.*

In order to analyze the impact of a relay on the gain in the key rate, we use the concept of *multiplexing gain* as a performance measure. In particular, the multiplexing gain of the key rate is the limit of the ratio of the key rate with relay cooperation to the key rate without relay cooperation as the signal-to-noise ratio (SNR) approaches infinity. Hence, it is equal to $\lim_{P\to\infty} R_{co}/R_s$, where R_{co} is given in (4.39) and R_s is given in (4.20).

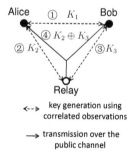

Figure 4.6: The key agreement protocol.

The following proposition provides the multiplexing gain for the proposed relay-assisted scheme.

Proposition 4.5.1. [18] *The multiplexing gain of the proposed scheme is* 2.

It is interesting to note that in the normal relay channel in which the relay helps to transmit information, the multiplexing gain is equal to one [26]. However, the multiplexing gain for secret key generation is two, which indicates that it is crucially important to exploit the presence of a relay in key generation (as opposed to information transmission), which provides a significant gain that doubles the rate of the key that can be generated from the wireless channel. It is also interesting to note that for a given amount of power P, we achieve a performance gain

$$\lim_{T \to \infty} R_{co}/R_s = 2. \tag{4.40}$$

This suggests that the proposed relay-assisted key generation scheme doubles the key rate even when the channel changes very slowly.

4.5.2 Relay-assisted Key Generation with Multiple Relays

The scheme discussed above can be easily extended to a network with N relays. Similarly to Section 4.5.1, relays are assumed to follow the designed transmission protocols, and not to leak information to Eve (i.e., not to collude with Eve). The generated key is required to be secure only from Eve, not necessarily from the relays.

As in Section 4.5.1, the key generation protocol includes channel estimation and key agreement steps. In the channel estimation step, Alice, Bob, and the relays take turns to broadcast their training sequences, and all other nodes estimate their corresponding channel gains. In the key agreement step, each pair of nodes agrees on one key based on their channel estimates using the Slepian–Wolf coding as illustrated in Figure 4.7. More specifically, Alice sends helper data to Bob, via which Alice and Bob agree on a key K_1. Then, each relay sends helper data to Alice and Bob, respectively, via which the relay agrees on a key $K_{A,i}$ with Alice and agrees on a key $K_{B,i}$ with Bob for $i = 1, \ldots, N$. Finally, each relay sends $K_{A,i} \oplus K_{B,i}$ to both Alice and Bob to allow them to recover all the keys. In the end, Alice and Bob concatenate $(K_1, (K_{A,1} \wedge K_{B,1}), \ldots, (K_{A,N} \wedge K_{B,N}))$ as the key, where $(K_i \wedge K_j)$ denotes either of K_i or K_j that has a smaller key rate.

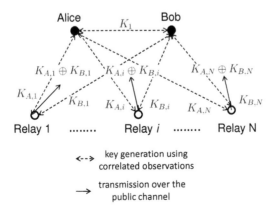

Figure 4.7: The key agreement protocol for N relays.

Using this approach, the achieved key rate is given by

$$R_{co,N} = \frac{1}{T}\Big\{ I(\tilde{h}_{1,A}; \tilde{h}_{1,B}) + \min\{I(\tilde{h}_{A1,A}; \tilde{h}_{A1,R}), I(\tilde{h}_{B1,B}; \tilde{h}_{B1,R})\}$$
$$+ \cdots + \min\{I(\tilde{h}_{AN,A}; \tilde{h}_{AN,R}), I(\tilde{h}_{BN,B}; \tilde{h}_{BN,R})\}\Big\}. \qquad (4.41)$$

Theorem 4.5.2. [18] *The multiplexing gain of having N relays is $N+1$, i.e.,*

$$\lim_{P \to \infty} R_{co,N}/R_s = N + 1, \qquad (4.42)$$

and this multiplexing gain is order-optimal.

The above theorem implies that the multiplexing gain of key generation scales linearly with the number of relay nodes.

4.5.3   Relay-oblivious Key Generation

An interesting scenario is also studied in [18], in which not only eavesdroppers but also relays are prevented from gaining any information about generated keys. That is, we wish to generate a relay-oblivious key while still benefiting from the relays. In this case, relays are still assumed to follow the designed transmission protocols, not to leak information to Eve (i.e., not to cooperate with Eve), and not to collude with each other, but are assumed to be curious and try to infer as much information about the generated key as possible.

We first describe the case with two relays (see Figure 4.8) as an example to illustrate the idea. The relay-oblivious key generation process has two main steps. In the first step, Alice, Bob, and the two relays follow the same protocol as developed in Section 4.5.2 to generate keys, as if there were no security constraint on the relays. In the second step, Alice and Bob distill a key that is unknown to the relays based on the keys already generated. More specifically, these nodes follow the same protocol as that in Section 4.5.2. Alice and Bob agree on a key K_1 from the correlated observations $(\tilde{h}_{1,A}, \tilde{h}_{1,B})$. Alice and relay 1 agree on a key K_2 from the correlated observations $(\tilde{h}_{2,A}, \tilde{h}_{2,R})$. Bob and relay 1 agree on a key K_3 from the correlated observations $(\tilde{h}_{3,B}, \tilde{h}_{3,R})$. Alice and relay 2 agree on a key K_4 from the correlated observations $(\tilde{h}_{4,A}, \tilde{h}_{4,R})$. Bob and relay 2 agree on a key K_5 from the correlated observations $(\tilde{h}_{5,B}, \tilde{h}_{5,R})$. Relay 1 broadcasts $K_2 \oplus K_3$ using the public channel. Relay 2 broadcasts $K_4 \oplus K_5$ using the public channel. We note that if we do not need to keep the key secret from each relay, we can use (K_1, K_2, K_4) as the key. However, if we have the additional privacy constraint, we have the additional key distillation step. Without loss of

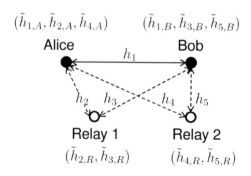

Figure 4.8: Observations at the nodes after training.

generality, assuming the length of K_2 is less than the length of K_3, and the length of K_4 is less than the length of K_5, then in the key distillation step, Alice and Bob concatenate $(K_1, K_2 \oplus K_4)$ as the key. It can easily be shown that each relay gains negligible information about the generated key $(K_1, K_2 \oplus K_4)$. Hence, the proposed scheme achieves the goal of generating relay-oblivious keys while exploiting the assistance of relays.

The extension to the general case with N relays is simple. Assuming that N is an even number, we then divide these N relays into $N/2$ pairs. Now, each pair of relays can run the same protocol as discussed above, and contribute a key that is unknown to the relays. A final key is obtained by concatenating these keys together. Based on this approach, the following corollary on the multiplexing gain in this case can be obtained.

Corollary 4.5.1. *The multiplexing gain of the rate of the key that is secure from relays is* $\lfloor N/2 \rfloor + 1$.

4.6 Key Agreement with the Presence of an Active Attacker

As mentioned in Section 5.1, it is assumed in many studies that Eve is *passive*, which means that she only overhears (does not transmit over) the channel and tries to infer information about the generated key. In reality, an eavesdropper can be an attacker who can initiate various active attacks to interrupt the key generation process. In this section, we introduce a model studied in [3], in which the eavesdropper is an active attacker.

4.6.1 Training Phase

In [3], the effects of active attacks are studied in the context of the joint source channel key agreement framework as discussed in Section 4.4. In particular, the optimal attacker strategy is characterized, and the achievable key rate given that the attacker applies the optimal strategy is obtained.

As shown in Figure 4.2, the key generation protocol has two phases: a training phase and a transmission phase. The active attacker can initiate an attack during both of these phases. We first characterize the attacker's optimal strategy for the training phase.

Suppose Alice sends a known sequence $\mathbf{S}_A$ of size $1 \times \alpha T_\tau$, with $0 < \alpha < 1$ during the training stage; then Bob receives

$$\mathbf{Y}_B = h\mathbf{S}_A + \mathbf{X}_{E1} + \mathbf{N}_B, \tag{4.43}$$

where $\mathbf{N}_B = [N_B(1), \cdots, N_B(\alpha T_\tau)]^T$ and $\mathbf{X}_{E1}$ is the attack signal transmitted by Eve. After that, Bob sends a known sequence $\mathbf{S}_B$ of size $1 \times (1-\alpha)T_\tau$ over the wireless channel, and Alice receives

$$\mathbf{Y}_A = h_1\mathbf{S}_B + \mathbf{X}_{E2} + \mathbf{N}_A, \tag{4.44}$$

where $\mathbf{N}_A = [N_A(1), \cdots, N_A((1-\alpha)T_\tau)]^T$ and again $\mathbf{X}_{E2}$ is the attack signal transmitted by Eve.

Following the protocol discussed in Section 4.4, Alice computes a statistic $\tilde{h}_{1,A}$ for $\mathbf{Y}_A$ via

$$\tilde{h}_{1,A} = \frac{\mathbf{S}_B^T}{||\mathbf{S}_B||^2}\mathbf{Y}_A = h_1 + \frac{\mathbf{S}_B^T}{||\mathbf{S}_B||^2}(\mathbf{X}_{E1} + \mathbf{N}_A). \tag{4.45}$$

Similarly, Bob computes a statistic $\tilde{h}_{1,B}$ for $\mathbf{Y}_B$ via

$$\tilde{h}_{1,B} = \frac{\mathbf{S}_A^T}{||\mathbf{S}_A||^2}\mathbf{Y}_B = h_{AB} + \frac{\mathbf{S}_A^T}{||\mathbf{S}_A||^2}(\mathbf{X}_{E2} + \mathbf{N}_B). \tag{4.46}$$

We use Γ_1 to denote $\mathbf{S}_B^T\mathbf{X}_{E1}/||\mathbf{S}_B||^2$, N_1 to denote $\mathbf{S}_B^T\mathbf{N}_A/||\mathbf{S}_B||^2$, Γ_2 to denote $\mathbf{S}_A^T\mathbf{X}_{E1}/||\mathbf{S}_A||^2$, and N_2 to denote $\mathbf{S}_A^T\mathbf{N}_B/||\mathbf{S}_A||^2$. Hence, (4.45) and (4.46) can be rewritten as

$$\tilde{h}_{1,A} = h_1 + \Gamma_1 + N_1, \tag{4.47}$$
$$\tilde{h}_{1,B} = h_1 + \Gamma_2 + N_2. \tag{4.48}$$

If the attacker is passive, as discussed in Section 4.4, $\tilde{h}_{1,A}$ and $\tilde{h}_{1,B}$ are jointly Gaussian random variables. However, when the attacker is active, the statistics of these two random variables depend on the attacker's strategy. Furthermore, Eve knows (Γ_1, Γ_2), which is correlated with the observations at Alice and Bob.

Alice and Bob will generate a key from these two correlated observations. As will be clear in the sequel, the proposed protocol will generate a key from $(\tilde{h}_{1,A}, \tilde{h}_{1,B})$ with a rate

$$R_s = \frac{1}{T}(I(\tilde{h}_{1,A} + Z; \tilde{h}_{1,B}) - I(\tilde{h}_{1,A} + Z; \Gamma_1, \Gamma_2)). \tag{4.49}$$

Here Z is a zero mean Gaussian random variable with variance σ_z^2, and is independent of other random variables of interest. Roughly speaking, $I(\tilde{h}_{1,A} + Z; \tilde{h}_{1,B})$ is the common randomness that both Alice and Bob share, and $I(\tilde{h}_{1,A} + Z; \Gamma_1, \Gamma_2)$ is the amount of information that Eve knows about the value of $\tilde{h}_{1,A}$. This is due to the fact that both $\tilde{h}_{1,A}$ and $\tilde{h}_{1,B}$ are related to the signal transmitted by Eve. Hence, the attacker will design her attack signal such that the mutual information between the observations at Alice and Bob is small, while the mutual information between the observations at Alice and the attack signal at Eve is large.

At the same time, Bob obtains an MMSE estimate $\hat{h}_{1,B}$ of the channel gain h_1 in the given coherence block. $\hat{h}_{1,B}$ will be treated as the true value of the channel gain in the second phase of the key agreement protocol. We can write $h_1 = \hat{h}_{1,B} + \bar{h}_1$, in which $\bar{h}_1$ is the estimation error. As will be clear in the sequel, the rate of the key that can be generated using the proposed protocol depends on the variance of $\bar{h}_1$, which will be denoted by σ_{est}^2. The larger the variance, the smaller the rate of the key.

Hence, Eve needs to design her attack signals $\mathbf{X}_{E1}$ and $\mathbf{X}_{E2}$ to simultaneously maximize σ_{est}^2 and minimize R_s. First, it is clear that the attacker should set $\mathbb{E}\{\Gamma_1\} = \mathbb{E}\{\Gamma_2\} = 0$. Assuming that Alice and Bob transmit with power P_τ during the training period, we have $||\mathbf{S}_B||^2 = (1-\alpha)T_\tau P_\tau$ and $||\mathbf{S}_A||^2 = \alpha T_\tau P_\tau$. Also, assuming that the attacker transmits at a power P_{E1} for $\mathbf{X}_{E1}$ and P_{E2} for $\mathbf{X}_{E2}$, respectively, then $\mathrm{Var}\{\Gamma_1\} = \sigma_1^2 = P_{E2}/P_\tau$ and $\mathrm{Var}\{\Gamma_2\} = \sigma_2^2 = P_{E1}/P_\tau$. Assuming that the correlation coefficient between Γ_1 and Γ_2 is ρ, we need to characterize the distribution of (Γ_1, Γ_2) that the attacker will adopt to maximize σ_{est}^2 and minimize R_s.

Theorem 4.6.1. [3] *Choosing (Γ_1, Γ_2) to be jointly Gaussian simultaneously minimizes R_s and maximizes σ_{est}^2. Furthermore, the optimal correlation coefficient between Γ_1 and Γ_2 is given by*

$$\rho_{opt} = \begin{cases} -\frac{\sigma_h^2}{\sigma_1\sigma_2}, & \text{if } \sigma_h^2 \le \sigma_1\sigma_2 \\ -1, & \text{otherwise}. \end{cases} \tag{4.50}$$

4.6.2 Key Generation Phase

As discussed in Section 4.4, after the training period of T_τ symbols, Alice will send two pieces of information to Bob via the wireless channel: (1) the information needed to distill a key from the correlated estimations $(\tilde{h}_{1,A}, \tilde{h}_{1,B})$ obtained in the first phase, which is public information and does not need to be kept secure; and (2) a new randomly generated key with a rate R_{ch}, which needs to be kept secure from the attacker. The total key rate will be $R_{ch} + R_s$.

Key generation from the correlated observations

We first look at the key distillation part, in which we generate a key from the correlated observations $(\tilde{h}_{1,A}, \tilde{h}_{1,B})$. Compared with the scenario discussed in Section 4.3, the attacker now possesses observations (Γ_1, Γ_2) that are correlated with the observations $(\tilde{h}_{1,A}, \tilde{h}_{1,B})$ at the legitimate users. Hence, this is a key generation problem with side-information at Eve as discussed in Section 4.2.3. Setting $U = \tilde{h}_{1,A} + Z$ in (4.11), with Z being $\mathcal{N}(0, \sigma_z^2)$ and independent of other random variables of interest, we obtain (4.49).

Key generation from the channel

Now, we look at how to send a newly generated key over the wireless channel. There are two main differences from the situation of Section 4.3: (1) the channel estimate is coarser due to the attack in the channel estimation stage; and (2) the attacker will send an attack signal in this stage.

More specifically, Bob still obtains an MMSE estimate $\hat{h}_{1,B}$ of the channel gain h_1 in the given coherence block,

$$\hat{h}_{1,B} = \frac{\sigma_h^2}{\sigma^2 + \sigma_2^2 + \alpha P_\tau T_\tau \sigma_h^2} \mathbf{S}_A^T \mathbf{Y}_B. \tag{4.51}$$

Bob will treat this as the true value of the channel gain. We can write $h_1 = \hat{h}_{1,B} + \bar{h}_1$, in which $\bar{h}_1$ is the estimation error. $\bar{h}_1$ is a zero mean Gaussian random variable with variance

$$\sigma_h^2 / (\sigma_h^2 \alpha P_\tau T_\tau + \sigma^2 + \sigma_2^2).$$

Now, when Alice transmits, Bob and Eve receive

$$Y_B = \hat{h}_1 X_A + \bar{h}_1 X_A + X_E + N_B, \tag{4.52}$$
$$\text{and}\quad Y_E = h_{AE} X_A + N_E. \tag{4.53}$$

Eve will choose her attack signal X_E to minimize R_{ch} specified by (4.24), which we reproduce here for the ease of presentation:

$$R_{ch} = \frac{T - T_\tau}{T} [I(X_A; Y_B | \hat{h}_{1,B}) - I(X_A; Y_E | h_{AE})]^+. \tag{4.54}$$

Obviously, the attacker will design X_E such that $I(X_A; Y_B | \hat{h}_{1,B})$ is minimized. Since Eve receives Y_E which is correlated with X_A, she can design X_E based on her knowledge of X_A.

To characterize the attacker's optimal attack strategy, we need a result from [27]. The result says that if h_1 is independent of X_A in the system and X_A is Gaussian, then even if Eve knows X_A completely, the optimal attack strategy of Eve is to send i.i.d. Gaussian noise that is independent of X_A. When one tries to use this result, caution should be exercised to satisfy this condition. As discussed in Section 4.3, X_A contains two pieces of information:

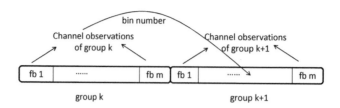

Figure 4.9: A scheme to avoid correlation between the channel gain and the transmitted codeword.

the number of the bin to which the channel gain belongs, and the newly generated key. That is, X_A is specified by the bin number i, which contains some information about the channel gain h_1. We can overcome this issue by using the scheme illustrated in Figure 4.9. More specifically, as discussed in Section 4.6.2, we divide the time into groups, each containing N symbol times (i.e., m fading blocks). In group k, Alice collects a vector of channel observations $\tilde{\mathbf{Y}}_A$, and determines the bin number i_k of this vector. Instead of transmitting i_k to Bob using the wireless channel during the kth group (which will introduce correlation between the channel gain and the codeword sent over the channel), we will transmit i_k over the $k+1$st block. With this idea, we can use the result of [27] and know that the optimal strategy of the attacker is to send i.i.d. Gaussian noise.

Suppose the powers used by Alice and Eve during this stage are P_d and P_{E3}, respectively; then (4.54) becomes

$$R_{ch} = \frac{T - T_\tau}{2T} \left[\mathbb{E}\left\{ \log\left(1 + \frac{\hat{h}_{1,B}^2 P_d}{\sigma^2 + P_{E3} + \sigma_{est}^2}\right) - \log\left(1 + \frac{h_{AE}^2 P_d}{\sigma^2}\right) \right\} \right]^+. \quad (4.55)$$

In summary, we have the following.

Theorem 4.6.2. [3] *Using a fading wireless channel, a key rate of*

$$R_{key} = \min_{P_{E1}, P_{E2}, P_{E3}} \max_{\alpha, P_\tau, T_\tau} \{R_s + R_{ch}\} \quad (4.56)$$

is achievable. Here, we require that

$$P_\tau T_\tau + P_d(T - T_\tau) \leq PT, \quad (4.57)$$
$$P_{E1}\alpha T_\tau + P_{E2}(1 - \alpha)T_\tau + P_{E3}(T - T_\tau) \leq P_E T. \quad (4.58)$$

At the same time, σ_z^2 should be chosen to satisfy the following condition:

$$\frac{I(U; \tilde{h}_{1,A}) - I(U; \tilde{h}_{1,B})}{T - T_\tau} \quad (4.59)$$

$$\leq \min\left\{ \mathbb{E}\left\{ \log\left(1 + \frac{\hat{h}_{1,B}^2 P_d}{\sigma^2 + P_{E3} + \sigma_{est}^2}\right) \right\}, \mathbb{E}\left\{ \log\left(1 + \frac{h_{AE}^2 P_d}{\sigma^2}\right) \right\} \right\}. \quad (4.60)$$

4.7 Conclusion

In this chapter, we have discussed how to generate secret keys via wireless fading channels. We have reviewed the basic information-theoretic models and results for key generation. We have then discussed the existing studies that apply these information-theoretic results to wireless channels. We have further discussed three new directions: the joint source-channel approach, the relay-assisted approach, and scenarios with active attackers. Some interesting topics for future research may include more sophisticated attack models, simultaneous key generation among nodes over multiple access and broadcast channels, and the impact of multiple antennas on key generation.

4.8 Acknowledgment

The work of L. Lai was supported by a National Science Foundation CAREER Award under Grant CCF-13-18980 and by the National Science Foundation under Grant CNS-13-21223. The work of Y. Liang was supported by a National Science Foundation CAREER Award under Grant CCF-10-26565 and by the National Science Foundation under Grants CCF-10-26566 and CNS-11-16932. The work of H. V. Poor was supported by the Office of Naval Research under Grant N00014-12-1-0767 and the National Science Foundation under Grant CCF-10-16671. The work of W. Du was supported by the National Science Foundation under Grant CNS-11-16932.

References

[1] R. Wilson, D. Tse, and R. A. Scholtz, "Channel identification: Secret sharing using reciprocity in ultrawideband channels," *IEEE Trans. Inform. Forensics and Security*, vol. 2, pp. 364–375, Sept. 2007.

[2] C. Ye, S. Mathur, A. Reznik, W. Trappe, and N. Mandayam, "Information-theoretic key generation from wireless channels," *IEEE Trans. Inform. Forensics and Security*, vol. 5, pp. 240–254, June 2010.

[3] L. Lai, Y. Liang, and H. V. Poor, "A unified framework for key agreement over wireless fading channels," *IEEE Trans. Inform. Forensics and Security*, vol. 7, pp. 480–490, Apr. 2012.

[4] T.-H. Chou, A. M. Sayeed, and S. C. Draper, "Minimum energy per bit for secret key acquisition over multipath wireless channels," in *Proc. IEEE Intl. Symposium on Inform. Theory* (Seoul, Korea), June 2009.

[5] T.-H. Chou, A. M. Sayeed, and S. C. Draper, "Impact of channel sparsity and correlated eavesdropping on secret key generation from multipath channel randomness," in *Proc. IEEE Intl. Symposium on Inform. Theory* (Austin, TX), June 2010.

[6] A. Sayeed and A. Perrig, "Secure wireless communications: Secret keys through multipath," in *Proc. IEEE Intl. Conf. on Acoustics, Speech, and Signal Processing* (Las Vegas, NV), Apr. 2008.

[7] S. Mathur, W. Trappe, N. Mandayam, C. Ye, and A. Reznik, "Radiotelephathy: Extracting a secret key from an unauthenticated wireless channel," in *Proc. ACM International Conference on Mobile Computing and Networking* (San Francisco, CA), Sept. 2008.

[8] K. Zeng, D. Wu, A. Chan, and P. Mohapatra, "Exploiting multiple antenna diversity for shared key generation in wireless networks," in *Proc. IEEE Conf. Computer Communications (Infocom)* (San Diego, CA), Mar. 2010.

[9] Q. Wang, H. Su, K. Ren, and K. Kim, "Fast and scalable secret key generation exploiting channel phase randomness in wireless networks," in *Proc. IEEE Conf. Computer Communications (Infocom)* (Shanghai, China), pp. 1422–1430, Apr. 2011.

[10] A. Khisti, S. N. Diggavi, and G. W. Wornell, "Secret-key agreement using asymmetry in channel state information," in *Proc. IEEE Intl. Symposium on Inform. Theory* (Seoul, Korea), June 2009.

[11] A. Khisti, "Interactive secret key generation over reciprocal fading channels," in *Proc. Allerton Conf. on Communication, Control, and Computing* (Monticello, IL), Oct. 2012.

[12] D. Tse and P. Viswanath, *Fundamentals of Wireless Communication.* Cambridge, UK: Cambridge University Press, May 2005.

[13] R. Ahlswede and I. Csiszár, "Common randomness in information theory and cryptography, Part I: Secret sharing," *IEEE Trans. Inform. Theory*, vol. 39, pp. 1121–1132, July 1993.

[14] S.-W. Ho, T. Chan, and C. Uduwerelle, "Error-free perfect-secrecy systems," in *Proc. IEEE Intl. Symposium on Inform. Theory*, (Saint-Petersburg, Russia), pp. 1613–1617, July-Aug. 2011.

[15] U. Maurer, "Secret key agreement by public discussion from common information," *IEEE Trans. Inform. Theory*, vol. 39, pp. 733–742, May 1993.

[16] I. Csiszár and P. Narayan, "Common randomness and secret key generation with a helper," *IEEE Trans. Inform. Theory*, vol. 46, pp. 344–366, Mar. 2000.

[17] I. Csiszár and P. Narayan, "Secrecy capacities for multiple terminals," *IEEE Trans. Inform. Theory*, vol. 50, pp. 3047–3061, Dec. 2004.

[18] L. Lai, Y. Liang, and W. Du, "Cooperative key generation in wireless networks," *IEEE Journal on Selected Areas in Communications*, vol. 30, pp. 1578–1588, Sept. 2012.

[19] A. D. Wyner, "The wire-tap channel," *Bell System Technical Journal*, vol. 54, pp. 1355–1387, Oct. 1975.

[20] J. N. Laneman, D. N. C. Tse, and G. W. Wornell, "Cooperative diversity in wireless networks: Efficient protocols and outage behavior," *IEEE Trans. Inform. Theory*, vol. 50, pp. 3062–3080, Dec. 2004.

[21] B. Hughes and P. Narayan, "Gaussian arbitrarily varying channels," *IEEE Trans. Inform. Theory*, vol. 33, pp. 267–284, Mar. 1987.

[22] I. Csiszár and P. Narayan, "Capacity of the Gaussian arbitrarily varying channel," *IEEE Trans. Inform. Theory*, vol. 37, pp. 18–26, Jan. 1991.

[23] P. K. Gopala, L. Lai, and H. El Gamal, "On the secrecy capacity of fading channels," *IEEE Trans. Inform. Theory*, vol. 54, pp. 4687–4698, Oct. 2008.

[24] M. Médard, "The effect upon channel capacity in wireless communications of perfect and imperfect knowledge of the channel," *IEEE Trans. Inform. Theory*, vol. 46, pp. 933–946, May 2000.

[25] B. Hassibi and B. M. Hochwald, "How much training is needed in multiple-antenna wireless links?," *IEEE Trans. Inform. Theory*, vol. 49, pp. 951–963, Apr. 2003.

[26] K. Azarian, H. El Gamal, and P. Schniter, "On the achievable diversity-multiplexing tradeoff in half-duplex cooperative channels," *IEEE Trans. Inform. Theory*, vol. 51, pp. 4152–4172, Dec. 2005.

[27] A. Kashyap, T. Basar, and R. Srikant, "Correlated jamming on MIMO Gaussian fading channels," *IEEE Trans. Inform. Theory*, vol. 50, pp. 2119–2123, Sept. 2004.

Chapter 5

Secrecy with Feedback

Xiang He
The Pennsylvania State University

Aylin Yener
The Pennsylvania State University

In this chapter, secure communication is studied for the case where the intended recipient of the message can also transmit signals. Several known strategies of utilizing this feedback capability to facilitate secure communication are described in the context of the Gaussian two-way wiretap channel model. Their respective achievable secrecy rates are computed. A converse on the theoretical transmission limit of confidential messages is described.

5.1 Introduction

Most communication links are bidirectional, i.e., the intended receiver of the message is also equipped with a transmitter. This capability could potentially help the sender of the message achieve secure communications that would otherwise be impossible in the presence of an eavesdropper. One can immediately see this if, for example, the intended receiver could reach the sender via some perfectly secure feedback link. In this case, the receiver could send a key to the message sender through the feedback link. The sender can then use the key to encrypt its message and send it through the forward link securely even if it is subject to eavesdropping.

In the real world, however, a secure feedback link is rare. More often than not, both the forward and feedback links are subject to eavesdropping. The question then becomes what the best strategy is to secure the transmitted messages in such a setting. There have been extensive efforts aiming to address this important question in recent years and many different strategies have been developed. In this chapter, we will review these developments.

Our discussion will be carried out in the context of the Gaussian two-way wiretap channel model [1,2]. We note that there are many related channel models that have been proposed and studied, which we will summarize at the end. The channel model is described in Section 5.2. From Section 5.3 to Section 5.6, we describe several different strategies to secure a confidential message and demonstrate how the secrecy rate is computed. At the end of this chapter, in Section 5.7, we briefly describe an outer bound for the Gaussian two-way wiretap channel and its implications.

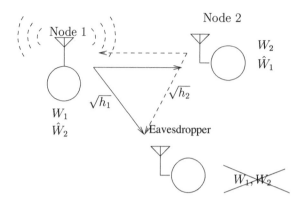

Figure 5.1: The Gaussian two-way wiretap channel.

5.2   The Gaussian Two-way Wiretap Channel

The Gaussian two-way wiretap channel model is shown in Figure 5.1. In this channel model, two nodes, Node 1 and 2, communicate with each other over two additive Gaussian channels, defined as:

$$Y_2 = X_2 + N_3, \tag{5.1}$$
$$Y_1 = X_1 + N_1. \tag{5.2}$$

The eavesdropper observes a noisy copy of the sum of their transmitted signals:

$$Z = \sqrt{h_1}X_1 + \sqrt{h_2}X_2 + N_2, \tag{5.3}$$

where $\sqrt{h_1}, \sqrt{h_2}$ are channel gains. $N_i, i = 1, 2, 3$ are independent Gaussian random variables with zero mean and unit variance, representing the channel noise.

X_1 and X_2 are transmitted signals from Node 1 and 2, respectively, and they are constrained in terms of average power. The power constraint of Node i, $i = 1, 2$, is P_i, i.e.,

$$\frac{1}{n}\sum_{k=1}^{n} E\left[X_{1,k}^2\right] \le P_1, \tag{5.4}$$

$$\frac{1}{n}\sum_{k=1}^{n} E\left[X_{2,k}^{\,2}\right] \le P_2. \tag{5.5}$$

In the most general form of this model, both nodes can send a confidential message to the other. Let W_k denote the message sent by Node k. The encoding functions used at the two nodes are allowed to be stochastic. Without loss of generality, we use M_j to model the local randomness in the encoding function used by Node j, $j = 1, 2$. At the ith channel use, the encoding function of Node 1 is defined as:

$$X_{1,i} = f_i(Y_2^{\,i-1}, W_1, M_1). \tag{5.6}$$

The encoding function of Node 2 is defined as

$$X_{2,i} = g_i(Y_1^{i-1}, W_2, M_2). \tag{5.7}$$

Note that with the introduction of M_j, $j = 1, 2$, we can define f_i, g_i as deterministic encoders. Also note that another way to define f_i is $X_{1,i} = f_i(X_1^{i-1}, Y_2^{\,i-1}, M_1)$. It is easy to see that this definition is equivalent to the definition given in (5.6).

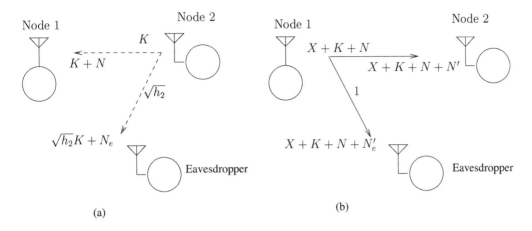

Figure 5.2: Achieving $R_1 > 0$ through Public Discussion: (a) the first channel use in a round; (b) the second channel use in a round.

Let n be the total number of channel uses. Node 2 must decode W_1 reliably from Y_1^n, X_2^n, M_2, W_2. Node 1 must decode W_2 reliably from Y_2^n, X_1^n, M_1, W_1. Let the decoding results be $\hat{W}_1$ and $\hat{W}_2$, respectively. Then we require

$$\lim_{n \to \infty} \Pr(W_j \neq \hat{W}_j) = 0, \quad j = 1, 2. \tag{5.8}$$

In addition, both messages must be kept secret from the eavesdropper. Hence we need

$$\lim_{n \to \infty} \frac{1}{n} I(W_1, W_2; Z^n) = 0 \tag{5.9}$$

The transmission rate pair (R_1, R_2) is computed as:

$$R_j = \lim_{n \to \infty} \frac{1}{n} H(W_j), \quad j = 1, 2. \tag{5.10}$$

The secrecy rate region is defined as all rate pairs $\{R_1, R_2\}$ for which (5.8) and (5.9) hold.

The benefit that the system can gain from Node 2 transmitting is most evident when $\sqrt{h_1} \geq 1$. In this case, if Node 2 is silent, the eavesdropper observes a higher received signal-to-noise ratio than Node 2 and it is not possible to achieve a positive secrecy rate R_1 from Node 1 to Node 2. In the following sections, we describe several strategies that achieve $R_1 > 0$ by taking advantage of Node 2's transmission capability.

5.3 Achieving Secrecy Using Public Discussion

The transmission strategy described in this section is developed in the spirit of what was proposed in [3] for a binary symmetric channel model and is generalized to Gaussian channels in [4]. For simplicity, in this section, we assume $\sqrt{h_1} = 1$, which implies $R_1 = 0$ if Node 2 does not transmit. We also assume the two nodes have the same power constraints $P_1 = P_2 = P$. We next show that it is possible to achieve a positive R_1 for any value of $\sqrt{h_2}$ for this configuration.

The transmission protocol is depicted in Figure 5.2 and is composed of multiple rounds. Each round is composed of two consecutive channel uses. During the first channel use, Node 2 sends a signal K to Node 1, which observes a noisy copy of K denoted by $K + N$, with N

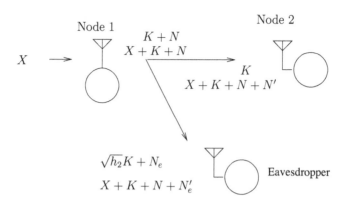

Figure 5.3: The wiretap channel equivalent to Figure 5.2, where one round of communication is viewed as one channel use.

being the channel noise. The eavesdropper observes $\sqrt{h_2}K + N_e$, where N_e is the noise. In the next channel use, Node 1 adds $K+N$ to its own signal X and transmits $X+K+N$. Node 2 observes $X + K + N + N'$ and the eavesdropper observes $X + K + N + N'_e$, where N', N'_e represent the additional noise in the observation. Therefore, at the end of each round of communication, Node 2 receives $K, X+K+N+N'$, and the eavesdropper observes $\sqrt{h_2}K + N_e, X+K+N+N'_e$. If K transmitted by Node 2 is an independent and identically distributed (i.i.d.) sequence, then the channel can be viewed as a memoryless wiretap channel shown in Figure 5.3, where each round of communication corresponds to one channel use in this equivalent wiretap channel. This allows us to apply the achievable secrecy rate results for the wiretap channel [5,6], which says there must exist a good codebook to achieve the secrecy rate $\max\{0, I(X;Y) - I(X;Z)\}$ for a wiretap channel $\Pr(Y;Z|X)$ with X, Y, Z being the inputs, the output received by the intended receiver, and the eavesdropper's observation, respectively. The achieved secrecy rate R_s is hence given by:

$$\max\{0, \frac{1}{2}[I(X; K, X + K + N + N') - I(X; \sqrt{h_2}K + N_e, X + K + N + N'_e)]\} \quad (5.11)$$

where $\frac{1}{2}$ reflects the fact that each round is composed of two channel uses.

Remark 5.3.1. *At this point, it is important to make the distinction between the nature of signals transmitted via K and X. Signals transmitted over K is an i.i.d. noise sequence. Signals transmitted via X need to convey the confidential message and hence cannot be an i.i.d noise sequence but instead are codewords from a predefined codebook. Reference [5] did not say how such a codebook can be constructed, but instead shows that among all codebooks which are composed of sequences sampled in an i.i.d. fashion from the so-called input distribution, at least one codebook must be good.*

We next evaluate (5.11). We choose X and K to have a Gaussian distribution with zero mean and variance P and let X and K be independent. Note that this independence is not just a design choice, but a necessity posed by the coding scheme, because the encoder at Node 1 does not rely on the signals it receives from Node 2 when it chooses the codeword to transmit via X. Since X and K are independent, we can drop K from $I(X; K, X + K + N + N')$ and rewrite (5.11) as

$$\max\{0, \frac{1}{2}[I(X; X + N + N') - I(X; \sqrt{h_2}K + N_e, X + K + N + N'_e)]\}. \quad (5.12)$$

We next evaluate (5.12). Let

$$A = \sqrt{h_2}K + N_e \tag{5.13}$$

$$B = K + N + N_e' \tag{5.14}$$

$$k = \frac{E[BA]}{E[A^2]} = \frac{\sqrt{h_2}P}{h_2P + 1} \tag{5.15}$$

Note that N, N', N_e, N_e' are independent Gaussian random variables with zero mean and unit variable. It is easy to verify that $B - kA$ and A are both Gaussian and are not correlated. This implies $B - kA$ and A are independent from each other. Therefore we can rewrite the term $I(X; \sqrt{h_2}K + N_e, X + K + N + N_e')$ in (5.12) as:

$$I(X; A, X + B) \tag{5.16}$$

$$= I(X; A, X + B - kA) \tag{5.17}$$

$$= I(X; X + B - kA) + I(X; A | X + B - kA) \tag{5.18}$$

$$\leq I(X; X + B - kA) + I(X, X + B - kA; A) \tag{5.19}$$

$$= I(X; X + B - kA) \tag{5.20}$$

$$= I(X; X + N + N_e' + (1 - k\sqrt{h_2})K - kN_e). \tag{5.21}$$

This means (5.12) is lower bounded by:

$$\max\{0, \frac{1}{2}[I(X; X + N + N') - I(X; X + N + N_e' + (1 - k\sqrt{h_2})K - kN_e)]\}. \tag{5.22}$$

Since $(1 - k\sqrt{h_2})K - kN_e$ is independent from $N + N_e'$, and $N + N'$ has the same variance as $N + N_e'$, we observe that the secrecy rate in (5.22) is always positive regardless of the value of h_2.

Despite the fact that the achieved secrecy rate is always positive, it should be noted that the secrecy rate does converge to zero when $\sqrt{h_2}$ goes to infinity. This is because

$$\lim_{\sqrt{h_2} \to \infty} k = 0, \tag{5.23}$$

$$\lim_{\sqrt{h_2} \to \infty} k\sqrt{h_2} = 1, \tag{5.24}$$

and hence the variance of $(1 - k\sqrt{h_2})K - kN_e$ converges to zero when $\sqrt{h_2}$ goes to infinity. This corresponds to the scenario where the eavesdropper is arbitrarily close to Node 2. In this case, the eavesdropper has a clear copy of the signal K sent from Node 2 and the secrecy rate decreases to zero as expected. In the next section, we shall introduce a different strategy that achieves a higher secrecy rate for this case.

5.4 Achieving Secrecy Using Cooperative Jamming

A strategy that can overcome the weakness of the previous section was proposed in References [1, 2, 7] and was termed "cooperative jamming" therein. Instead of letting Node 2 transmit first followed by a transmission from Node 1, in this scheme, Node 1 and Node 2 transmit simultaneously. The signals sent by Node 2 jam the eavesdropper, preventing it from receiving signals transmitted by Node 1, hence the terminology "making Node 2 the *cooperative jammer*." The achievable secrecy rate of this strategy depends on whether Node 2 can transmit and receive signals at the same time, i.e., whether it is a full duplex node or a half duplex node. Each case is described in a separate section below.

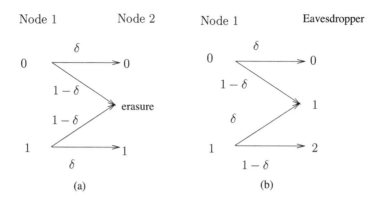

Figure 5.4: The equivalent wiretap channel when Node 1 and Node 2 use on-off transmission strategy. (a) Channel between Node 1 and Node 2. (b) Channel between Node 1 and the eavesdropper.

5.4.1 Full-duplex Node

In this case, Node 2 transmits an independent identically distributed (i.i.d.) Gaussian sequence, denoted by J, with power P_2 at each channel use. Node 1 sends X. Node 2 receives $X + N$, while the eavesdropper receives $X + \sqrt{h_2}J + N_e$. Let $C(x) = \frac{1}{2}\log_2(1 + x)$. The achieved secrecy rate is hence given by [2]:

$$\max\left\{0, I(X; X + N) - I(X; X + \sqrt{h_2}J + N_e)\right\} \tag{5.25}$$

$$= \max\left\{0, C(P_1) - C\left(\frac{P_1}{1 + h_2 P_2}\right)\right\} \tag{5.26}$$

The achieved secrecy rate given by (5.26) approaches zero when $\sqrt{h_2}$ goes to zero, which corresponds to the case when the cooperative jammer cannot reach the eavesdropper. Note that this is different from the secrecy rate in (5.12), which approaches zero when $\sqrt{h_2}$ goes to infinity.

5.4.2 Half-duplex Node

Interestingly, even when Node 2 can not transmit and receive simultaneously, a positive secrecy rate is still achievable through cooperative jamming [8, 9].

For clarity of exposition, we assume $\sqrt{h_2} = 1$. To simplify the computation of mutual information terms, we also make the approximation that the received signal-to-noise ratios at Node 1, Node 2, and the eavesdroppers are sufficiently high such that we can ignore the channel noise.

In this case, Nodes 1 and 2 use an on-off transmission strategy: let Node 1 use a binary input distribution over zero and one, and $\Pr(X = 1) = 1/2$. Node 1 uses this input distribution to generate its codebook. During each channel use, Node 2 chooses to transmit 1 with the probability $1 - \delta$. Otherwise it transmits 0. Hence if Node 1 transmits 0 during a channel use, the eavesdropper receives 0 with probability δ, and receives 1 with probability $1 - \delta$. If Node 1 transmits 1 during a channel use, the eavesdropper receives 1 with probability δ, and receives 2 with probability $1 - \delta$. Node 2 cannot receive when it transmits hence it observes an erasure channel with erasure probability $1 - \delta$. The equivalent channel is shown in Figure 5.4. The confusion at the eavesdropper happens when only one node transmits. In this case, if Node 1 and Node 2 control their received signal-to-noise ratio (SNR) at the

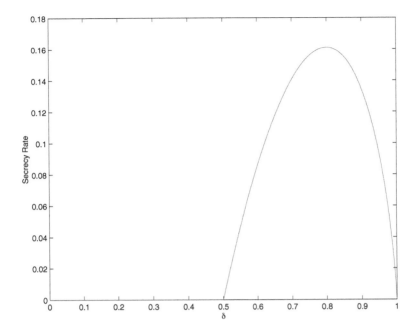

Figure 5.5: Secrecy rate achieved by (5.28)

eavesdropper to be close to each other, the eavesdropper is uncertain where the transmission comes from, which provides an opportunity to achieve secrecy.

Define the notation $H(p_1, p_2) = -\sum_{i=1}^{2} p_i \log_2(p_i)$ and $H(p_1, p_2, p_3) = -\sum_{i=1}^{3} p_i \log_2(p_i)$. Let Y, Z denote the signals received by Node 2 and the eavesdropper. The achieved secrecy rate is then evaluated to be

$$\max\{0, I(X;Y) - I(X;Z)\} \tag{5.27}$$

$$= \max\{0, \delta - (H(\frac{\delta}{2}, \frac{1-\delta}{2}, \frac{1}{2}) - H(\delta, 1-\delta))\}. \tag{5.28}$$

The rate achieved by (5.28) is illustrated in Figure 5.5. By letting $\delta = 3/4$, it is easy to verify that the rate above is positive.

5.5 Achieving Secrecy through Discussion and Jamming

Based on the discussion in the previous sections, it is clear that neither schemes introduced in the previous sections achieve the largest possible secrecy rate for all channel gain configurations. The achievable scheme in [9] combines these schemes, which we shall describe in this section.

5.5.1 Jamming with Codewords

Recall that in the scheme in Section 5.4.1, Node 2 jams the eavesdropper with an i.i.d. Gaussian sequence. It is also possible to jam the eavesdropper using sequences from a codebook. This strategy was used in [2, 10, 11], which we shall describe below.

Let n be the codeword length. Let δ_n be positive and converge to zero when n goes to infinity. We use $\bar{R}_k > 0$ to denote the rate of the codebook used by Node k. This

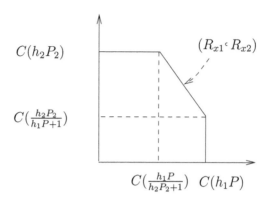

Figure 5.6: The bin size should be chosen such that the rate of the subcodebook in each bin should operate at the dominant face of the multiple access channel observed by the eavesdropper.

means the codebook used by Node k is composed of $2^{n(\bar{R}_k - \delta_n)}$ sequences sampled in an i.i.d. fashion from a given input distribution. Here the input distribution is chosen to be a zero mean Gaussian with variance P_k. Each time a new sequence is generated, it is labeled with two labels (i_k, j_k). i_k takes values from $1, ..., 2^{n(\bar{R}_k - R_{xk})}$, and j_k takes values from $1, ..., 2^{n(R_{xk} - \delta_n)}$. The labeling operation produces a useful structure in the codebook: the codewords with the same label i_k form a subcodebook by itself. Each subcodebook contains $2^{nR_{xk}}$ codewords which are also sampled in an i.i.d. fashion from the same input distribution.

Node k encodes the message W_k to a codeword as follows: the encoder considers all codewords with label $i_k = W_k$. Among these $2^{nR_{xk}}$ codewords, it chooses one codeword according to a uniform distribution.

$\bar{R}_k$ is chosen such that the transmitted codeword can be decoded by the other node reliably. Therefore

$$\bar{R}_k \leq C(P_k), \quad k = 1, 2. \tag{5.29}$$

By the construction of the codebook, we require

$$R_{xk} \leq \bar{R}_k, \quad k = 1, 2. \tag{5.30}$$

R_{xk} is chosen such that

$$R_{xk} \leq C(h_k P_k), \quad k = 1, 2, \tag{5.31}$$
$$R_{x1} + R_{x2} = C(h_1 P_1 + h_2 P_2). \tag{5.32}$$

The intuition is that the rates of the subcodebooks formed by codewords with the given label i_1, i_2, which is $(R_{x1} - \delta_n, R_{x2} - \delta_n)$, should be close to the rate limits at which eavesdroppers can decode. In this case, as illustrated in Figure 5.6, it should operate close to the boundary of the capacity region of the multiple access channel observed by the eavesdropper. This property later is used in proving the message is secured against the eavesdropper.

Let $\mathcal{C}$ denote the codebooks used by the two nodes. We next show that there must exist a good codebook to achieve the secrecy rate pair $(\bar{R}_1 - R_{x1}, \bar{R}_2 - R_{x2})$.

$$H(W_1, W_2 | Z^n, \mathcal{C}) \tag{5.33}$$
$$= H(W_1, W_2, X_1^n, X_2^n | Z^n, \mathcal{C}) - H(X_1^n, X_2^n | Z^n, W_1, W_2, \mathcal{C}). \tag{5.34}$$

Recall that given W_k, X_k^n is sampled from a subcodebook with rate R_{xk}. R_{x1}, R_{x2} is within the MAC region observed by the eavesdropper. Hence, given W_1, W_2, the eavesdropper should be able to decode X_1^n, X_2^n from Z^n. From Fano's inequality, this implies

$$H\left(X_1^n, X_2^n | Z^n, W_1, W_2, \mathcal{C}\right) \leq n\delta_n. \tag{5.35}$$

This means (5.34) is lower bounded by

$$H\left(X_1^n, X_2^n | Z^n, \mathcal{C}\right) - n\delta_n \tag{5.36}$$
$$=H\left(X_1^n, X_2^n | \mathcal{C}\right) - I\left(X_1^n, X_2^n; Z^n | \mathcal{C}\right) - n\delta_n \tag{5.37}$$
$$\geq H\left(X_1^n, X_2^n | \mathcal{C}\right) - I\left(\mathcal{C}, X_1^n, X_2^n; Z^n\right) - n\delta_n \tag{5.38}$$
$$=H\left(X_1^n, X_2^n | \mathcal{C}\right) - I\left(\mathcal{C}; Z^n | X_1^n, X_2^n\right) + I\left(X_1^n, X_2^n; Z^n\right) - n\delta_n. \tag{5.39}$$

Given $\{X_1^n, X_2^n\}$, Z^n is independent from the codebook being used. Hence $I\left(\mathcal{C}; Z^n | X_1^n, X_2^n\right) = 0$ and we can rewrite (5.39) as

$$H\left(X_1^n, X_2^n | \mathcal{C}\right) + I\left(X_1^n, X_2^n; Z^n\right) - n\delta_n \tag{5.40}$$
$$=n(R_1 + R_2) - nI\left(X_1, X_2; Z\right) - 2n\delta_n \tag{5.41}$$
$$=n\left(R_1 + R_2 - R_{x1} - R_{x2}\right) - 2n\delta_n \tag{5.42}$$
$$=H\left(W_1, W_2 | \mathcal{C}\right) - 2n\delta_n. \tag{5.43}$$

Hence we have

$$\lim_{n\to\infty} \frac{1}{n} I(W_1, W_2; Z^n | \mathcal{C}) = 0. \tag{5.44}$$

Let $p_{ek} = \Pr(W_k \neq \hat{W}_k)$ denote the probability of decoding errors for message W_k. Define $E[A|B] = \sum_b \Pr(B = b)E[A|B = b]$, where $E[A|B = b]$ is the conditional expectation of the random variable A conditioned on $B = b$. Then from [12], it follows that

$$\lim_{n\to\infty} E[p_{ek}|\mathcal{C}] = 0. \tag{5.45}$$

Let $\bar{e}_k$ denote the average transmission power of Node k computed by the energy of each codeword per channel use averaged over the codebook:

$$\bar{e}_k = \frac{1}{Q_k} \sum_{i=1}^{Q_k} \frac{1}{n} \sum_{j=1}^{n} |x_{i,j}|^2 \tag{5.46}$$

where Q_k is the total number of codewords in the codebook used by Node k. $x_{i,j}$ is the jth component of the ith codeword. Then from the codebook construction, we have $E[\bar{e}_k|\mathcal{C}] = P_k$ and $\lim_{n\to\infty} E[\bar{e}_k^2|\mathcal{C}] = P_k^2$. Therefore

$$\lim_{n\to\infty} E\left[(\bar{e}_k - P_k)^2 | \mathcal{C}\right] = 0. \tag{5.47}$$

This implies that

$$\lim_{n\to\infty} E[p_{e1} + p_{e2}|\mathcal{C}] + E\left[(\bar{e}_k - P_k)^2 | \mathcal{C}\right] \tag{5.48}$$

$$+ \frac{1}{n} I(W_1, W_2; Z^n | \mathcal{C}) = 0. \tag{5.49}$$

Hence there must exist a sequence of codebooks $\{\mathcal{C}\}$ with increasing codelength n, for which both $\frac{1}{n}I(W_1, W_2; Z^n)$ and p_{e1}, p_{e2} vanish and $\bar{e}_k$ converges to the transmission power target P_k when n goes to infinity. This means W_1, W_2 is secure and can be transmitted reliably.

Remark 5.5.1. *To satisfy any given power constraint $\bar{P}_k$, we only need to set P_k as $\bar{P}_k - \varepsilon_k$ for some positive but arbitrarily small ε_k. Then there must exist a code length n_0 such that for any $n > n_0$, we have $\bar{e}_k < \bar{P}_k$.*

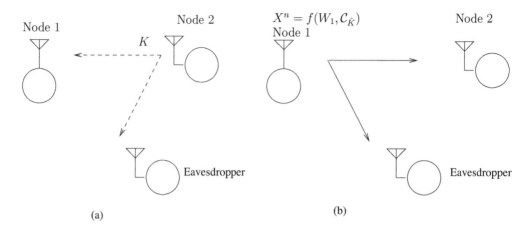

Figure 5.7: Achieving $R_1 > 0$ through secret key generation: (a) first phase: Node 2 sends a secret key K to Node 1. (b) Second phase: Node 1 decodes K as $\hat{K}$ and uses the $\hat{K}$th codebook $\mathcal{C}_{\hat{K}}$ to encode W_1.

5.5.2 Secrecy through Key Generation

The public discussion scheme used in [9] is slightly different from the one described in Section 5.3, in that Node 1 tries to decode the signals from Node 2 instead of adding the noisy signals it received to its own transmission signals directly. To facilitate explanation, in this section we describe a simplified version of it, which first appeared in [4].

As shown in Figure 5.7, the scheme is composed of two phases. Assume the first phase spans over n_1 channel uses and the second phase spans over n_2 channel uses. Let $n = n_1 + n_2$. Let $\alpha_1 = n_1/(n_1 + n_2)$ and $\alpha_2 = 1 - \alpha_1$. During the first phase, Node 2 sends a secret message K which takes values from $\{1, ..., 2^{n_1 R_2}\}$ to Node 1 using a good codebook for the wiretap channel it observes, with average power P_2. At the same time, Node 1 jams the eavesdropper with Gaussian noise P_1 as shown in Section 5.4.1. Hence

$$0 \leq R_2 \leq \max\{0, C(P_2) - C(\frac{h_2 P_2}{1 + h_1 P_1})\}. \tag{5.50}$$

Node 1 decodes K and during the second phase uses it to encrypt part of its message as follows: Node 1 generates $2^{n_1 R_2}$ codebooks instead of one. Each of them is sampled in an i.i.d. fashion from a Gaussian distribution with zero mean and variance P_1. The codeword with each codebook is then labeled with two labels (i, j) as described in the previous section, where i takes values from $1, ...2^{n_2 R_1}$ and j takes values from $1, ..., 2^{n_2(R_x - \delta_n)}$. The codebook is labeled by k, which takes values from $1, ..., 2^{n R_2}$. We choose R_x such that $R_x \geq 0$ and

$$R_x + \frac{\alpha_1}{\alpha_2} R_2 = C(\frac{h_1 P}{h_2 P_2 + 1})). \tag{5.51}$$

Note that due to the contribution of the secret key, the required randomness in the codebook of Node 1, represented by R_x, decreases from $C(\frac{h_1 P}{h_2 P_2 + 1}))$ to $C(\frac{h_1 P}{h_2 P_2 + 1})) - \frac{\alpha_1}{\alpha_2} R_2$.

Node 1 encodes a confidential message as follows: each time, Node 1 chooses a codebook based on the value of $\hat{K}$, where $\hat{K}$ is the decoding result from phase 1. Then it considers all codewords in this codebook whose label $i = W_1$, and chooses one according to a uniform distribution. This codeword is then transmitted during the second phase.

If Node 1 decodes K correctly, then $\hat{K} = K$. Then Node 2 knows which codebook is used by Node 1. This limits the number of possible codewords transmitted by Node 1 to

$2^{n(R_1+R_x-\delta_n)}$. In order for the codeword to be decoded by Node 2 reliably, we require

$$0 \leq R_1 + R_x \leq C(P_1). \qquad (5.52)$$

The transmission rate for the message W_1 is therefore given by $\alpha_2 R_1$. In order to satisfy the constraints (5.50), (5.51), and (5.52), $\alpha_2 R_1$ is given by:

$$\alpha_2[C(P) - [C(\frac{h_1 P_1}{h_2 P_2 + 1})) - \frac{\alpha_1}{\alpha_2}[C(P_2) - C(\frac{h_2 P_2}{h_1 P_1 + 1})\}]^+]^+]^+ \qquad (5.53)$$

where $[x]^+ = \max\{x, 0\}$. We next show that W_1 transmitted with such a scheme is secure. The proof is adapted from [4].

The following notation is used in the remainder of the proof: $\bar{x}$ denotes any signal x which is related to the second phase. Otherwise, the signal is related to the first phase. Let $\mathcal{C}$ denote the codebook used by the two nodes. The equivocation rate is then bounded as follows:

$$H\left(W_1|Z^{n_1}, \bar{Z}^{n_2}, \mathcal{C}\right)$$
$$=H\left(\bar{X}^{n_2}, W_1|Z^{n_1}, \bar{Z}^{n_2}, \mathcal{C}\right) - H\left(\bar{X}^{n_2}|W_1, Z^{n_1}, \bar{Z}^{n_2}, \mathcal{C}\right) \qquad (5.54)$$
$$\geq H\left(\bar{X}^{n_2}, W_1|Z^{n_1}, \bar{Z}^{n_2}, \mathcal{C}\right) - n_2 \varepsilon \qquad (5.55)$$

where in (5.55) we upper bound the term $H\left(\bar{X}^{n_2}|W_1, Z^{n_1}, \bar{Z}^{n_2}, \mathcal{C}\right)$ with $n_2 \varepsilon$ where $\varepsilon > 0$ and $\lim_{n \to \infty} \varepsilon = 0$. This is because given the message W_1, for a fixed codebook, the eavesdropper needs to consider at most $2^{n(R_x+R_2)}$ codewords, which is a subcodebook composed of i.i.d. sequences sampled from a Gaussian distribution. Because of (5.51), the eavesdropper can decode the codeword transmitted by Node 1 if it knows W_1. Hence (5.55) follows from Fano's inequality.

We next write (5.55) as:

$$H\left(W_1|Z^{n_1}, \bar{Z}^{n_2}, \bar{X}^{n_2}, \mathcal{C}\right) + H\left(\bar{X}^{n_2}|Z^{n_1}, \bar{Z}^{n_2}, \mathcal{C}\right) - n_2 \varepsilon \qquad (5.56)$$
$$=H\left(\bar{X}^{n_2}|Z^{n_1}, \bar{Z}^{n_2}, \mathcal{C}\right) - n_2 \varepsilon \qquad (5.57)$$
$$=H\left(\bar{X}^{n_2}|Z^{n_1}, \bar{Z}^{n_2}, \mathcal{C}\right) - H\left(\bar{X}^{n_2}, \mathcal{C}\right) + H\left(\bar{X}^{n_2}|\mathcal{C}\right) - n_2 \varepsilon \qquad (5.58)$$
$$=H\left(\bar{X}^{n_2}|\mathcal{C}\right) - I\left(\bar{X}^{n_2}; Z^{n_1}, \bar{Z}^{n_2}|\mathcal{C}\right) - n_2 \varepsilon \qquad (5.59)$$
$$=H\left(\bar{X}^{n_2}|\mathcal{C}\right) - I\left(\bar{X}^{n_2}; \bar{Z}^{n_2}|\mathcal{C}\right) - I\left(\bar{X}^{n_2}; Z^{n_1}|\bar{Z}^{n_2}, \mathcal{C}\right) - n_2 \varepsilon. \qquad (5.60)$$

The third term in (5.60) can then be bounded as follows:

$$I\left(\bar{X}^{n_2}; Z^{n_1}|\bar{Z}^{n_2}, \mathcal{C}\right) \qquad (5.61)$$
$$=h\left(Z^{n_1}|\bar{Z}^{n_2}, \mathcal{C}\right) - h\left(Z^{n_1}|\bar{Z}^{n_2}, \bar{X}^{n_2}, \mathcal{C}\right) \qquad (5.62)$$
$$=h\left(Z^{n_1}|\bar{Z}^{n_2}, \mathcal{C}\right) - h\left(Z^{n_1}|\bar{X}^{n_2}, \mathcal{C}\right) \qquad (5.63)$$
$$\leq h\left(Z^{n_1}|\bar{Z}^{n_2}, \mathcal{C}\right) - h\left(Z^{n_1}|\bar{X}^{n_2}, K, \mathcal{C}\right) \qquad (5.64)$$
$$=h\left(Z^{n_1}|\bar{Z}^{n_2}, \mathcal{C}\right) - h\left(Z^{n_1}|K, \mathcal{C}\right) \qquad (5.65)$$
$$=h\left(Z^{n_1}|\bar{Z}^{n_2}, \mathcal{C}\right) - h\left(Z^{n_1}|\mathcal{C}\right) - h\left(Z^{n_1}|K, \mathcal{C}\right) + h\left(Z^{n_1}|\mathcal{C}\right) \qquad (5.66)$$
$$=I\left(Z^{n_1}; K|\mathcal{C}\right) - I\left(Z^{n_1}; \bar{Z}^{n_2}|\mathcal{C}\right) \qquad (5.67)$$
$$\leq I\left(Z^{n_1}; K|\mathcal{C}\right) \leq n_1 \varepsilon. \qquad (5.68)$$

Substituting (5.68) into (5.60), we have

$$H\left(W_1|Z^{n_1}, \bar{Z}^{n_2}, \mathcal{C}\right) \qquad (5.69)$$

$$\geq H\left(\bar{X}^{n_2}|\mathcal{C}\right) - I\left(\bar{X}^{n_2}; \bar{Z}^{n_2}|\mathcal{C}\right) - (n_1 + n_2)\varepsilon. \tag{5.70}$$

Then following similar derivation to the previous section, $I\left(\bar{X}^{n_2}; \bar{Z}^{n_2}|\mathcal{C}\right)$ is upper bounded by $n_2 I(X;Z) = n_2 C(\frac{h_1 P_1}{h_2 P_2 + 1}) = n_2(R_x + R_2)$. $H(\bar{X}^{n_2}|\mathcal{C})$, is given by $n_2(R_1 + R_x + R_2)$. Hence

$$H\left(W_1|Z^{n_1}, \bar{Z}^{n_2}, \mathcal{C}\right) \geq n_2(R_1 + R_x + R_2) - n_2(R_x + R_2) = n_2 R_1. \tag{5.71}$$

Hence

$$\lim_{n\to\infty} \frac{1}{n} H\left(W_1|Z^{n_1}, \bar{Z}^{n_2}, \mathcal{C}\right) = \alpha_2 R_1 = \lim_{n\to\infty} \frac{1}{n} H\left(W_1\right). \tag{5.72}$$

Following similar derivation to the previous section, it can be shown that there exists a codebook $\mathcal{C}$ which secures W_1.

5.5.3   Block Markov Coding Scheme

Reference [9] uses block Markov coding along with the key generation scheme in Section 5.5.2 so that the message transmission overlaps with the key generation process to avoid the rate loss due to time sharing factor α. We describe its achievable scheme in this section.

Each node divides its power into two parts: Node 1 uses power P_n' to transmit Gaussian noise, and uses power P_c' to transmit codewords. Node 2 uses power P_n to transmit Gaussian noise, and uses power P_c to transmit codewords. This power splitting strategy makes sure the achieved rates reap the benefits of the strategy in Section 5.4.1. To satisfy average power constraint, we require:

$$0 \leq P_c' + P_n' \leq P_1, \tag{5.73}$$

$$0 \leq P_c + P_n \leq P_2. \tag{5.74}$$

The construction of the codebook is the same as described in Section 5.5.1, except that as in Section 5.5.2, Node 1 will use multiple codebooks. The actual codebook it uses will be determined by a secret key it received from Node 2 earlier: Node 1 generates 2^{nR_K} codebooks. Each codebook contains 2^{nR_1} subcodebooks. Each subcodebook contains $2^{n(R_{x1} - \delta_n)}$ codewords. Each codeword is sampled in an i.i.d. fashion from a Gaussian distribution with zero mean and variance P_c'.

Node 2 uses one codebook. Each codebook contains 2^{nR_K} subcodebooks. Each subcodebook contains $2^{n(R_{x2} - \delta_n)}$ codewords. Each codeword is sampled in an i.i.d. fashion from a Gaussian distribution with zero mean and variance P_c.

The transmission is divided into several blocks. During each block, as shown in Figure 5.8, Node 1 chooses a codebook based on the secret key it receives from node 2 in the previous block. It then chooses the subcodebook in this codebook based on the value of the secret message W_1. A codeword is chosen randomly from this subcodebook and transmitted by Node 1. At the same time, Node 2 sends a new secret key to Node 1, whose value is chosen between $\{1, ..., 2^{nR_K}\}$ according to a uniform distribution.

Let K_t, W_t be the secret key and the message transmitted during the tth block. Then to make sure $\lim_{n\to\infty} \frac{1}{n} I(K_t, W_t; Z_t^n) = 0$, as in Section 5.5.1, we require:

$$R_K + R_{x1} + R_{x2} = C\left(\frac{h_1 P_c' + h_2 P_c}{1 + h_1 P_n' + h_2 P_n}\right) \tag{5.75}$$

$$R_K + R_{x1} \leq C\left(\frac{h_1 P_c'}{1 + h_1 P_n' + h_2 P_n}\right) \tag{5.76}$$

$$X_{1,t}^n = f(W_t, \mathcal{C}_{\hat{K}_{t-1}})$$

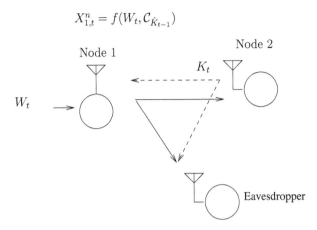

Figure 5.8: Achieving $R_1 > 0$ through secret key generation and cooperative jamming: during the tth block, Node 2 sends a secret key K_t to Node 1. Node 1 decodes K_{t-1} from the previous phase as $\hat{K}_{t-1}$ and uses the $\hat{K}_{t-1}$th codebook $\mathcal{C}_{\hat{K}}$ to encode W_t.

$$R_{x2} \leq C(\frac{h_2 P_2}{1 + h_1 P_n' + h_2 P_n}) \tag{5.77}$$

$$R_{x1} \geq 0, R_{x2} \geq 0, R_k \geq 0. \tag{5.78}$$

The difference is that R_{x1} is replaced with $R_K + R_{x1}$.

In order for Node 2 to be able to decode W_1, we require $R_1 \geq 0$ and

$$R_1 + R_{x1} \leq C(\frac{P_c'}{1 + P_n'}). \tag{5.79}$$

In order for Node 1 to decode W_1, we require:

$$R_K + R_{x2} \leq C(\frac{P_c}{1 + P_n}). \tag{5.80}$$

The achievable secrecy rate is given by R_1 such that the constraints (5.73) and (5.74) through (5.75)-(5.80) are satisfied. Reference [9] shows that this can be further simplified to

$$R_1 \leq \max\{0, \min\{C(\frac{P_c'}{1 + P_n'}), C(\frac{P_c'}{1 + P_n'}) + C(\frac{P_c}{1 + P_n}) - C(\frac{h_1 P_c' + h_2 P_c}{1 + h_1 P_n' + h_2 P_n})\}\}. \tag{5.81}$$

We refer the interested readers to [9] for the complete derivation that leads to (5.81). Here we shall show this for an example given in [4]: $\sqrt{h_1} = \sqrt{2}, \sqrt{h_2} = 1, P_2 = 1$. With this configuration, the rate achieved by the strategy in (5.26) in Section 5.4.1 is always zero, regardless of the value of P. Reference [4] shows that by using a strategy similar to the one in Section 5.3, a positive secrecy rate is achievable. We next evaluate the secrecy rate achievable with the scheme in this section.

We first show that (5.81) is an upper bound on the secrecy rate expression by considering a subset of the constraints that R_1 needs to satisfy. Then we maximize the upper bound expression by optimizing over the power allocation, which simplifies this upper bound expression. Finally, we show this maximal upper bound is attainable by choosing R_k, R_{x1}, R_{x2}, R_s, which satisfies all constraints.

The fact that (5.81) is an upper bound on the secrecy rate expression can be seen from (5.75), (5.79), and (5.80). Adding (5.79) and (5.80) and subtracting the equality (5.75) from both sides, we get the second term inside the minimum in (5.81). The first term comes from (5.79).

We next maximize this upper bound. Fix the value of $P'_c + P'_n$ and $P_c + P_n$. For $\sqrt{h_1} = \sqrt{2}$, $\sqrt{h_2} = 1$, it can be verified that $C(\frac{P'_c}{1+P'_n}) + C(\frac{P_c}{1+P_n}) - C(\frac{h_1 P'_c + h_2 P_c}{1+h_1 P'_n + h_2 P_n})$ is a decreasing function in P_n. This means this bound is maximized when $P_n = 0$. On the other hand, when $P_n = 0$, $C(\frac{P'_c}{1+P'_n}) + C(\frac{P_c}{1+P_n}) - C(\frac{h_1 P'_c + h_2 P_c}{1+h_1 P'_n + h_2 P_n})$ is an increasing function in P'_n. Since $C(\frac{P'_c}{1+P'_n})$ is a decreasing function in P'_n, the right hand side of (5.81) is maximized when exactly one of the following two cases take place:

$$C(\frac{P'_c}{1 + P'_n}) = C(\frac{P'_c}{1 + P'_n}) + C(\frac{P_c}{1 + P_n}) - C(\frac{h_1 P'_c + h_2 P_c}{1 + h_1 P'_n + h_2 P_n}) \tag{5.82}$$

or

$$P'_n = P_1 \text{ and } C(\frac{P'_c}{1 + P'_n}) > C(\frac{P'_c}{1 + P'_n}) + C(\frac{P_c}{1 + P_n}) - C(\frac{h_1 P'_c + h_2 P_c}{1 + h_1 P'_n + h_2 P_n}). \tag{5.83}$$

The second case implies $R_1 = 0$. Hence we only need to consider the first case. This means, if $R_1 > 0$, then the right hand side of (5.81) is maximized when (5.82) is satisfied and $P_n = 0$, which implies $P_c = P_2$ and $P'_c + P'_n = P_1$.

We next show a feasible solution when R_K, R_{x1}, R_{x2}, R_1 exists for this power allocation. We choose R_{x2} so that the constraint (5.77) is active:

$$R_{x2} = C(\frac{h_2 P_2}{1 + h_1 P'_n}) \tag{5.84}$$

and choose $R_{x1} = 0$. In order to satisfy (5.75), we choose R_K as

$$R_K = C(\frac{h_1 P'_c + h_2 P_2}{1 + h_1 P'_n + h_2 P_n}) - R_{x2} = C(\frac{h_1 P'_c}{1 + h_1 P'_n + h_2 P_2}). \tag{5.85}$$

It can be verified that all constraints are satisfied. In particular, due to (5.82) and $R_{x1} = 0$, (5.75) and (5.80) are satisfied simultaneously. This implies R_1 can be as large as

$$R_1 = C(\frac{P'_c}{1 + P'_n}) \tag{5.86}$$

when (5.82) has a solution for P'_n when $P'_c = P_1 - P'_n$, $P_c = P_2$, $P_n = 0$. Otherwise $R_1 = 0$.

This result has the following operation meaning: Node 1 splits its power and uses part of it to jam the eavesdropper. It then chooses a codebook based on the secret key it receives from the previous block. Node 1 then transmits the codeword from this codebook based on the value of the confidential message. No further randomness is used in Node 1's encoder since $R_{x1} = 0$. Node 2 does not transmit any Gaussian noise since $P_n = 0$. During each block, it randomly generates a secret key and transmits a codeword chosen randomly from all the codewords in its codebook whose label i equals the secret key value.

5.6   When the Eavesdropper Channel States Are Not Known

All of the achievable strategies in the previous sections rely on the knowledge of the eavesdropper channel state information, as evident in that the rate expressions depend on the eavesdropper's channel gains $\sqrt{h_1}, \sqrt{h_2}$, and the security of these strategies will be compromised if the actual values of $\sqrt{h_1}, \sqrt{h_2}$ are different from the expected values. For example, we had pointed out that the secrecy rate achieved in Section 5.3 decreases to zero when $\sqrt{h_2}$ goes to infinity. Hence if we underestimate the value of $\sqrt{h_2}$, which means the eavesdropper observes a higher received signal to noise ratio from Node 2 than we expected, then Node 1 could transmit the message at a rate higher than the secrecy rate supported by the coding scheme, resulting in a security breach. For the cooperative jamming scheme described in Section 5.4.1, the security compromise happens when we overestimate $\sqrt{h_2}$, which means the noise sent by Node 2 cannot jam the eavesdropper effectively. For the scheme in Section 5.4.2, miscalculating $\sqrt{h_1}$ and $\sqrt{h_2}$ means it is easier for the eavesdropper to differentiate which node is transmitting. Overall, none of these schemes is robust against deviations in eavesdropper channel gains. This brings the question whether the secrecy is still achievable for the Gaussian two-way wiretap channel if $\sqrt{h_1}\sqrt{h_2}$ is only known to the eavesdropper herself. Interestingly, recent progress in this direction [13, 14] showed that a positive secrecy rate is still achievable in this setting. We note that the methods of computing secrecy rate such as in [5] cannot be applied to this setting, and new tools have been recently developed [15]. Here, we shall describe the protocol proposed in [13] and refer the interested readers to [13–15] for its secrecy rate computation and secrecy analysis.

The communication is divided to three stages:

1. The first stage takes n channel uses. In this stage, Nodes 1 and 2 transmit i.i.d. Gaussian random sequences with zero mean. Let P denote its variance. Let J_i denote the signal transmitted by node $i, i = 0, 1$ during this stage.

2. The second stage also takes n channel uses. In this stage, only Node 1 transmits. Node 1 generates a binary i.i.d. sequence T^n, such that $\Pr(T_i = 0) = 1/2, i = 1, ..., n$. Observe that Node 1 at this moment has the knowledge of

   (a) J_0^n, which is the sequence it transmitted during the first stage, and
   (b) $J_1^n + N_{c1}^n$, which is the sequence it received during the first stage, where N_{c1}^n is the channel noise.

   Node 1 then constructs a jamming sequence J_r^n, such that

   $$J_{r,i} = \begin{cases} J_{0,i}, & T_i = 0 \\ J_{1,i} + N_{c1,i}, & T_i = 1 \end{cases} \qquad (5.87)$$

   The transmitter at Node 1 takes input V^n and transmits

   $$V_i + J_{r,i} + N_{J,i} \qquad (5.88)$$

   during the ith channel use in the second stage, where $\{N_J^n\}$ is an i.i.d. Gaussian sequence with zero mean and unit variance.

   The confidential message W shall be encoded in V^n through a stochastic encoder.

3. During the third stage, only Node 1 transmits. In this stage, Node 1 broadcasts T^n as a public message to Node 2.

Note that this scheme is similar to the public discussion scheme in Section 5.3, in that Node 1 also adds a noise sequence to its signals V^n before sending out. The difference is how this noise sequence is generated. In Section 5.3, the noise sequence is sent from Node 2 via K. Hence the eavesdropper can counteract by trying to achieve a good reception from Node 2 in order to estimate this noise sequence. Here the noise sequence is mixed from two noise sequences which are transmitted from Nodes 1 and 2 simultaneously. The way that this mixture is performed is only determined afterwards by T^n. The eavesdropper cannot get a good reception from Node 1 and Node 2 simultaneously, hence she cannot get a good estimate of this noise sequence, which provides an opportunity to achieve secrecy.

5.7 Converse

So far, we presented achievable rates with bidirectional links. The remainder of this chapter is devoted to deriving outer bounds on the secrecy rate region of the two-way wiretap channel.

5.7.1 Outer Bounds

The proof for following bound can be found in [4].

Theorem 5.7.1. *The secrecy capacity region of the channel model in Figure 5.1 is bounded by*

$$\cup_{\Pr(X_1, X_2)} \{(R_1, R_2) : (5.90)\ (5.91)\ (5.92)\ \text{holds}\} \tag{5.89}$$

$$0 \leq R_1 \leq I(X_1; Y_1) \tag{5.90}$$
$$0 \leq R_2 \leq I(X_2; Y_2) \tag{5.91}$$
$$R_1 + R_2 \leq \min \left\{ \begin{array}{l} I(X_1; Y_1 | Z, X_2) + I(X_2; Z, Y_2 | X_1), \\ I(X_2; Y_2 | Z, X_1) + I(X_1; Z, Y_1 | X_2) \end{array} \right\}. \tag{5.92}$$

For a deterministic binary wiretap channel, Theorem 5.7.1 leads to the equivocation capacity region, as shown by the following theorem:

Theorem 5.7.2. *When X_1, X_2 are binary and $Y_1 = X_1 \oplus X_2, Y_2 = X_2 \oplus X_1, Z = X_1 \oplus X_2$, the secrecy capacity region is given by*

$$R_j \geq 0, j = 1, 2 \tag{5.93}$$
$$R_1 + R_2 \leq 1. \tag{5.94}$$

Proof. The achievability follows from [1, Theorem 2]. The converse follows from Theorem 5.7.1. The sum rate bound specializes as follows:

$$I(X_1; Y_1 | Z, X_2) + I(X_2; Y_2, Z | X_1) \tag{5.95}$$
$$= I(X_1; X_1 | X_1 \oplus X_2, X_2) + I(X_2; X_2, X_1 \oplus X_2 | X_1) \tag{5.96}$$
$$= I(X_1; X_1 | X_1, X_2) + I(X_2; X_2, X_1 | X_1) \tag{5.97}$$
$$= I(X_2; X_2, X_1 | X_1) \tag{5.98}$$
$$\leq H(X_2) \leq 1. \tag{5.99}$$

$\square$

We next consider the Gaussian channel.

Theorem 5.7.3. *The secrecy capacity region of the Gaussian two-way wiretap channel is included in*

$$0 \le R_1 \le C(P) \tag{5.100}$$

$$0 \le R_2 \le C(P_2) \tag{5.101}$$

$$R_1 + R_2 \le \min \left\{ \begin{array}{l} \inf_{\sigma^2 \ge 0} C\left(\frac{P(1+\sigma^2)}{(h_1 P + 1 + \sigma^2)}\right) + C\left(\frac{P_2(h_2 + 1 + \sigma^2)}{1+\sigma^2}\right) \\ \inf_{\sigma^2 \ge 0} C\left(\frac{P_2(1+\sigma^2)}{(h_2 P_2 + 1 + \sigma^2)}\right) + C\left(\frac{P(h_1 + 1 + \sigma^2)}{1+\sigma^2}\right) \end{array} \right\}. \tag{5.102}$$

Proof. Define N_4 as a Gaussian random variable such that $N_4 \sim \mathcal{N}(0, \sigma^2)$ and is independent from $N_i, i = 1, 2, 3$. We consider a channel where the eavesdropper receives $Z + N_4$ and derive an outer bound for this new channel.

To prove the theorem, we first show $I(X_1; Y_1)$, $I(X_1; Y_1 | Z, X_2)$, $I(X_2; Y_2, Z | X_1)$, $I(X_2; Y_2)$, $I(X_2; Y_2 | Z, X_2)$, and $I(X_1; Z, Y_1 | X_2)$ are maximized simultaneously when X_1 and X_2 are independent, $X_1 \sim \mathcal{N}(0, P)$, and $X_2 \sim \mathcal{N}(0, P_2)$.

Due to the symmetry of the channel model, we only need to show $I(X_1; Y_1)$, $I(X_1; Y_1 | Z, X_2)$, and $I(X_2; Y_2, Z | X_1)$ are maximized by this distribution.

The case of $I(X_1; Y_1 | Z, X_2)$ was shown in the proof of Theorem 5.7.4.

For $I(X_2; Y_2, Z | X_1)$, we have:

$$I(X_2; Y_2, Z | X_1) \tag{5.103}$$

$$= I\left(X_2; X_2 + N_3, \sqrt{h_2} X_2 + N_2 + N_4 | X_1\right) \tag{5.104}$$

$$= h\left(\sqrt{h_2} X_2 + N_2 + N_4, X_2 + N_3 | X_1\right) - h(N_2 + N_4, N_3) \tag{5.105}$$

$$\le h\left(\sqrt{h_2} X_2 + N_2 + N_4, X_2 + N_3\right) - h(N_2 + N_4, N_3). \tag{5.106}$$

Hence $I(X_2; Y_2, Z | X_1)$ is maximized when X_1 and X_2 are independent, $X_1 \sim \mathcal{N}(0, P)$, and $X_2 \sim \mathcal{N}(0, P_2)$. The theorem then is a consequence of Theorem 5.7.1 when evaluated at this input distribution. □

Recall in some achievability strategy, signals from Node 2 are not used by Node 1 for encoding, which is the case in Section 5.4. The next result upper bounds the secrecy rate R_1 achievable by this class of strategies [4, 16].

Theorem 5.7.4. *When Y_2 is a constant, that is, Y_2 is ignored by Node 1, the secrecy rate R_1 is upper bounded by*

$$\max_{\Pr(X_1, X_2)} \min\{I(X_1; Y_1), I(X_1; Y_1 | Z, X_2) + I(X_2; Z | X_1)\}. \tag{5.107}$$

For the Gaussian bound, this bound becomes

$$\inf_{\sigma^2 \ge 0} C\left(\frac{P(1+\sigma^2)}{(h_1 P + 1 + \sigma^2)}\right) + C\left(\frac{h_2 P_2}{1 + \sigma^2}\right). \tag{5.108}$$

Proof. We refer the interested readers to [4, 16] for the proof of (5.107). Here we provide the evaluation of (5.107), from which we prove (5.108).

Recall that Z is the signal received by the eavesdropper. We next consider a channel where the eavesdropper receives $Z + N_4$, where N_4 was defined in the proof for Theorem 5.7.3.

Since $Z + N_4$ is a degraded version of Z, we can find an upper bound of the original channel by deriving an upper bound for this new channel. This upper bound is found by applying the bound (5.107).

We next prove that all terms in the upper bound (5.107) is maximized when X_1, X_2 are independent and each has a Gaussian distribution with zero mean and maximum possible variance: $I(X_1; Y_1)$ is obviously maximized by this distribution. For the other two terms, we have:

$$I(X_1; Y_1 | X_2, Z) \tag{5.109}$$

$$= I\left(X_1; X_1 + N_1 | X_2, \sqrt{h_1} X_1 + \sqrt{h_2} X_2 + N_2 + N_4\right) \tag{5.110}$$

$$= h\left(X_1 + N_1 | X_2, \sqrt{h_1} X_1 + \sqrt{h_2} X_2 + N_2 + N_4\right) - h(N_1 | N_2 + N_4) \tag{5.111}$$

$$\leq h\left(X_1 + N_1 | \sqrt{h_1} X_1 + N_2 + N_4\right) - h(N_1 | N_2 + N_4) \tag{5.112}$$

and

$$I(X_2; Z | X_1) \tag{5.113}$$

$$= I\left(X_2; \sqrt{h_2} X_2 + N_2 + N_4 | X_1\right) \tag{5.114}$$

$$= h\left(\sqrt{h_2} X_2 + N_2 + N_4 | X_1\right) - h(N_2 + N_4) \tag{5.115}$$

$$\leq h\left(\sqrt{h_2} X_2 + N_2 + N_4\right) - h(N_2 + N_4). \tag{5.116}$$

Equations (5.112) and (5.116) show that the second term in (5.107) are maximized when X_1 and X_2 are independent. Moreover, (5.112) is known to be maximized when X_1 has a Gaussian distribution with the maximum possible variance; see [17]. Equation (5.116) is also maximized when X_2 has a Gaussian distribution with the maximum possible variance. Hence we have shown the optimal input distribution for X_1, X_2 is an independent Gaussian distribution. For this distribution, it can be verified the second term in (5.107) becomes (5.108).

This concludes the proof of the theorem.   □

5.7.2   Discussion

We next describe the implications of the bounds we just derived.

The next result shows that when the transmission powers of Nodes 1 and 2 are proportional to each other, the simple strategy in Section 5.4.1 is actually close to being optimal.

Theorem 5.7.5. *When $P_2 = kP$, k is a positive constant, and $h_j \neq 0, j = 1, 2$, the loss in secrecy rates when received signals are not used to compute transmitting signals at Node j, $j = 1, 2$ is bounded by a constant, which is only a function of h_1 and h_2.*

Proof. The proof is given in Section 5.9.   □

Figure 5.9 illustrates the case described in Theorem 5.7.5. In this figure, we fix $P = P_2$, increase P, and compare the achievable sum secrecy rate and its upper bound. It is observed that the gap between these two does not increase with P.

On the other hand, when P is not increasing linearly with respect to P_2, the simple strategy in Section 5.4.1 could incur a significant loss in the secrecy rate. Figure 5.10 plots the achievable secrecy rate R_1 computed according to Section 5.5.3. Also plotted in the figure is an upper bound on R_1 found without utilizing Y_2, i.e., without feedback in the model.

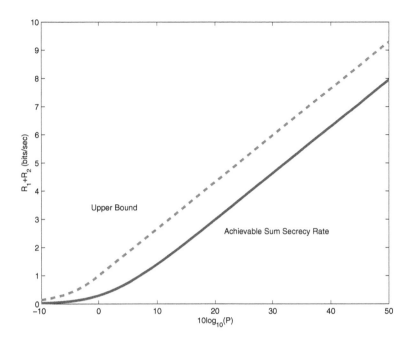

Figure 5.9: Comparison of the sum secrecy rate when $P = P_2$, $h_1 = h_2 = 1$, $\rho = \eta = 0$. The sum secrecy rate computed with (5.117), when Y_2 is ignored at Node 1. The upper bound on the sum rate is computed with (5.102) [4].

Figure 5.10 shows that the achieved secrecy rate with feedback can exceed the upper bound without feedback by an amount that increases with power, a fact we shall state formally in Theorem 5.7.6. This means that *it is impossible to achieve the same secrecy rate by designing the encoder at Node 2 if Node 1 ignores the feedback signal Y_2.*

Next, the following theorem states the fact illustrated in Figure 5.10.

Theorem 5.7.6. *Even in the case where cooperative jamming is possible ($h_j \neq 0, j = 1, 2$), when P is not linearly increasing with P_2, ignoring Y_2 at Node 1 can lead to unbounded loss in the secrecy rate.*

Proof. The proof is given in Section 5.10. □

5.8 Conclusion

In this chapter, we discussed several different schemes to achieve secrecy in the Gaussian two-way wiretap channel. We also described an outer bound on the secrecy capacity region. By comparing the outer bound with the achievable rates, we found that the benefits of feedback signals to secrecy is more pronounced when the power of each of the two transmitters do not increase in proportion to the other.

Our coverage of channel models with feedback for secret message transmission is by no means complete. In addition to the two-way wiretap channel discussed in this chapter, a large body of works consider multiple nodes having access to correlated random sources and that are connected with a public discussion link. These nodes then use the public discussion link to generate a secret key from the correlated random sources (see [3,18–20] for example) and references therein. Another channel model which is closely related to the Gaussian two-way wiretap channel is called the two-way relay channel with an untrusted relay channel,

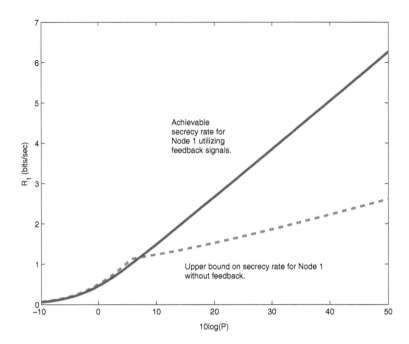

Figure 5.10: Comparison of the sum secrecy rate when $P_2 = P^{1/4}$, $h_1 = h_2 = 1$, $\rho = \eta = 0$, $\alpha = 0.5$. Achievable secrecy rate at Node 1 is computed according to Section 5.5.2. Upper bound on the secrecy rate at Node 1 when Y_2 is ignored is computed with Theorem 5.7.4 [4].

which was studied in [4, 21]. For this model, where the relay is an eavesdropper and could potentially perform a Byzantine attack, References [4, 21] derive an achievable secrecy rate and a converse and proves that any Byzantine attack can be reliably detected respectively. A Gaussian wiretap channel where the sender observes the signals received by the intended receiver was considered in [22, 23] and it was shown that the achieved secrecy rate could equal the channel capacity without an eavesdropper in certain cases. A wiretap channel with secure rate-limited feedback was studied in [24] and its secrecy capacity was found therein.

5.9   Proof of Theorem 5.7.5

The proof is from [4]. Since received signals are not used to compute transmitting signals at Node j, $j = 1, 2$, we let $\alpha = 1$ in (5.53). In this case, when $P = kP_2$, we can achieve R_1 when $R_2 = 0$:

$$R_1 = C\left(P\right) - C\left(\frac{h_1}{h_2 k + 1/P}\right), \tag{5.117}$$

Similarly, we can achieve R_2 when $R_1 = 0$:

$$R_2 = C\left(kP\right) - C\left(\frac{h_2 k}{h_1 + 1/P}\right). \tag{5.118}$$

The sum rate bound given by Theorem 5.7.3 is upper bounded by:

$$\min\left\{C\left(\frac{P}{h_1 P + 1}\right) + C\left((h_2 + 1)\, kP\right), C\left(\frac{kP}{h_2 kP + 1}\right) + C\left((h_1 + 1)\, P\right)\right\}. \tag{5.119}$$

To prove Theorem 5.7.5, it is sufficient to show that both R_1 and R_2 are within constant gaps of (5.119), as we show below:

$$C\left(\frac{P}{h_1 P + 1}\right) + C\left((h_2 + 1)\,kP\right) - R_1 \tag{5.120}$$

$$=C\left(\frac{P}{h_1 P + 1}\right) + C\left((h_2 + 1)\,kP\right) - C\left(P\right) + C\left(\frac{h_1}{h_2 k + 1/P}\right) \tag{5.121}$$

$$\leq C\left(\frac{P}{h_1 P + 1}\right) + C\left((h_2 + 1)\,kP\right) - C\left(P\right) + C\left(\frac{h_1}{h_2 k}\right) \tag{5.122}$$

$$\leq C\left(\frac{1}{h_1}\right) + C\left((h_2 + 1)\,kP\right) - C\left(P\right) + C\left(\frac{h_1}{h_2 k}\right) \tag{5.123}$$

$$=C\left(\frac{1}{h_1}\right) + \frac{1}{2}\log_2\left(\frac{1 + (h_2 + 1)\,kP}{1 + P}\right) + C\left(\frac{h_1}{h_2 k}\right) \tag{5.124}$$

$$\leq C\left(\frac{1}{h_1}\right) + \frac{1}{2}\log_2\left(\frac{1 + \max\{1,(h_2 + 1)\,k\}\,P}{1 + P}\right) + C\left(\frac{h_1}{h_2 k}\right) \tag{5.125}$$

$$\leq C\left(\frac{1}{h_1}\right) + \frac{1}{2}\log_2\left(\max\{1,(h_2 + 1)\,k\}\right) + C\left(\frac{h_1}{h_2 k}\right). \tag{5.126}$$

For R_2, we have:

$$C\left(\frac{P}{h_1 P + 1}\right) + C\left((h_2 + 1)\,kP\right) - R_2 \tag{5.127}$$

$$=C\left(\frac{P}{h_1 P + 1}\right) + C\left((h_2 + 1)\,kP\right) - C\left(kP\right) + C\left(\frac{h_2 k}{h_1 + 1/P}\right) \tag{5.128}$$

$$\leq C\left(\frac{P}{h_1 P + 1}\right) + C\left((h_2 + 1)\,kP\right) - C\left(kP\right) + C\left(\frac{h_2 k}{h_1}\right) \tag{5.129}$$

$$\leq C\left(\frac{1}{h_1}\right) + C\left((h_2 + 1)\,kP\right) - C\left(kP\right) + C\left(\frac{h_2 k}{h_1}\right) \tag{5.130}$$

$$=C\left(\frac{1}{h_1}\right) + \frac{1}{2}\log_2\left(\frac{1 + (h_2 + 1)\,kP}{1 + kP}\right) + C\left(\frac{h_2 k}{h_1}\right) \tag{5.131}$$

$$\leq C\left(\frac{1}{h_1}\right) + \frac{1}{2}\log_2\left(h_2 + 1\right) + C\left(\frac{h_2 k}{h_1}\right). \tag{5.132}$$

Hence we have proved the Theorem.

5.10 Proof of Theorem 5.7.6

The proof is from [4]. To prove this theorem, we only need show that it is possible to achievable a secrecy rate for Node 1 that exceeds the upper bound given by Theorem 5.7.4. Consider the case when $h_1 = h_2 = 1$. Then by evaluating (5.108) at $\sigma^2 = 0$ and $\sigma^2 \to \infty$ with $\rho = \eta = 0$, we find the secrecy rate R_1 is bounded by

$$\min\{C(P), C(P_2) + 0.5\} \tag{5.133}$$

when Y_2 is ignored by Node 1. Choose P_2 and P such that

$$C(P_2) + 0.5 < 0.4C(P). \tag{5.134}$$

For this power configuration, from (5.133), we observe that R_1 is upper bounded by $0.4C(P)$.

Let the α in (5.53) be 0.5. R_1 then becomes:

$$0.5C\left(P\right) - 0.5\left[C\left(\frac{P}{P_2+1}\right) - C\left(P_2\right) + C\left(\frac{P_2}{P+1}\right)\right]^+. \tag{5.135}$$

A sufficient condition for $R_1 = 0.5C(P)$ is that

$$C\left(\frac{P}{P_2+1}\right) + C\left(\frac{P_2}{P+1}\right) > C\left(P_2\right). \tag{5.136}$$

It can be verified that this condition is equivalent to

$$\frac{\left(\frac{P}{P_2+1}+1\right)^2}{P+1} > 1. \tag{5.137}$$

A sufficient condition for it to hold is:

$$\frac{\left(\frac{P}{P_2+1}+1\right)^2}{\left(\sqrt{P}+1\right)^2} > 1 \tag{5.138}$$

which means

$$\sqrt{P} > P_2 + 1. \tag{5.139}$$

Choose $P_2 = P^{1/4}$. For sufficiently large P, both (5.134) and (5.139) can be fulfilled. In this case, the achievable rate is $0.5C(P)$, which is greater than the upper bound $0.4C(P)$. The difference is $0.1C(P)$, which is not a bounded function of P. Hence we have proved the theorem.

References

[1] E. Tekin and A. Yener. Achievable Rates for Two-Way Wire-Tap Channels. In *International Symposium on Information Theory*, June 2007.

[2] E. Tekin and A. Yener. The General Gaussian Multiple Access and Two-Way Wire-Tap Channels: Achievable Rates and Cooperative Jamming. *IEEE Transactions on Information Theory*, 54(6):2735–2751, June 2008.

[3] U. M. Maurer. Secret Key Agreement by Public Discussion from Common Information. *IEEE Transactions on Information Theory*, 39(3):733–742, May 1993.

[4] X. He and A. Yener. The Role of Feedback in Two-Way Secure Communication. Submitted to IEEE Transactions on Information Theory, November 2009, available online at http://arxiv.org/abs/0911.4432, revised May 2012.

[5] I. Csiszár and J. Körner. Broadcast Channels with Confidential Messages. *IEEE Transactions on Information Theory*, 24(3):339–348, May 1978.

[6] S. Leung-Yan-Cheong and M. Hellman. The Gaussian Wire-tap Channel. *IEEE Transactions on Information Theory*, 24(4):451–456, July 1978.

[7] E. Tekin and A. Yener. The Gaussian Multiple Access Wire-tap Channel: Wireless Secrecy and Cooperative Jamming. In *Information Theory and Applications Workshop*, January 2007.

[8] L. Lai, H. El Gamal, and H. V. Poor. The Wiretap Channel with Feedback: Encryption over the Channel. *IEEE Transactions on Information Theory*, 54(11):5059–5067, November 2008.

[9] A. El-Gamal, O. O. Koyluoglu, M. Youssef, and H. El-Gamal. The Two Way Wiretap Channel: Theory and Practice. Submitted to the IEEE Transactions on Information Theory, June, 2010, available online at http://arxiv.org/abs/1006.0778.

[10] O. Koyluoglu and H. El-Gamal. Cooperative binning and channel prefixing for secrecy in interference channels. Submitted to IEEE Transaction on Information Theory, 2009.

[11] X. He and A. Yener. Providing Secrecy with Structured Codes: Tools and Applications to Gaussian Two-user Channels. Submitted to IEEE Transactions on Information Theory, July 2009, in revision, available online at http://arxiv.org/abs/0907.5388.

[12] T. M. Cover and J. A. Thomas. *Elements of Information Theory*. Wiley-Interscience, New York, 2006.

[13] X. He and A. Yener. Secrecy When the Eavesdropper Controls its Channel States. In *IEEE International Symposium on Information Theory*, July 2011.

[14] X. He and A. Yener. Gaussian Two-way Wiretap Channel with an Arbitrarily Varying Eavesdropper. In *IEEE Global Telecommunication Conference, Workshop on Physical Layer Security*, December 2011.

[15] X. He and A. Yener. MIMO Wiretap Channels with Arbitrarily Varying Eavesdropper Channel States. Submitted to the IEEE Transactions on Information Theory, July, 2010, available online at http://arxiv.org/abs/1007.4801.

[16] M. Bloch. Channel Scrambling for Secrecy. In *IEEE International Symposium on Information Theory*, June 2009.

[17] A. Khisti and G. Wornell. Secure Transmission with Multiple Antennas-I: The MISOME Wiretap Channel. *IEEE Transactions on Information Theory*, 56(7):3088–3104, July 2010.

[18] I. Csiszár and P. Narayan. Secrecy capacities for multiple terminals. *IEEE Transactions on Information Theory*, 50(12):3047–3061, 2004.

[19] A. Khisti, S. N. Diggavi, and G. W. Wornell. Secret-key Agreement with Channel State Information at the Transmitter. *IEEE Transactions on Information Forensics and Security*, 6(3):672–681, September 2011.

[20] S. Watanabe and Y. Oohama. Secret Key Agreement from Correlated Gaussian Sources by Rate Limited Public Communication. *IEICE transactions on Fundamentals of Electronics, Communications and Computer Sciences*, 93(11):1976–1983, 2010.

[21] X. He and A. Yener. Strong Secrecy and Reliable Byzantine Detection in the Presence of an Untrusted Relay. To appear in IEEE Transactions on Information Theory, submitted in March 2010, available online at http://arxiv.org/abs/1004.1423.

[22] D. Gunduz, D. R. Brown III, A. Goldmith, and H. V. Poor. Secure Communication in the Presence of Feedback. In *Joint Workshop on Coding and Communications*, October 2008.

[23] D. Gunduz, D. R. Brown III, and H. V. Poor. Secure Communication with Feedback. In *IEEE International Symposium on Information Theory and its Applications*, December 2008.

[24] E. Ardetsanizadeh, M. Franceschetti, T. Javidi, and Y. H. Kim. Wiretap Channel with Secure Rate-limited Feedback. *IEEE Transactions on Information Theory*, 55(12):5353–5361, 2009.

Chapter 6

MIMO Signal Processing Algorithms for Enhanced Physical Layer Security

Amitav Mukherjee
Hitachi America Ltd.

S. Ali A. Fakoorian
University of California Irvine

Jing Huang
University of California Irvine

Lee Swindlehurst
University of California Irvine

The use of physical layer methods for improving the security of wireless links has recently become the focus of a considerable research effort. Such methods can be used in combination with cryptography to enhance the exchange of confidential messages over a wireless medium in the presence of unauthorized eavesdroppers, or they can be used to enable secrecy in the absence of shared secret keys through the use of coding strategies, jamming or beamforming. Indeed, one of the driving forces behind the recent emergence of physical layer techniques for security is the push toward adding extra degrees of freedom in the form of multiple antennas at the transmitter and receiver of the link. Multiple-input multiple-output (MIMO) wireless systems have been extensively studied during the past two decades, and their potential gains in throughput, diversity, and range have been well quantified. MIMO approaches are now an integral part of the WiFi and 4G standards in use today. It is not surprising that MIMO architectures are useful in improving wireless security as well, since they can provide focused transmit selectivity of both information and noise toward desired and undesired receivers. In this chapter, we discuss a number of different ways that physical layer security can be achieved in wireless networks with MIMO links. We will focus primarily on signal processing related issues (e.g., beamforming, power control, resource allocation) that enable reliable reception at intended recipients and minimize data leakage to eavesdroppers, and we will consider a variety of different MIMO settings including point-to-point, broadcast, interference, and multi-hop networks. We cannot offer an exhaustive survey of such methods in just a single chapter; instead, we present a few representative approaches that illustrate the gains that multiple antennas provide. Additional and more complete treatments of the physical layer security problem can be found in [1–4] as well as other chapters in this book.

6.1   Introduction

The availability of multiple antennas in multiuser networks is akin to a double-edged sword from the security point of view: the enhanced transmission quality of the legitimate users is offset by the increased interception capabilities of unauthorized receivers. Consequently, MIMO signal processing techniques must be carefully designed in order to ensure both the reliability of the desired communication links and the degradation of those for the eavesdroppers. The chapter commences with a description of transceiver optimization algorithms for the conventional three-user MIMO wiretap channel with perfect channel state information (CSI). As with MIMO techniques in general, the availability and accuracy of CSI is crucial to success, and plays a major role in achieving secure communications. Thus, we subsequently examine robust techniques that compensate for partial or inaccurate CSI in the MIMO wiretap channel. A key secrecy-enhancing technique in scenarios with partial CSI is the embedding of "artificial noise" or jamming signals together with the confidential message, the objective being to selectively degrade the signal quality at any unintended receiver.

Generalizing to multiuser wiretap scenarios opens new avenues for study, since resource allocation among multiple legitimate users must be balanced with potentially limited cooperation between them. Two prime examples are the MIMO wiretap channel with external cooperative jammers or helpers, and the broadcast MIMO wiretap network with multiple receivers. For the former example, we describe the optimal precoding and power allocation operations at the helping jammer in order to maximally suppress the information decodable at the eavesdroppers, without information being exchanged with the transmitter. In the latter example, we present linear and non-linear MIMO downlink precoding algorithms that are categorized by the level of eavesdropper CSI assumed. Subsequently, we study MIMO interference networks with secrecy constraints as a more involved wiretap network scenario, since in this application the design of MIMO precoders requires that the confidential information exchanged between pairs is suppressed at cochannel terminals without causing excessive degradation to the secrecy sum rate or secure degrees of freedom.

Finally, we examine MIMO multi-hop networks where intermediary relays play a pivotal role in establishing secure one-way or two-way links. In some cases, the relay(s) acts as an enabler in enhancing secrecy, for example by acting as an external jammer if a direct source-to-destination link is available. Alternatively, the relay itself may be an unauthenticated node from which the exchanged messages must be kept secret. For each case, we examine MIMO relay optimization procedures and their robust counterparts, depending upon the CSI assumed to be known.

6.2   Physical Layer Security

6.2.1   Signal Processing Aspects

An early information-theoretic treatment of physical layer security by Wyner [5] in 1975 demonstrated that perfect secrecy is achievable in the wiretap channel without the use of encryption keys, provided that the eavesdropper's channel is of lower quality than the desired receiver. In general, approaches to physical layer security in the information theory literature have primarily focused on code design and bounds for achievable secrecy capacities/regions. Such analyses frequently involve idealized assumptions of perfectly known global CSI, random coding arguments, Gaussian inputs, and so on. The signal processing perspective on physical layer security then naturally pertains to optimal and near-optimal transceiver design in situations where these assumptions may or may not hold. As a result, the optimization of power allocation algorithms together with the design of spatial transmit/receive filters under perfect and imperfect channel state information form the major thrust of signal processing research in this area.

6.2.2 Secrecy Performance Metrics

Secret communication problems arise in multi-terminal networks comprising a minimum of three nodes: the legitimate transmitter and its intended receiver, and an unauthorized interloper commonly referred to as an eavesdropper. This three-terminal system is referred to as the wiretap channel [1]. Several of the most frequently used secrecy metrics are introduced below, while their formal descriptions are relegated to subsequent sections.

Secrecy Rate—In the wiretap channel, the secrecy rate is a transmission rate that can be reliably supported on the primary channel, but which is undecodable on the eavesdropper's channel. For Gaussian channels, it is calculated as the difference between the mutual information on the primary and eavesdropper's channels. Secrecy capacity is achieved when the secrecy rate is maximized. When multiple communication links are present, for example as in broadcast or interference channels, then one is typically interested in defining the achievable secrecy rate or secrecy capacity regions or the aggregate secrecy sum rate or sum capacity.

Secrecy Outage Probability—The secrecy outage probability (SOP) represents the probability that a certain target secrecy rate is not achieved for a given communication link. The SOP characterizes the likelihood of simultaneously reliable and secure data transmission, and is most often employed in situations where only statistical CSI about the eavesdropper is available.

Secret Diversity/Multiplexing Gain—Secret diversity and multiplexing gains for wiretap channels can be defined similarly to their counterparts in conventional MIMO systems [6]. The secret diversity gain is the asymptotic rate of decrease with signal-to-noise ratio (SNR) in probability of error at the desired receiver when subject to secrecy constraints, while the secret multiplexing gain (or degrees of freedom) is the asymptotic rate of increase with SNR in the secrecy rate. The secret diversity/multiplexing trade-off (DMT) captures the interplay between these competing metrics.

Secret Key Rate—The secret key rate quantifies the rate at which legitimate users can agree upon a shared key sequence by exchanging messages over a public channel that is observable to eavesdroppers.

6.2.3 The Role of CSI

Realization of the gains promised by multiple antenna transceivers hinges in large part on the ability of the system to obtain and exploit accurate CSI. In the wiretap channel, this includes not only CSI for the primary link, but also possibly CSI for the eavesdroppers.

Complete CSI—Intuitively, the greatest level of security can be achieved when the legitimate transmitters have complete knowledge of the wireless channels to all receivers, including the eavesdroppers. Such information may be available if the potential eavesdropper is an active network node, and has previously communicated with the transmitter. The knowledge of global CSI can be exploited to design MIMO transmit precoders that minimize the information leaked into eavesdropper channels, or to accurately direct jamming signals toward the eavesdroppers, as described in the following sections.

Partial CSI—In many cases of practical interest, the eavesdroppers may be passive and their instantaneous channels are not known at the transmitters. If the statistical distribution of the eavesdropper's channel is known, then the transmit signals can be designed to optimize an ergodic secrecy metric. If nothing is known about the eavesdropper, measures can still be taken to improve secrecy, although in such cases the benefits of such measures are difficult to quantify. Going one step further, the CSI of the intended receiver itself may only be known partially due to estimation error or limited feedback, in which case robust or worst-case secrecy optimization methods must be developed.

6.3 MIMO Wiretap Channels

A MIMO wiretap channel consists of a transmitter (Alice), a legitimate receiver (Bob), and an eavesdropper (Eve) equipped with $N_T, N_R,$ and N_E antennas, respectively. A general representation for the signal received by the legitimate receiver is

$$\mathbf{y}_b = \mathbf{H}_b \mathbf{x}_a + \mathbf{n}_b, \qquad (6.1)$$

while the received signal at the eavesdropper is

$$\mathbf{y}_e = \mathbf{H}_e \mathbf{x}_a + \mathbf{n}_e, \qquad (6.2)$$

where $\mathbf{x}_a \in \mathbb{C}^{N_T \times 1}$ is the transmitted signal with covariance $E\left\{\mathbf{x}_a \mathbf{x}_a^H\right\} = \mathbf{Q}_x$, $\mathbf{H}_b \in \mathbb{C}^{N_R \times N_T}, \mathbf{H}_e \in \mathbb{C}^{N_E \times N_T}$ are the complex MIMO channel matrices associated with Bob and Eve, respectively, and $\mathbf{n}_b, \mathbf{n}_e$ represent additive noise at the two receivers. The above scenario with multiple antennas at all nodes is occasionally referred to as the MIMOME (multiple-input multiple-output multiple-eavesdropper) channel [7].

When the additive noise is Gaussian, a Gaussian input $\mathbf{x}_a$ is the optimal choice for achieving the secrecy capacity under an average power constraint on the transmit covariance matrix [8]–[11]:

$$C_s = \max_{\mathbf{Q}_x, \mathrm{Tr}(\mathbf{Q}_x) \leq P} [I\left(\mathbf{X}_a; \mathbf{Y}_b\right) - I\left(\mathbf{X}_a; \mathbf{Y}_e\right)] \qquad (6.3)$$

where $I(\cdot; \cdot)$ denotes mutual information, and $\mathbf{X}_a, \mathbf{Y}_a, \mathbf{Y}_e$ are the random variables whose realizations follow the system model in (6.1)–(6.2). For Gaussian input signaling, this is equivalent to

$$C_s = \max_{\mathbf{Q}_x, \mathrm{Tr}(\mathbf{Q}_x) \leq P} \log \det \left(\mathbf{I} + \mathbf{H}_b \mathbf{Q}_x \mathbf{H}_b^H\right) - \log \det \left(\mathbf{I} + \mathbf{H}_e \mathbf{Q}_x \mathbf{H}_e^H\right), \qquad (6.4)$$

assuming for simplicity that the noise is spatially white and unit-variance at both receivers.

Secret-key agreement over wireless channels is another promising application of physical layer security since the noise, interference, and fading affecting wireless communications provide a convenient source of randomness. A secret-key rate is informally defined as the ratio between the number of key bits k obtained at the end of a key-distillation strategy and the number of noisy channel uses n required to obtain it. Assuming the eavesdropper and receiver observe independent signals and are under an average power constraint, the MIMO secret-key capacity is [12,13]

$$C_k = \max_{\mathbf{Q}_x, \mathrm{Tr}(\mathbf{Q}_x) \leq P} I\left(\mathbf{X}_a; \mathbf{Y}_b | \mathbf{Y}_e\right). \qquad (6.5)$$

For Gaussian input signaling, C_k has a form similar to (6.4) with an equivalent channel $\tilde{\mathbf{H}}_b^H \tilde{\mathbf{H}}_b = \mathbf{H}_b^H \mathbf{H}_b + \mathbf{H}_e^H \mathbf{H}_e$ introduced into the first log-determinant term on the RHS.

6.3.1 Complete CSI

For the case where the transmitter possesses instantaneous CSI for both the desired receiver and the eavesdropper, a precise characterization of the secrecy capacity and the corresponding transmit covariance $\mathbf{Q}_x$ for the MIMO wiretap channel under an average power constraint is unknown. Solutions can be found in special cases; for example, in the so-called MISOME case where $N_R = 1, N_T, N_E > 1$, the optimal solution is based on transmit beamforming: $\mathbf{Q}_x = P\boldsymbol{\psi}_m\boldsymbol{\psi}_m^H$, where $\boldsymbol{\psi}_m$ is the unit-norm generalized eigenvector corresponding to the largest generalized eigenvalue λ_m of

$$(\mathbf{I} + P\,\mathbf{h}_b^H\mathbf{h}_b)\boldsymbol{\psi}_m = \lambda_m(\mathbf{I} + P\mathbf{H}_e^H\mathbf{H}_e)\boldsymbol{\psi}_m. \tag{6.6}$$

Other special cases have been investigated, but the general solution remains an open problem.

An alternative power constraint was considered by Bustin et al. [14], who reexamined the MIMO wiretap channel by exploiting the derivative relationship between mutual information and mean-squared error to provide a closed-form expression for the capacity-achieving $\mathbf{Q}_x$ under a matrix power covariance $\mathbf{Q}_x \preceq \mathbf{S}$. For this constraint, the MIMO secrecy capacity was shown to be given by

$$C_{sec}(\mathbf{S}) = \sum_{i=1}^{\lambda} \log \alpha_i \tag{6.7}$$

where α_i are the generalized eigenvalues of the pencil

$$(\mathbf{S}^{\frac{1}{2}}\mathbf{H}_b^H\mathbf{H}_b\mathbf{S}^{\frac{1}{2}} + \mathbf{I}, \quad \mathbf{S}^{\frac{1}{2}}\mathbf{H}_e^H\mathbf{H}_e\mathbf{S}^{\frac{1}{2}} + \mathbf{I}) \tag{6.8}$$

that are greater than one. The matrix power constraint is a more narrow constraint that places considerable limits on the per-antenna power and transmit correlation structure [44]. The average power constraint is a much less restrictive constraint that provides considerable additional flexibility in increasing the secrecy rate of the MIMO wiretap channel. In principle, at least, the secrecy capacity of the MIMO wiretap channel under the average power constraint could be found via an exhaustive search over the set $\{\mathbf{S} : \mathbf{S} \succeq \mathbf{0}, \mathrm{Tr}(\mathbf{S}) \le P\}$ [41, Lemma 1], [44]:

$$\mathcal{C}_{sec}(P_t) = \max_{\mathbf{S} \succeq \mathbf{0}, \mathrm{Tr}(\mathbf{S}) \le P_t} C_{sec}(\mathbf{S}), \tag{6.9}$$

where for any given semidefinite $\mathbf{S}$, $\mathcal{C}_{sec}(\mathbf{S})$ should be computed as in (6.7).

A closed-form solution is possible in certain special cases, for example when $\mathbf{Q}_x$ is known to be full rank [15, 16], or in the high-SNR regime [7]. In this latter case, it was shown that the asymptotically optimal solution for high SNR can be found by decomposing the system into parallel channels based upon the generalized singular value decomposition (GSVD) of the matrix pair $(\mathbf{H}_b, \mathbf{H}_e)$. The optimal power allocation for GSVD precoding was derived in [17], and is shown to empirically achieve the MIMO secrecy capacity in (6.4). The GSVD precoding approach is also applicable to MIMO secret-key establishment due to the close similarity between C_s and C_k.

An example of the secrecy rate performance of various transmission strategies for the MIMO wiretap channel is shown in Figure 6.1. The GSVD scheme assumes instantaneous knowledge of the eavesdropper channel $\mathbf{H}_e$, the artificial noise scheme requires the statistics of $\mathbf{H}_e$, and the relatively poor performance of waterfilling on the main channel is also shown for the case where no information is available regarding $\mathbf{H}_e$. In this example, the power allocation given in [17] is observed to essentially achieve the secrecy capacity of the wiretap channel.

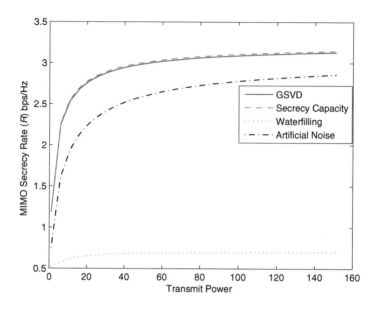

Figure 6.1: The MIMO secrecy rates of GSVD beamforming [8, 17], artificial noise [9], and waterfilling over the main channel, $N_T = N_E = 3, N_R = 2$.

6.3.2 Partial CSI

When only CSI for the primary channel is available at the transmitter, an alternative scheme for the MIMO wiretap channel is to transmit an *artificial* or synthetic noise signal $\mathbf{z}'$ in conjunction with the information signal. The artificial noise signal is generally designed to be orthogonal to the intended receiver such that only the eavesdropper is jammed [7–9], although in certain cases secrecy rates can be improved by allowing some noise leakage to the receiver [18]. An example of an artificial noise transmission strategy is

$$\mathbf{x}_a = \mathbf{T}\mathbf{z} + \mathbf{T}'\mathbf{z}', \qquad (6.10)$$

where $\mathbf{T}, \mathbf{T}'$ are the $N_T \times d$, $N_T \times (N_T - d)$ precoding matrices for the $d \times 1$ information vector $\mathbf{z}$ and uncorrelated $(N_T - d) \times 1$ jamming signal $\mathbf{z}'$, respectively. These signals can be made orthogonal at the intended receiver by choosing $\mathbf{T}$ and $\mathbf{T}'$ as disjoint sets of the right singular vectors of $\mathbf{H}_b$ [9, 22].

A frequent assumption in the literature is that only the statistical distribution of $\mathbf{H}_e$ is known to the transmitter, along with perfect knowledge of $\mathbf{H}_b$. In such scenarios, the artificial noise injection strategy as first suggested by Goel and Negi [9] remains the best known secure transmission strategy, with approaches for optimizing the number of spatial dimensions and power allocated to the artificial noise ranging from an exhaustive search to optimizing an approximate expression for the ergodic secrecy rate. For the MISOME channel in this scenario, rank-1 beamforming without artificial noise has been shown to be the optimal strategy [19]. An alternative assumption regarding imperfect eavesdropper CSI is that the error in knowledge of $\mathbf{H}_e$ is norm-bounded [20, 21]. In [20], the authors studied the optimal transmit strategy by exploiting the relationship between the secrecy MISO channel and the cognitive radio MISO channel. The authors in [21] considered the MISO channel with and without an external helper, and obtained robust beamforming/jamming solutions via numerical methods.

The use of artificial noise can be problematic when the channel to the intended receiver is also not known perfectly, since the noise can leak into the desired receiver's signal. The imperfect CSI at the transmitter can be written as

$$\hat{\mathbf{H}}_b = \mathbf{H}_b + \Delta\mathbf{H}_b$$

where $\Delta\mathbf{H}_b$ is the unknown channel perturbation matrix. The two common approaches for modeling $\Delta\mathbf{H}_b$ are (1) to assume it follows a zero-mean complex Gaussian matrix distribution with a known covariance matrix $\mathbf{C}_\Delta$ [22], and (2) a deterministic model where the "size" of the perturbation lies within known bounds, e.g., $\|\Delta\mathbf{H}_b\|_F \leq \varepsilon$ [23]. In [22], the Gaussian perturbation model is considered. The authors propose a rank-1 beamforming strategy, compute the degradation in receiver SINR due to noise leakage based on SVD perturbation theory, and construct robust transmit/receive beamformers to recover the loss in SINR. In [23] on the other hand, the deterministic perturbation case is investigated, and the nonconvex secrecy rate maximization problem is numerically solved by applying a semidefinite relaxation based on a worst-case channel perturbation. The authors of [24] consider the case where $\Delta\mathbf{H}_b$ arises due to limited feedback from the receiver, and devise optimal power allocation and feedback rules to constrain the level of artificial noise leakage.

As an extreme case of partial CSI, consider the scenario where neither the realization nor the distribution of the eavesdropper channel is known to the legitimate parties. Information-theoretic solutions have been focused on the design of robust coding schemes that guarantee a minimum secrecy rate irrespective of the eavesdropper's channel state. A far simpler method, albeit without such a guarantee, is to devote just enough resources to achieve a desired signal-to-interference-and-noise ration (SINR) or rate to the intended receiver, and allocate any unused power/spatial resources to the generation of orthogonal artificial noise [22, 25]. Interestingly, the conventional waterfilling strategy is not necessarily the optimal approach for achieving a desired rate on the main channel, and can be outperformed by rank-minimization methods that yield greater unused spatial dimensions for artificial noise [25].

MIMO secret-key rates under assumptions of partial CSI are comparatively less well studied. In [26], the transmitter knows $\mathbf{H}_b$ but assumes $\mathbf{H}_e$ is drawn arbitrarily from a known finite set of possible matrices. This scenario naturally leads to a max–min approach to optimize the worst-case secret-key rate; it is shown that a saddlepoint exists and the optimal transmit covariance matrix is subsequently characterized.

It is important to note in closing that the vast majority of prior work on the MIMO wiretap channel assumes the use of Gaussian input signals for analytical simplicity. Moving to a model with inputs drawn from a finite alphabet immediately renders analyses more intractable due to the lack of simple closed-form expressions for mutual information. In most cases with discrete inputs, we must resort to numerical methods for MIMO transceiver optimization [28].

6.4   MIMO Wiretap Channel with an External Helper

It was shown in [29] that for a wiretap channel without feedback, a nonzero secrecy capacity can only be obtained if the eavesdropper's channel is of lower quality than that of the intended recipient. If such a situation does not exist endogenously, one solution is to solicit the assistance of external nodes such as relays or *helpers* in facilitating the transmission of confidential messages from the source to the destination. While the helper could in principle assist with the transmission of the confidential messages themselves, the computational cost associated with this approach may be prohibitive and there are difficulties associated with the coding and decoding schemes at both the helper and the intended receiver. Alternatively, the helper can simply transmit noise-like jamming signals independent of the source message, to confuse the eavesdropper and increase the range of channel conditions under which secure communications can take place [30, 32, 33]. This so-called *cooperative jamming* (CJ) or noise-forwarding approach is analogous to the artificial noise strategy described previously for the MIMOME channel.

Prior work on CJ assuming single-antenna nodes includes [30], which considers multiple single-antenna users communicating with a common receiver (i.e., the multiple access channel) in the presence of an eavesdropper, and the optimal transmit power allocation that achieves the maximum secrecy sum-rate is obtained. The work of [30] shows that any user prevented from transmitting based on the obtained power allocation can help increase the secrecy rate for other users by transmitting artificial noise to the eavesdropper via cooperative jamming. In [32], a MISO source–destination system in the presence of multiple helpers and multiple eavesdroppers is considered, where the helpers can transmit weighted jamming signals to degrade the eavesdropper's ability to decode the source. While the objective is to select the weights so as to maximize the secrecy rate under a total power constraint, or to minimize the total power under a secrecy rate constraint, the results in [32] yield suboptimal weights for both single and multiple eavesdroppers, due to the assumption that the jamming signal must be nulled at the destination. The noise-forwarding scheme of [33] requires that the interferer's codewords be decoded by the intended receiver. A generalization of [30, 32] and [33] is proposed in [34], in which the helper's codewords do not have to be decoded by the receiver. In [35], a CJ scenario is considered with an arbitrary number of antennas at the transmitters but with only single-antenna receivers, and it is shown that beamforming is the optimal strategy at both the main transmitter and the helper. A closed-from solution is obtained for the optimal channel input at the transmitter side based on the CJ signal chosen by the helper. Also, [35] analytically shows that sending CJ in the null space of $\mathbf{G}_b$ is a near-optimal solution.

While all the previous work considers SISO or MISO scenarios, [36] and [37] propose solutions for the MIMO case. In [36], multiple cooperative jammers were studied, wherein the jammers aligned their interference to lie within a prespecified "jamming subspace" at the receiver, but the dimensions of the subspace and the power allocation were not optimized. In [37], necessary and sufficient conditions are obtained to have a jamming signal which does not reduce the mutual information between the source and the legitimate receiver. Thus the jamming signal only degrades the mutual information between the main transmitter and the eavesdropper, and as a result, the achievable secrecy rate improves.

Figure 6.2 shows the maximum achievable secrecy rate for the proposed algorithm in [37] versus the helper's transmit power P_h. In this figure, it is assumed that the total average power $P_t + P_h = 110$, where P_t is the power assigned to the source. While channels are assumed to be quasistatic flat Rayleigh fading and independent of each other, direct and cross channels have i.i.d. entries distributed as $\mathcal{CN}(0, \sigma_d^2)$ and $\mathcal{CN}(0, \sigma_c^2)$, respectively. The figure considers a situation in which $\sigma_c > \sigma_d$, or in other words where the channel between the transmitter and the intended receiver is weaker than the channel between the transmitter and the eavesdropper, and the channel between the helper and the intended receiver is weaker than the channel between the helper and the eavesdropper. The arrow in the figure shows the secrecy capacity without the helper ($P_h = 0$). The figure shows that a helper with just a single antenna can provide a dramatic improvement in secrecy rate with very little power allocated to the jamming signal; in fact, the optimal rate is obtained when P_h is less than 2% of the total available transmit power. If the number of antennas at the helper increases, a much higher secrecy rate can be obtained, but at the expense of allocating more power to the helper and less to the signal for the desired user.

While many cases assume the external helper knows the CSI of the eavesdropper, in some

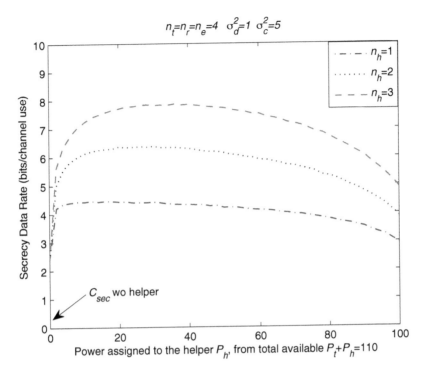

Figure 6.2: Comparison of the achievable secrecy data rate for the MIMO Gaussian wiretap channel with and without a helper versus P_h for a different number of antennas at the helper, $P_t + P_h = 110$, assuming the eavesdropper's channels are stronger than those of the receiver ($\sigma_d^2 = 1, \sigma_c^2 = 5$).

scenarios perfect CSI may be difficult to obtain due to the nature of the eavesdropper, and thus robust beamforming/precoding strategies are needed to guarantee efficient cooperative jamming. The authors in [21] considered optimizing the worst-case secrecy performance for a MISO wiretap channel with a helper where the source and the helper only possess imperfect CSI for the eavesdropper and the channel error is norm-bounded by some known constraints. By converting the resulting nonconvex maximin problem into a quasiconvex problem, [21] derived optimal transmit covariance matrices for both the source and the helper under the zero-forcing constraint that the jamming signals are nulled at the legitimate receivers, and a Quality-of-Service constraint that a minimum receive SINR at the legitimate receiver is imposed.

6.5 MIMO Broadcast Channel

The original wiretap channel, as proposed by Wyner [5], is a form of broadcast channel (BC) where the source sends confidential messages to the destination while the messages should be kept as secret as possible from the other receiver(s)/eavesdropper(s). Csiszár and Körner extended this work to the case where the source sends common information to both the destination and the eavesdropper, and confidential messages are sent only to the destination [29]. The secrecy capacity region for the case of a BC with parallel independent subchannels is considered in [39] and the optimal source power allocation that achieves the boundary of the secrecy capacity region is derived. A special case of the MIMO Gaussian broadcast channel with common and confidential messages is considered in [40], where the common message is intended for both receivers but the confidential message is intended only for receiver 1, and must be kept secret from receiver 2. A matrix characterization of the secrecy capacity region is established by first splitting receiver 1 into two virtual receivers and then enhancing only the virtual receiver that decodes the confidential message. It should be noted that the notion of an enhanced broadcast channel was first introduced in [41] to characterize the capacity region of the conventional Gaussian MIMO broadcast channel without secrecy constraints.

Prior work has considered the discrete memoryless broadcast channel with two confidential messages sent to two receivers, where each receiver acts as an eavesdropper for the other. This problem has been addressed in [42], where inner and outer bounds for the secrecy capacity region were established. Further work in [43] studied the MISO Gaussian case under the average power constraint, and [44] considered the general MIMO Gaussian case under the matrix power constraint.

For the two-user BC where each user is to receive own confidential message, it was shown in [44] that, under a matrix input power-covariance constraint, both confidential messages can be simultaneously communicated at their respective maximum secrecy rates as if over two separate MIMO Gaussian wiretap channels. In other words, under the matrix power constraint $\mathbf{S}$, the secrecy capacity region $\{R_1, R_2\}$ is rectangular. This interesting result is obtained using secret dirty-paper coding (S-DPC), and the corner point rate (R_1^*, R_2^*) of the secrecy capacity region under the matrix constraint $\mathbf{S}$ can be calculated as [44, Theorem 3],

$$R_1^* = \log |\mathbf{\Lambda}_1|$$
$$R_2^* = -\log |\mathbf{\Lambda}_2| \tag{6.11}$$

where the diagonal matrices $\mathbf{\Lambda}_1$ and $\mathbf{\Lambda}_2$ contain the generalized eigenvalues of

$$\left(\mathbf{S}^{\frac{1}{2}} \mathbf{H}_b^H \mathbf{H}_b \mathbf{S}^{\frac{1}{2}} + \mathbf{I}, \quad \mathbf{S}^{\frac{1}{2}} \mathbf{H}_e^H \mathbf{H}_e \mathbf{S}^{\frac{1}{2}} + \mathbf{I} \right)$$

that are respectively greater than or less than one.

The secrecy capacity region of the MIMO Gaussian broadcast channels with both confidential and common messages under the matrix power constraint is characterized in [45] and [46], where the transmitter has two independent confidential messages and a common message. The achievability is obtained using secret dirty-paper coding, while the converse is proved by using the notion of channel splitting [45]. Secure broadcasting with more than two receivers has been considered in [47–50] and references therein. The secrecy capacity region for the two-legitimate receiver case is characterized by Khandani et al. [48] using enhanced channels, and for an arbitrary number of legitimate receivers by Ekrem and Ulukus [49] who use relationships between the minimum mean square error and mutual information, and between the Fisher information and the differential entropy to provide the converse proof. Liu et al. [50] considered the secrecy capacity regions of the degraded vector Gaussian MIMO broadcast channel with layered confidential messages, and presented a vector generalization of Costa's entropy power inequality to provide the converse proof. The role of artificial noise for jamming unintended receivers in multiuser downlink channels was investigated, for example, in [51, 52].

It should be noted that, under the average power constraint, there is not a computable secrecy capacity expression for the general MIMO broadcast channel case. However, optimal solutions based on linear precoding have been found. For example, in [53], a linear precoding scheme is proposed for a general MIMO BC under the matrix covariance constraint $\mathbf{S}$. Conditions are derived under which the proposed linear precoding approach is optimal and achieves the same secrecy rate region as S-DPC. Then this result is used to derive a closed-form suboptimal algorithm based on linear precoding for an average power constraint. In [54], GSVD-based beamforming is used for the MIMO Gaussian BC to simultaneously diagonalize the channels. The GSVD creates a set of parallel independent

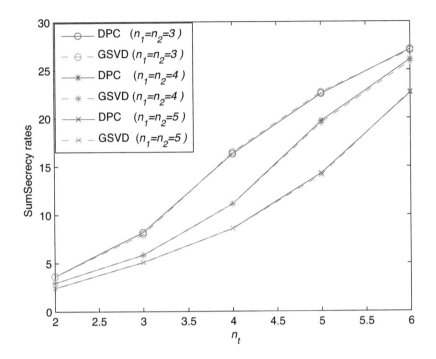

Figure 6.3: Comparison of the sum-secrecy rate of GSVD and S-DPC versus N_T, for different numbers of antennas $n_1 = n_2$ at the two receivers.

subchannels between the transmitter and the receivers, and it suffices for the transmitter to use independent Gaussian codebooks across these subchannels. In particular, the confidential message W_1 for receiver 1 is sent only over the subchannels for which the output at receiver 2 is a degraded version of the output at receiver 1. These subchannels correspond to the generalized singular values that are larger than one. On the other hand, the confidential message W_2 for receiver 2 is sent only over those subchannels for which the output at receiver 1 is a degraded version of the output at receiver 2. These subchannels correspond to the generalized singular values that are less than one. The optimal power allocation for these subchannels is derived in [54] to maximize the sum-secrecy rate for a given fraction α and $1 - \alpha$ of the total power allocated to W_1 and W_2, respectively. The Pareto boundary of the suboptimal secrecy rate region is then achieved by sweeping through all values of $\alpha \in \{0, 1\}$.

Figure 6.3 compares the average over 1,000 channel realizations of the sum-secrecy rate of the GSVD-based beamforming approach and the optimal S-DPC methods, when N_T varies from two to six and for different numbers of antennas at the receivers. As the figure shows, the performance of the proposed GSVD-based beamforming approach is essentially identical to that of the optimal S-DPC, while requiring considerably less computation.

6.6 MIMO Interference Channel

The interference channel (IFC) refers to the case where multiple communication links are simultaneously active in the same time and frequency slot, and hence potentially interfere

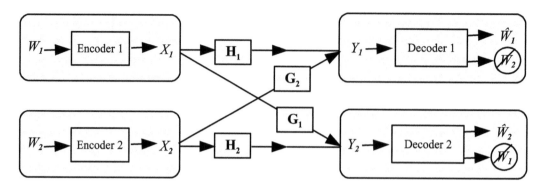

Figure 6.4: System model for the two-user MIMO interference channel with confidential messages.

with each other. A two-user IFC is depicted in Figure 6.4. A special application of the IFC with secrecy constraints is addressed in [55], where the message from only one of the transmitters is considered confidential. The more general case was studied in [42] where, in the absence of a common message, the authors imposed a perfect secrecy constraint and obtained inner and outer bounds for the perfect secrecy capacity region of a two-user discrete memoryless IFC.

For multiuser networks, a useful metric that captures the scaling behavior of the sum secrecy rate R_Σ as the transmit SNR, ρ, goes to infinity is secret multiplexing gain or degrees of freedom (DoF), which can be defined as

$$\eta = \lim_{\rho \to \infty} \frac{R_\Sigma(\rho)}{\log(\rho)}. \tag{6.12}$$

Calculation of the number of secure DoF for the K-user Gaussian IFC ($K \geq 3$) has been addressed in [56] and [57], where it was shown that under very strong interference, positive secure DoFs are achievable for each user in the network. For the case of a K-user SISO Gaussian interference channel with confidential messages, where each node has one antenna and each transmitter needs to ensure the confidentiality of its message from all nonintended receivers, a secure DoF of

$$\eta = \frac{K-2}{2K-2} \tag{6.13}$$

is almost surely achievable for each user [56]. The achievability of this result is obtained by interference alignment and channel extension. Moreover, for the case of a K-user SISO Gaussian interference channel with an external eavesdropper, each user can achieve

$$\eta = \frac{K-2}{2K}. \tag{6.14}$$

While the above references [55] through [57] assume single-antenna nodes, it should be noted that the secrecy *capacity* region of a two-user multiple antenna Gaussian interference channel, even for the MISO case, is still unknown. Indeed, from [42, Theorem 2], any rate pair

$$0 \leq R_1 \leq I(V_1; Y_1) - I(V_1; Y_2|V_2)$$
$$0 \leq R_2 \leq I(V_2; Y_2) - I(V_2; Y_1|V_1) \tag{6.15}$$

over all $p(V_1, V_2, X_1, X_2, Y_1, Y_2) = p(V_1)p(V_2)p(X_1|V_1)p(X_2|V_2)p(Y_1, Y_2|X_1, X_2)$ is achievable, where the independent precoding auxiliary random variables V_1 and V_2 represent two independent stochastic encoders. To ensure information-theoretic secrecy, the bound for the achievable rate R_1 includes a penalty term $I(V_1; Y_2|V_2)$, which is the conditional mutual information of receiver 2's eavesdropper channel assuming that receiver 2 can first decode its own information. Thus, the achievable rates in (6.15) correspond to the worst-case scenario, i.e., where the intended messages are decoded treating the interference as noise, but the eavesdropped message is decoded without any interference (which is assumed to be subtracted before) [42, 58].

In Jorswieck et al. [58], studied the achievable secrecy rates of a two-user MISO interference channel, where each single-antenna receiver acts as an eavesdropper for the other link, and each transmitter only sends its information signal. In [60] and [61], a two-user MIMO Gaussian IFC is investigated where each node has arbitrary number of antennas:

$$
\begin{aligned}
\mathbf{y}_1 &= \mathbf{H}_1\mathbf{x}_1 + \mathbf{G}_2\mathbf{x}_2 + \mathbf{n}_1 \\
\mathbf{y}_2 &= \mathbf{H}_2\mathbf{x}_2 + \mathbf{G}_1\mathbf{x}_1 + \mathbf{n}_2 .
\end{aligned}
\tag{6.16}
$$

Based on different assumptions about the CSI available at the transmitter, several cooperative and noncooperative transmission schemes are described, and their achievable secrecy rate regions are derived.

In one noncooperative game, it is assumed that each transmitter i knows not only its own direct channel $\mathbf{H}_i$, but also the cross-channel $\mathbf{G}_i$ to the potential eavesdropper. When both $\mathbf{H}_i$ and $\mathbf{G}_i$ are available to transmitter i, a reasonable precoding scheme is obtained by simply treating the IFC as two parallel wiretap channels and using the GSVD method to design the precoders for each transmitter. Thus, transmitter 1 constructs $\mathbf{x}_1$ as

$$
\mathbf{x}_1 = \mathbf{A}_1\mathbf{s}_1, \quad \mathbf{s}_1 \sim \mathcal{CN}(\mathbf{0}, \mathbf{P}_1)
\tag{6.17}
$$

where $\mathbf{A}_1$ is obtained from the GSVD of the pair $(\mathbf{H}_1, \mathbf{G}_1)$, each nonzero element of the vector $\mathbf{s}_1$ represents an independently encoded Gaussian codebook symbol that is beamformed with the corresponding column of the matrix $\mathbf{A}_1$, and $\mathbf{P}_1$ is a positive semidefinite diagonal matrix representing the power allocated to each data stream [61]. A similar description applies for transmitter 2.

In the cooperative solution, [61] assumes that each transmitter i knows not only its own channel $\mathbf{H}_i$ and the cross-channel $\mathbf{G}_i$, but also the subspaces span($\mathbf{G}_j\mathbf{A}_j\mathbf{P}_j$) and span($\mathbf{H}_j\mathbf{A}_j\mathbf{P}_j$), which represent the subspaces in which transmitter j's information signal lies when it reaches receiver i and j, respectively. It is assumed that the transmitters exchange information about these subspaces with each other. For the cooperative scheme, transmitter i devotes a fraction $1 - \alpha_i$ of its power ($0 \le \alpha_i \le 1$) to transmit artificial noise. Mathematically, the transmitted signal vector for user i is given by:

$$
\mathbf{x}_i = \mathbf{A}_i\mathbf{s}_i + \mathbf{T}_i\mathbf{z}_i \quad (i \ne j \quad i, j \in \{1, 2\})
\tag{6.18}
$$

where $\mathbf{s}_i \sim \mathcal{CN}(\mathbf{0}, \mathbf{P}_i)$, $\mathbf{z}_i \sim \mathcal{CN}(\mathbf{0}, \mathbf{Q}_{zi})$, and

$$
\mathbf{Q}_{zi} = \frac{(1 - \alpha_i)p_i}{\mathrm{Tr}(\mathbf{T}_i\mathbf{T}_i^H)}.
\tag{6.19}
$$

The elements of the vector $\mathbf{z}_i$ represent synthetic noise symbols broadcast by transmitter i along the column vectors of the beamforming matrix $\mathbf{T}_i$. Again, $\mathbf{A}_i$, the precoder for the information signal, is determined using the GSVD of the pair $(\mathbf{H}_i, \mathbf{G}_i)$ as in the noncooperative GSVD approach, with power loading matrix $\mathbf{P}_i$, but with the updated power constraint $\mathrm{Tr}(\mathbf{A}_i\mathbf{P}_i\mathbf{A}_i^H) \le \alpha_i P_i$.

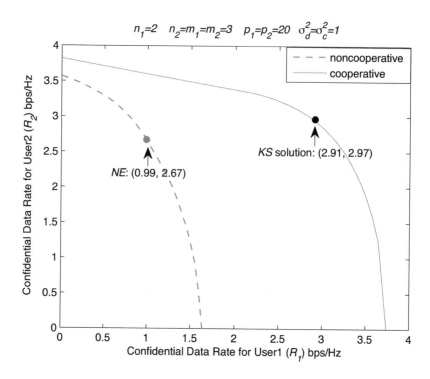

Figure 6.5: System model for the two-user MIMO interference channel with confidential messages.

In [61], a closed-form solution is proposed to construct $\mathbf{T}_i$, by considering the fact that there are three possible goals that, for example, transmitter 1 could consider when cooperatively designing $\mathbf{T}_1$ to broadcast artificial noise[1]:

1. Eliminate the impact of the artificial noise on its own information signal at Rx1: $\mathrm{span}(\mathbf{H}_1\mathbf{A}_1\mathbf{P}_1) \perp \mathrm{span}(\mathbf{H}_1\mathbf{T}_1)$,

2. Eliminate the impact of the artificial noise on the information signal from Tx2 at Rx2: $\mathrm{span}(\mathbf{H}_2\mathbf{A}_2\mathbf{P}_2) \perp \mathrm{span}(\mathbf{G}_1\mathbf{T}_1)$, and

3. Align the artificial noise with Tx2's information signal at Rx1: $\mathrm{span}(\mathbf{G}_2\mathbf{A}_2\mathbf{P}_2) \cong \mathrm{span}(\mathbf{H}_1\mathbf{T}_1)$.

The last two goals are entirely altruistic, and aid Tx2 in improving its secrecy rate. A subspace fitting method that minimizes a weighted sum of the errors in achieving goals 1 through 3 is explained in [61].

Two-user IFC scenarios with confidential messages are also interesting to consider from game-theoretic perspective. In [62], a cognitive radio scenario is considered where only the message from the primary transmitter is considered confidential and must be kept secret at the cognitive receiver. While the primary transmitter and receiver employ a single-user wiretap channel encoder and decoder, respectively, the strategy of the cognitive transmitter is power allocation on the nonsecure cognitive message and the noise signal. For this scenario, the Nash Equilibrium (NE) is established for the case that all nodes have a single antenna.

---

[1]Note that we do not include the goal of using the artificial noise to hide its own information signal at Rx2; this is because we are using GSVD precoding for this purpose, which is known to be more effective than artificial noise [7].

In [63], a two-user one-sided interference channel with confidential messages is considered, in which one transmitter/receiver pair is interference-free. For this scenario, the NE for the binary deterministic channel is determined. In [58], an iterative algorithm is proposed to compute the NE point for the MISO Gaussian IFC, where each transmitter wishes to maximize its own utility function. The authors of [59] obtain a closed-form solution for the NE point where each multiantenna transmitter desires to maximize the *difference* between its secrecy rate and the secrecy rate of the other link. In [61], a game-theoretic formulation is adopted for an arbitrary two-user MIMO Gaussian interference channel, where the transmitters find an operating point that balances network performance and fairness using the so-called Kalai-Smorodinsky (K-S) bargaining solution [61]. Figure 6.5 shows the achievable secrecy rate regions of the proposed schemes in [61], along with the NE from the noncooperative GSVD approach, and the K-S rate point for the cooperative GSVD and artificial noise alignment method. In this figure, n_i is the number of antennas at transmitter i and m_i is the number of antennas at receiver i. The benefit of transmitter cooperation in the IFC network is clearly evident from the secrecy point of view.

6.7 MIMO Relay Wiretap Networks

Physical layer security in relay networks can be considered as a natural extension to methods for secure transmission in MIMO networks. The scenarios considered in MIMO relay wiretap networks can be classified into two main categories: *trusted* and *untrusted* relay wiretap networks. The trusted relay network refers to a traditional scenario where the relays and destinations are all legitimate users. However, in an untrusted relay wiretap network, the relay itself is a potential eavesdropper, although it may still be able to offer assistance for cooperative transmission. For these models, the secrecy capacity and achievable secrecy rate bounds have been investigated for various types of relay-eavesdropper channels, and many cooperative strategies, based on those for conventional relay systems, have been proposed.

6.7.1 Relay-aided Cooperation

In trusted relay wiretap networks, the relays can play various roles with external eavesdroppers. They may act purely as traditional relays while utilizing help from other nodes to ensure security, they may also act as both relaying components as well as cooperative jamming partners to enhance the secure transmission, or they can assume the role of stand-alone helpers to facilitate jamming unintended receivers.

One-way Relays

A typical model of a one-way relay channel with an external eavesdropper is investigated in [33], where the four-terminal network is introduced and an outer bound on the optimal rate-equivocation region is derived. The authors also propose a noise-forwarding strategy where the full-duplex relay sends codewords independent of the secret message to confuse the eavesdropper. A more practical two-hop half-duplex relay wiretap channel is studied in [32], where several cooperative schemes are proposed for a two-hop multiple-relay network, and the corresponding relay weights are derived aiming to maximize the achievable secrecy rate, under the constraint that the link between the source and the relay is not protected from eavesdropping.

A more general case where cooperative jamming strategies guarantee secure communication in both hops without using external helpers is studied in [64], as illustrated in Figure 6.6. In the proposed cooperative jamming strategies, the source and destination nodes act as temporary helpers to transmit jamming signals during transmission phases in

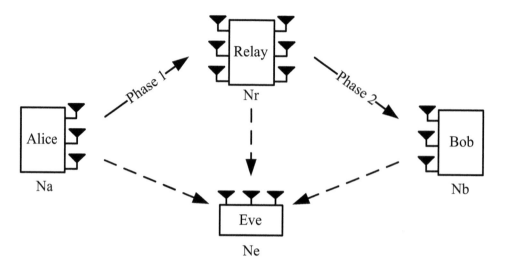

Figure 6.6: Two-hop MIMO relay network with external eavesdropper.

which they are normally inactive. In general, the signals transmitted by Alice in the first phase may contain both information and jamming signals, and Bob may also transmit jamming signals at the same time. Thus the signals received by the relay and Eve in the first phase will be given by (the expressions for the second phase are similar in form)

$$\mathbf{y}_r = \mathbf{H}_{ar}(\mathbf{T}_a \mathbf{z}_a + \mathbf{T}'_a \mathbf{z}'_a) + \mathbf{H}_{br}\mathbf{T}'_b\mathbf{z}'_b + \mathbf{n}_r \tag{6.20}$$

$$\mathbf{y}_{e1} = \mathbf{H}_{ae}(\mathbf{T}_a \mathbf{z}_a + \mathbf{T}'_a \mathbf{z}'_a) + \mathbf{H}_{be}\mathbf{T}'_b\mathbf{z}'_b + \mathbf{n}_{e1} \tag{6.21}$$

where $\mathbf{z}_a$ is the data signal vector, $\mathbf{z}'_a$ and $\mathbf{z}'_b$ are jamming signal vectors transmitted by Alice and Bob, respectively, and $\mathbf{T}'_a$ and $\mathbf{T}'_b$ are the corresponding transmit beamformers. The case where both $\mathbf{z}'_a \neq \mathbf{0}$ and $\mathbf{z}'_b \neq \mathbf{0}$ is defined as fully cooperative jamming (FCJ), while if either of them is zero, it is referred to as partially cooperative jamming (PCJ). One possible example of PCJ is for Alice to utilize the conventional GSVD transmit strategy, while Bob implements jamming in a reverse GSVD fashion, i.e., Bob in phase 1 considers Eve as the intended receiver of the jamming and wants to avoid leaking interference signals to the relay. The performance in terms of secrecy rate of this approach is depicted in Figure 6.7.

For the case where the relay acts as a helper, an interesting approach is to split the transmission time into two phases. In the first phase, the transmitter and the intended receiver both transmit independent artificial noise signals to the helper nodes. The helper nodes and the eavesdropper receive different weighted versions of these two signals. In the second stage, the helper nodes simply replay a weighted version of the received signal, using a publicly available sequence of weights. At the same time, the transmitter sends its secret message, while also canceling the artificial noise at the intended receiver [9].

Two-way Relays

In [30], a two-way wiretap channel is considered, in which both the source and the receiver transmit information over the channel to each other in the presence of a wiretapper. Achievable rates for the two-way Gaussian channel are derived. In addition, a cooperative jamming scheme that utilizes the potential jammers is shown to be able to further increase the secrecy sum rate. In [65], it is shown that using feedback for encoding is essential in Gaussian full-duplex two-way wiretap channels, while feedback can be ignored in the Gaussian half-duplex two-way relay channel with untrusted relays. More recently, secure transmission strategies

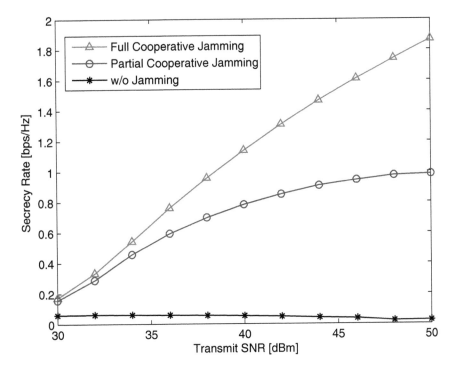

Figure 6.7: Secrecy rate vs. transmit SNR. Each node has four antennas. The distance between the source and destination is 800m with target rate $R_t = 2$bps/Hz.

have been studied for multiantenna two-way relay channels with network coding in the presence of eavesdroppers [66], [67]. By applying the analog network-coded relaying protocol, the end nodes exchange messages in two time slots. In this scenario, the eavesdropper has a significant advantage since it obtains two observations of the transmitted data compared to a single observation at each of the end nodes. As a countermeasure, in each of the two communication phases the transmitting nodes jam the eavesdropper, either by optimally using any available spatial degrees of freedom, or with the aid of external helpers.

6.7.2 Untrusted Relaying

For the untrusted relay case, the relay node acts both as an eavesdropper and a intermediate helper, i.e., the eavesdropper is colocated with the relay node. This type of relay may belong to a heterogeneous network without the same security clearance as the source and destination nodes. The source desires to use the relay to communicate with the destination, but at the same time intends to shield the message from the relay. This type of model was first studied in [68] for the general relay channel. Coding problems of the relay-wiretap channel are studied under the assumption that some transmitted messages are confidential to the relay, and deterministic and stochastic rate regions are explicitly derived in [69–71], which show that cooperation from the untrusted relay is still essential for achieving a nonzero secrecy rate. A more symmetric case is discussed in [72], where both the source and the relay send their own private messages while keeping them secret from the destination. Assuming a half-duplex amplify-and-forward (AF) protocol, another effective countermeasure in this case is to have the destination jam the relay while it is receiving data from the source. This intentional interference can then be subtracted out by the destination from the signal it receives via the relay.

In [73], the authors consider the joint source/relay beamforming design problem for secrecy rate maximization, through a one-way untrusted MIMO relay. For the two-way untrusted relay case, [74] proposes an iterative algorithm to solve for the joint beamformer optimization problem. In realistic fading channels, the secrecy outage probability (SOP) is more meaningful compared with the ergodic secrecy rate, which is ill-defined under finite delay constraints. Thus [75] focuses on the SOP of the AF relaying protocol, which is chosen due to its increased security vis-à-vis decode-and-forward relaying and has lower complexity compared to compress-and-forward approaches. The SOP is a criterion that indicates the fraction of fading realizations where a secrecy rate R can be supported, and also provides a security metric when the source and destination have no CSI for the eavesdropper [76].

In general for untrusted AF relaying, when the number of antennas at the source and the destination is fixed, more antennas at the unauthenticated relay will facilitate a more powerful wiretapping ability and thus the secrecy performance will be degraded. Therefore, an alternative scenario is to consider implementing antenna selection at the relay (e.g., when the relay has only one RF chain) combined with cooperative jamming, in order to suppress the relay's wiretapping ability while still providing diversity gain for the legitimate destination. One example is the case where the relay has no knowledge of the channel to Bob, which applies to the case where Bob transmits no training data to the relay, and jams only when Alice is transmitting so the relay cannot collect interference information [75]. In such cases, the relay selects the receive antenna with the largest channel gain during the first hop, and then uses either the same or some random antenna for transmission during the second hop. Note that the first hop antenna selection uses the same scheme as in traditional AF relaying where the best antenna pair is chosen for the first hop. With this scheme, the SOP is shown to be a nonincreasing function of the number of antennas at the relay, which indicates a possible solution for using multiple antennas at an untrusted relay.

6.8   Conclusions

We have surveyed a number of different methods that have been used to exploit the availability of multiple antennas in enhancing the security of wireless networks at the physical layer. Several different applications were considered, including MIMO wiretap, broadcast, and interference networks, as well as the use of external helpers for cooperative jamming. We have seen that multiple antennas provide considerable flexibility in the form of spatial degrees of freedom that can be exploited for steering signals of interest away from eavesdroppers and toward desired users, and doing the opposite for jamming signals. While performance limits (e.g., secrecy capacity regions, optimal solutions, etc.) have been derived for certain special cases, we have highlighted a number of open problems that require additional attention by researchers in the field. We hope that this chapter can serve as a helpful "jumping-off" point for those interested in investigating some of these problems.

References

[1] Y.-S. Shiu, S. Chang, H.-C. Wu, S. C.-H. Huang, and H.-H. Chen, "Physical layer security in wireless networks: A tutorial," *IEEE Wireless Comm.*, pp. 66–74, Apr. 2011.

[2] M. Bloch and J. Barros, *Physical-Layer Security: From Information Theory to Security Engineering*, Cambridge University Press, 2011.

[3] *Securing Wireless Communications at the Physical Layer*, R. Liu and W. Trappe, Editors, Springer, 2010.

[4] Y. Liang, H. V. Poor, and S. Shamai (Shitz), *Information Theoretic Security*, Now Publishers Inc., 2009.

[5] A. D. Wyner, "The wire-tap channel," *Bell Sys. Tech. Journ.*, vol. 54, pp. 1355–1387, 1975.

[6] M. Yuksel and E. Erkip, "Diversity-multiplexing tradeoff for the multiple-antenna wiretap channel," *IEEE Trans. Wireless Commun.*, vol. 10, no. 3, pp. 762–771, Mar. 2011.

[7] A. Khisti and G. Wornell, "Secure transmission with multiple antennas II: The MI-MOME wiretap channel," *IEEE Trans. Inf. Theory*, vol. 56, no. 11, pp. 5515–5532, Nov. 2010.

[8] A. Khisti and G. Wornell, "Secure transmission with multiple antennas I: The MIS-OME wiretap channel," *IEEE Trans. Inf. Theory*, vol. 56, no. 7, pp. 3088–3104, July 2010.

[9] S. Goel and R. Negi, "Guaranteeing secrecy using artificial noise," *IEEE Trans. Wireless Commun.*, vol. 7, no. 6, pp. 2180–2189, June 2008.

[10] F. Oggier and B. Hassibi, "The secrecy capacity of the MIMO wiretap channel," *IEEE Trans. Inf. Theory*, vol. 57, no. 8, pp. 4961–4972, Aug. 2011.

[11] T. Liu and S. Shamai, "A note on the secrecy capacity of the multiple-antenna wiretap channel," *IEEE Trans. Inf. Theory*, vol. 55, no. 6, pp. 2547–2553, June 2009.

[12] T. F. Wong, M. Bloch, and J. M. Shea, "Secret sharing over fast-fading MIMO wiretap channels," *EURASIP J. on Wireless Communications and Networking*, vol. 2009.

[13] F. Renna, M. Bloch, and N. Laurenti, "Semi-blind key agreement over MIMO fading channels," in *Proc. IEEE International Conference on Communications*, Kyoto, Japan, June 2011.

[14] R. Bustin, R. Liu, H. V. Poor, and S. Shamai, "An MMSE approach to the secrecy capacity of the MIMO Gaussian wiretap channel," *EURASIP J. Wireless Communications and Networking*, 2009.

[15] S. A. A. Fakoorian and A. L. Swindlehurst, "Full Rank Solutions for the MIMO Gaussian Wiretap Channel with an Average Power Constraint," *IEEE Trans. Signal Process.*, 2013 (to appear).

[16] S. Loyka and C. D. Charalamboust, "On Optimal Signaling over Secure MIMO Channels," in *Proc. IEEE International Symposium on Information Theory (ISIT)*, Boston, 2012.

[17] S. A. A. Fakoorian and A. L. Swindlehurst, "Optimal power allocation for GSVD-based beamforming in the MIMO wiretap channel," in *Proc. IEEE International Symposium on Information Theory (ISIT)*, Boston, 2012.

[18] S.-H. Lai, P.-H. Lin, S.-C. Lin, and H.-J. Su, "On optimal artificial-noise assisted secure beamforming for the multiple-input multiple-output fading eavesdropper channel," in *Proc. IEEE WCNC 2012*, pp. 513–517.

[19] S. Shafiee and S. Ulukus, "Achievable rates in Gaussian MISO channels with secrecy constraints," in *Proc. IEEE International Symposium on Information Theory (ISIT)*, pp. 2466–2470, June 2007.

[20] L. Zhang, Y.-C. Liang, Y. Pei, and R. Zhang, "Robust beamforming design: From cognitive radio MISO channels to secrecy MISO channels," in *Proc. IEEE Global Telecommunications Conf. (GLOBECOM)*, Nov. 2009.

[21] J. Huang and A. L. Swindlehurst, "Robust Secure Transmission in MISO Channels Based on Worst-Case Optimization," *IEEE Trans. Signal Process.*, vol. 60, no. 4, pp. 1696–1707, Apr. 2012.

[22] A. Mukherjee and A. L. Swindlehurst, "Robust beamforming for secrecy in MIMO wiretap channels with imperfect CSI," *IEEE Trans. Signal Process.*, vol. 59, no. 1, pp. 351–361, Jan. 2011.

[23] Q. Li and W.-K. Ma, "Optimal and robust transmit designs for MISO channel secrecy by semidefinite programming," *IEEE Trans. Signal Process.*, vol. 59, no. 8, pp. 3799–3812, Aug. 2011.

[24] S.-C. Lin, T.-H. Chang, Y.-L. Liang, Y.-W. P. Hong, and C.-Y. Chi, "On the impact of quantized channel feedback in guaranteeing secrecy with artificial noise: The noise leakage problem," *IEEE Trans. Wireless Commun.*, vol. 10, no. 3, pp. 901–915, Mar. 2011.

[25] A. Mukherjee and A. L. Swindlehurst, "Fixed-rate power allocation strategies for enhanced secrecy in MIMO wiretap channels," *Proc. of 10th IEEE International Workshop on Signal Processing Advances for Wireless Communications (SPAWC)*, pp. 344–348, Perugia, Italy, June 2009.

[26] A. Wolf and E. Jorswieck, "Maximization of worst-case secret key rates in MIMO systems with eavesdropper," in *Proc. IEEE GLOBECOM*, 2011.

[27] J. Li and A. Petropulu, "Ergodic secrecy rate for multiple-antenna wiretap channels with Rician fading," *IEEE Trans. Inform. Forens. Security*, vol. 6, no. 3, pp. 861–867, Sep. 2011.

[28] Y. Wu, C. Xiao, Z. Ding, X. Gao, and S. Jin, "Linear precoding for finite-alphabet signaling over MIMOME wiretap channels," *IEEE Trans. Veh. Tech.*, vol. 61, no. 6, pp. 2599–2612, July 2012.

[29] I. Csiszàr and J. Körner, "Broadcast channels with confidential messages," *IEEE Trans. Inf. Theory*, vol. 24, no. 3, pp. 339–348, May 1978.

[30] E. Tekin and A. Yener, "The general Gaussian multiple access and two-way wire-tap channels: Achievable rates and cooperative jamming," *IEEE Trans. Inf. Theory*, vol. 54, no. 6, pp. 2735-2751, Jun. 2008.

[31] L. Dong, Z. Han, A. P. Petropulu, and H. V. Poor, "Cooperative jamming for wireless physical layer security," in *Proc. of IEEE Workshop on Statistical Signal Processing*, Cardiff, Wales, U.K., 2009

[32] L. Dong, Z. Han, A. P. Petropulu, and H. V. Poor, "Improving wireless physical layer security via cooperating relays," *IEEE Trans. Signal Proc.*, vol. 58, no. 3, pp. 1875–1888, Mar. 2010.

[33] L. Lai and H. El Gamal, "The relay-eavesdropper channel: Cooperation for secrecy," *IEEE Trans. Inf. Theory*, vol. 54, no. 9, pp. 4005–4019, Sep. 2008.

[34] X. Tang, R. Liu, P. Spasojevic, and H. V. Poor, "The Gaussian wiretap channel with a helping interferer," in *Proc. IEEE Int. Symp. Inf. Theory*, Toronto, ON, Canada, Jul. 2008.

[35] S. A. A. Fakoorian and A. L. Swindlehurst, "Secrecy capacity of MISO Gaussian wiretap channel with a cooperative jammer," in *IEEE SPAWC Workshop*, San Francisco, June 2011.

[36] J. Wang and A. L. Swindlehurst, "Cooperative jamming in MIMO ad hoc networks," in *Proc. Asilomar Conf. on Signals, Systems and Computers*, pp. 1719–1723, Nov. 2009.

[37] S. A. A. Fakoorian and A. L. Swindlehurst, "Solutions for the MIMO Gaussian wiretap channel with a cooperative jammer," *IEEE Trans. Signal Proc.*, vol. 59, no. 10, pp. 5013–5022, Oct. 2011.

[38] E. MolavianJazi, M. Bloch, and J. N. Laneman, "Arbitrary jamming can preclude secure communication," in *Proc. Allerton Conf. Communications, Control, and Computing*, Monticello, IL, Sept. 2009.

[39] Y. Liang, H. V. Poor, and S. Shamai, "Secure communication over fading channels," *IEEE Trans. Inf. Theory*, vol. 54, no. 6, pp. 2470-2497, June 2008.

[40] H. D. Ly, T. Liu, and Y. Liang, "Multiple-input multiple-output Gaussian broadcast channels with common and confidential messages," *IEEE Trans. Inf. Theory*, vol. 56, no. 11, pp. 5477–5487, Nov. 2010.

[41] H. Weingarten, Y. Steinberg, and S. Shamai, "The capacity region of the Gaussian multiple-input multiple-output broadcast channel," *IEEE Trans. Inf. Theory*, vol. 52, no. 9, pp. 3936-3964, Sep. 2006.

[42] R. Liu, I. Maric, P. Spasojevic, and R. D. Yates, "Discrete memoryless interference and broadcast channels with confidential messages: Secrecy rate regions," *IEEE Trans. Inf. Theory*, vol. 54, no. 6, pp. 2493–2512, June 2008.

[43] R. Liu and H. V. Poor, "Secrecy capacity region of a multiple-antenna Gaussian broadcast channel with confidential messages," *IEEE Trans. Inf. Theory*, vol. 55, no. 3, pp. 1235–1249, Mar. 2009.

[44] R. Liu, T. Liu, H. V. Poor, and S. Shamai, "Multiple-input multiple-output Gaussian broadcast channels with confidential messages," *IEEE Trans. Inf. Theory*, vol. 56, no. 9, pp. 4215–4227, Sep. 2010.

[45] R. Liu, T. Liu, H. V. Poor, and S. Shamai (Shitz), "MIMO Gaussian broadcast channels with confidential and common messages," in *Proc. IEEE Int. Symp. Information Theory*, Texas, U.S.A., June 2010, pp. 2578–2582.

[46] E. Ekrem and S. Ulukus,"Capacity region of Gaussian MIMO broadcast channels with common and confidential messages," *IEEE Trans. Inf. Theory*, vol. 58, no. 9, pp. 5669–5680, Sep. 2012.

[47] A. Khisti, A. Tchamkerten, and G. Wornell, "Secure broadcasting over fading channels," *IEEE Trans. Inf. Theory*, vol. 54, no. 6, pp. 2453–2469, June 2008.

[48] G. Bagherikaram, A. S. Motahari and A. K. Khandani, "The secrecy capacity region of the Gaussian MIMO broadcast channel," submitted to *IEEE Trans. Inf. Theory*, March 2009, available at http://arxiv.org/PScache/arxiv/pdf/0903/0903.3261v2.pdf.

[49] E. Ekrem and S. Ulukus, "The secrecy capacity region of the Gaussian MIMO multi-receiver wiretap channel," *IEEE Trans. Inf. Theory*, vol. 57, no. 4, pp. 2083–2114, Apr. 2011.

[50] R. Liu, T. Liu, H. V. Poor, and S. Shamai, "A vector generalization of Costa's entropy-power inequality with applications," *IEEE Trans. Inf. Theory*, vol. 56, no. 4, pp. 1865–1879, Apr. 2010.

[51] A. Mukherjee and A. L. Swindlehurst, "Utility of beamforming strategies for secrecy in multiuser MIMO wiretap channels," in *Proc. of Forty-Seventh Allerton Conf.*, Oct. 2009.

[52] W. Liao, T. Chang, W. Ma, and C. Chi, "Joint transmit beamforming and artificial noise design for QoS discrimination in wireless downlink," in *Proc. IEEE ICASSP*, pp. 2561–2565, Dallas, Mar. 2010.

[53] S. A. A. Fakoorian and A. L. Swindlehurst, "On the Optimality of linear precoding for secrecy in the MIMO broadcast channel," *IEEE Journal on Selected Areas on Communications*, 2013 (to appear).

[54] S. A. A. Fakoorian and A. L. Swindlehurst, "Dirty paper coding versus linear GSVD-based precoding in MIMO broadcast channel with confidential messages," in *Proc. IEEE GLOBECOM*, 2011.

[55] Y. Liang, A. Somekh-Baruch, H. V. Poor, S. Shamai, and S. Verdu, "Capacity of cognitive interference channels with and without confidential messages," *IEEE Trans. Inf. Theory*, vol. 55, no. 2, pp. 604–619, 2009.

[56] O. O. Koyluoglu, H. El Gamal, L. Lai, and H. V. Poor, "Interference alignment for secrecy," *IEEE Trans. Inf. Theory*, vol. 57, no. 6, June 2011.

[57] X. He and A. Yener, "K-user interference channels: Achievable secrecy rate and degrees of freedom," in *Proc. IEEE Information Theory Workshop on Networking and Information Theory, ITW'09*, Volos, Greece, June 2009, pp. 336–340.

[58] E. A. Jorswieck and R. Mochaoura, "Secrecy rate region of MISO interference channel: Pareto boundary and non-cooperative games," in *Proc. Int'l ITG Workshop on Smart*

Antennas (WSA), 2009.

[59] S. A. A. Fakoorian and A. L. Swindlehurst, "Competing for secrecy in the MISO interference channel," *IEEE Tran. Sig. Proc.*, vol. 61, no. 1, Jan. 2013.

[60] S. A. A. Fakoorian and A. L. Swindlehurst, "MIMO interference channel with confidential messages: Game theoretic beamforming designs," in *Proc. Asilomar Conf. on Signals, Systems, and Computers*, Nov. 2010.

[61] S. A. A. Fakoorian and A. L. Swindlehurst, "MIMO interference channel with confidential messages: Achievable secrecy rates and beamforming design," *IEEE Trans. on Inf. Forensics and Security*, vol. 6, no. 3, Sep. 2011.

[62] E. Toher, O. O. Koyluoglu, and H. El Gamal, "Secrecy games over the cognitive channel," in *Proc. IEEE Int. Symp. Information Theory* Austin, Texas, Aug. 2010, pp. 2637–2641.

[63] J. Xie and S. Ulukus, "Secrecy games on the one-sided interference channel," in *Proc. IEEE Int. Symp. Information Theory*, Saint-Petersburg, Russia, Aug. 2011.

[64] J. Huang and A. L. Swindlehurst, "Cooperative jamming for secure communications in MIMO relay networks," *IEEE Trans. Signal Process.*, vol. 59, no. 10, pp. 4871–4884, Oct. 2011.

[65] X. He and A. Yener, "On the role of feedback in two-way secure communication," in *Proc. 42nd Asilomar Conf. Signals, Systems and Computers*, pp. 1093–1097, Oct. 2008.

[66] A. Mukherjee and A. L. Swindlehurst, "Securing multi-antenna two-way relay channels with analog network coding against eavesdroppers," in *Proc. 11th IEEE SPAWC*, Jun. 2010.

[67] S. Al-Sayed and A. Sezgin, "Secrecy in Gaussian MIMO bidirectional broadcast wiretap channels: Transmit strategies," in *Proc. 44th Asilomar Conf. Signals, Systems and Computers*, Nov. 2010.

[68] Y. Oohama, "Capacity theorems for relay channels with confidential messages," in *Proc. IEEE ISIT*, pp. 926-930, Jun. 2007.

[69] X. He and A. Yener, "Cooperation with an untrusted relay: A secrecy perspective," *IEEE Trans. Inf. Theory*, vol. 56, no. 8, pp. 3807–3827, Aug. 2010.

[70] X. He and A. Yener, "On the equivocation region of relay channels with orthogonal components," in *Proc. Forty-First Asilomar Conf.*, pp. 883–887, Nov. 2007.

[71] X. He and A. Yener, "The role of an untrusted relay in secret communication," in *Proc. IEEE ISIT*, pp. 2212–2216, Jul. 2008.

[72] E. Ekrem and S. Ulukus, "Secrecy in cooperative relay broadcast channels," in *Proc. IEEE ISIT*, pp. 2217–2221, Jul. 2008.

[73] C. Jeong, I.-M. Kim, and D. I. Kim, "Joint secure beamforming design at the source and the relay for an amplify-and-forward MIMO untrusted relay system," *IEEE Trans. Signal Process.*, vol. 60, no. 1, pp. 310–325, Jan. 2012.

[74] J. Mo, M. Tao, Y. Liu, B. Xia, and X. Ma, "Secure beamforming for MIMO two-way transmission with an untrusted relay," in *Proc. IEEE WCNC*, to appear, Apr. 2013.

[75] J. Huang, A. Mukherjee, and A. L. Swindlehurst, "Secure communication via an untrusted non-regenerative relay in fading channels," *IEEE Trans. Signal Process.*, to appear, 2013.

[76] M. Bloch, J. Barros, M. Rodrigues, and S. McLaughlin, "Wireless information-theoretic security," *IEEE Trans. Inf. Theory*, vol. 54, no. 6, pp. 2515–2534, Jun. 2008.

Chapter 7

Discriminatory Channel Estimation for Secure Wireless Communication

Y.-W. Peter Hong
National Tsing Hua University

Tsung-Hui Chang
National Taiwan University of Science and Technology

Through a review of the so-called *discriminatory channel estimation* (DCE) scheme [1,2], this chapter introduces a way to enhance physical layer secrecy from a channel estimation perspective. The DCE scheme is a novel training scheme that is aimed at providing discrimination between the channel estimation performances at a legitimate receiver (LR) and an unauthorized receiver (UR). This effectively enlarges the difference between the signal-to-noise ratios (SNRs) at the two receivers and leaves more room for secrecy coding or transmission schemes in the data transmission phase. A key feature of DCE designs is the insertion of artificial noise (AN) in the training signal to degrade the channel estimation performance at UR. To do so, AN must be placed in a carefully chosen subspace to minimize its effect on LR. However, this requires preliminary knowledge of the channel at the transmitter, which can be difficult to achieve without benefiting the channel estimation at UR as well. To achieve this task, two DCE schemes were proposed in the literature, namely, the feedback-and-retraining [1] and the two-way training [2] based DCE schemes. Both of these schemes require a preliminary training stage to provide the transmitter with a rough estimate of LR's channel and an AN-assisted training stage to enable channel estimation at LR while degrading channel estimation at UR. These schemes as well as methods to optimally allocate power among pilot signals and AN are reviewed in this chapter.

7.1 Introduction

With the proliferation of wireless devices, secrecy has become an important but challenging issue in the design of wireless communication systems. In particular, in wireless systems, signals are broadcast through the wireless medium and can be easily intercepted or eavesdropped by unauthorized receivers. These issues have been addressed in the past using encryption and decryption techniques in the higher layers of the network protocol stack. However, this approach requires reliable secret-key exchange and management that can be difficult to achieve in distributed or highly dynamic environments. This has motivated studies on the so-called physical layer secrecy [3–11], which utilize properties of the physical wireless channel to discriminate the reception performances at the legitimate receiver (LR) and the unauthorized receiver (UR). This can be achieved using both coding and signal processing techniques, e.g., in [3–7] and [9–11], respectively.

Most studies in the literature on physical layer secrecy, e.g., [3–11], focus on the design of data transmission schemes while often assuming that perfect channel state information is available at all terminals. However, in practice, channel knowledge is typically obtained through training and channel estimation, and the channel estimation quality has a significant impact on the data reception performance. For instance, low channel estimation quality leads to poor effective channel condition at the corresponding receiver and, thus, poor reception performance in the data transmission phase [12]. This motivates the design of training procedures to discriminate the channel estimation performance at the two receivers and, in this way, enhance the achievable secrecy performance in the physical layer [13–15]. The design is referred to as the *discriminatory channel estimation* (DCE) scheme [1,2]. From an information-theoretic viewpoint [3–6], this leaves more room for secrecy coding and leads to a higher achievable secrecy rate (i.e., the rate achievable between the transmitter and LR subject to the secrecy constraint). From a signal processing perspective, this prohibits UR from achieving a detection probability as low as that of LR.

The key to achieving discriminatory channel estimation is the insertion of AN in the training signal to degrade UR's channel estimation performance. However, to minimize the interference on LR, AN must be placed mostly in the null space of LR's channel. This requires preliminary channel knowledge at the transmitter, which is difficult to obtain in the channel estimation phase without benefiting the channel estimation at UR as well. Two DCE schemes were proposed in the literature, namely, the feedback-and-retraining [1] and the two-way training [2] based DCE schemes. In the feedback-and-retraining scheme, the transmitter first emits a preliminary training signal to assist LR in obtaining a rough estimate of its channel. Then, LR feeds back this rough estimate to the transmitter, who then emits a new training signal with artificial noise (AN) embedded in the null space of the estimated channel matrix. The power utilized in the preliminary training must be low enough to prevent UR from obtaining a good channel estimate. However, this also limits the channel estimation performance at LR. To overcome this issue, multiple stages of feedback and retraining are used to gradually refine transmitter's knowledge of the channel and to more accurately place AN in the desired subspace. In the two-way training scheme, preliminary training is sent by LR to facilitate channel estimation directly at the transmitter. If the channel is reciprocal, the transmitter can obtain knowledge of the forward channel from an estimate of the reverse channel and can immediately emit an AN-assisted training afterwards to achieve the desired performance discrimination. If the channel is nonreciprocal, an additional round-trip training is required to provide the transmitter with knowledge of the forward channel. The feedback-and-retraining scheme is easily extendable to cases with multiple LRs, but requires large training overhead when stringent constraints are given on UR's channel estimation performance. The two-way training scheme has a smaller overhead, but is limited by LR's ability to participate in the training process.

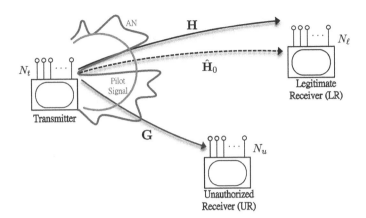

Figure 7.1: Diagram of a wireless MIMO system consisting of a multiantenna transmitter, a multiantenna legitimate receiver (LR), and a multiantenna unauthorized receiver (UR).

7.2 Discriminatory Channel Estimation—Basic Concept

Let us consider a wireless multiple-input multiple-output (MIMO) system that consists of a transmitter, a legitimate receiver (LR), and an unauthorized receiver (UR)[1]. They are equipped with N_t, N_ℓ, and N_u antennas, respectively, as shown in Figure 7.1. Throughout this chapter, we assume that $N_t > N_\ell$. (For the case where $N_t < N_\ell$, training can be designed such that channels corresponding to subsets of antennas with size less than N_t are estimated in turn at LR. However, details will not be examined in this chapter.) Let $\mathbf{H}$ and $\mathbf{G}$ be the channel matrices from the transmitter to LR and UR, respectively. The entries of $\mathbf{H}$ and $\mathbf{G}$ are assumed to be independent and identically distributed (i.i.d.) zero-mean complex Gaussian random variables with variances σ_h^2 and σ_g^2, respectively (which are denoted by $\mathcal{CN}(0, \sigma_h^2)$ and $\mathcal{CN}(0, \sigma_g^2)$, respectively). The channel fading coefficients are assumed to remain constant during each transmission block, which consists of a training phase and a data transmission phase. The main objective of the DCE scheme is to discriminate the channel estimation performances at LR and UR, i.e., to allow a good estimate of $\mathbf{H}$ at LR but a bad estimate of $\mathbf{G}$ at UR.

A general DCE scheme consists of two stages, a preliminary training stage and an AN-assisted training stage. In the preliminary training stage, a rough estimate of the channel between the transmitter and LR, i.e., $\mathbf{H}$, is to be acquired at the transmitter. In the AN-assisted training stage, a pilot signal plus AN is emitted by the transmitter to enable channel estimation at LR while simultaneously disrupting the channel estimation at UR. The AN placement is determined based on the channel knowledge obtained at the transmitter in the preliminary training stage. With channel knowledge, AN can be placed in the null space of the estimated channel from the transmitter to LR to minimize its effect on LR.

Suppose that an estimate of $\mathbf{H}$, denoted by $\widehat{\mathbf{H}}$, is obtained at the transmitter in the preliminary training stage. Then, in the AN-assisted training stage, AN can be placed in the null space of the estimated channel matrix and, thus, the training signal can be expressed

---

[1] For ease of presentation, we will focus on the scenario with one LR and one UR throughout the paper. The presented methods, nevertheless, can be easily extended to the scenario with multiple LRs and URs as shown in [1].

as

$$\mathbf{X} = \sqrt{\frac{PT}{N_t}} \mathbf{C}_t + \mathbf{A} \mathbf{K}_{\widehat{\mathbf{H}}}^H, \tag{7.1}$$

where $\mathbf{C}_t \in \mathbb{C}^{T \times N_t}$ is the pilot matrix with $\mathrm{Tr}(\mathbf{C}_t^H \mathbf{C}_t) = N_t$, P is the pilot signal power, and T is the training length. Here, $\mathbf{A} \in \mathbb{C}^{T \times (N_t - N_\ell)}$ is the AN matrix whose entries are i.i.d. $\mathcal{CN}(0, \sigma_a^2)$ and are statistically independent of the channels and noises at all terminals. $\mathbf{K}_{\widehat{\mathbf{H}}} \in \mathbb{C}^{N_t \times (N_t - N_\ell)}$ is a matrix whose columns form an orthonormal basis for the left null space of $\widehat{\mathbf{H}}$, that is, $\mathbf{K}_{\widehat{\mathbf{H}}}^H \widehat{\mathbf{H}} = \mathbf{0}_{(N_t - N_\ell) \times N_\ell}$ and $\mathbf{K}_{\widehat{\mathbf{H}}}^H \mathbf{K}_{\widehat{\mathbf{H}}} = \mathbf{I}_{N_t - N_\ell}$ (i.e., the $(N_t - N_\ell) \times (N_t - N_\ell)$ identity matrix). Notice that AN is superimposed on top of the pilot matrix and is placed in the left null space of $\widehat{\mathbf{H}}$ to minimize its interference on LR. The received signals at LR and UR are respectively given by

$$\mathbf{Y} = \sqrt{\frac{PT}{N_t}} \mathbf{C}_t \mathbf{H} + \mathbf{A} \mathbf{K}_{\widehat{\mathbf{H}}}^H \mathbf{H} + \mathbf{W}, \tag{7.2}$$

$$\mathbf{Z} = \sqrt{\frac{PT}{N_t}} \mathbf{C}_t \mathbf{G} + \mathbf{A} \mathbf{K}_{\widehat{\mathbf{H}}}^H \mathbf{G} + \mathbf{V}, \tag{7.3}$$

where $\mathbf{W} \in \mathbb{C}^{T \times N_\ell}$ and $\mathbf{V} \in \mathbb{C}^{T \times N_u}$ are the additive white Gaussian noise (AWGN) matrices at LR and UR, respectively, with entries being i.i.d. $\mathcal{CN}(0, \sigma_w^2)$ and $\mathcal{CN}(0, \sigma_v^2)$, respectively. Because $\widehat{\mathbf{H}} = \mathbf{H} + \Delta \mathbf{H}$ and $\mathbf{K}_{\widehat{\mathbf{H}}}^H \widehat{\mathbf{H}} = \mathbf{0}$, (7.2) can be rewritten as

$$\mathbf{Y} = \sqrt{\frac{PT}{N_t}} \mathbf{C}_t \mathbf{H} - \mathbf{A} \mathbf{K}_{\widehat{\mathbf{H}}}^H \Delta \mathbf{H} + \mathbf{W}. \tag{7.4}$$

Then, with the received signals, LR and UR can update their estimates of $\mathbf{H}$ and $\mathbf{G}$, respectively, with the hope that AN will cause significant degradation to the estimation performance at UR while having limited effect on LR. The challenge of designing the DCE scheme lies mostly in the channel acquisition at the transmitter and the optimal power allocation between pilot signals and AN. The feedback-and-retraining and the two-way DCE schemes as well as their respective performances are described in detail in the ensuing sections.

7.3   DCE via Feedback and Retraining

The first DCE scheme that we introduce is the feedback-and-retraining DCE scheme proposed in [1] and as illustrated in Figure 7.2. In the preliminary training stage, the transmitter emits a pure pilot signal to enable a rough channel estimate at LR, as done in conventional MIMO training schemes [16,17]. Then, by having LR feedback its preliminary channel estimate to the transmitter, another training signal with AN embedded in the left null space of LR's channel can then be emitted to improve the channel estimate at LR while preventing UR from obtaining a good channel estimate. The power used for preliminary training should on the one hand be low enough to limit UR's channel estimation performance, but should on the other hand be set high enough to allow for better channel estimation performance at LR.

7.3.1   Two-Stage Feedback-and-Retraining

Let us first consider the basic two-stage feedback-and-retraining DCE scheme that utilizes a preliminary training stage for channel acquisition at the transmitter and one feedback-and-retraining stage for channel quality discrimination.

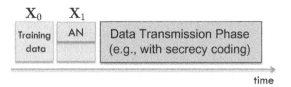

Figure 7.2: Illustration of the transmission phases in a two-stage feedback-and-retraining DCE scheme.

Stage 0 (Channel Acquisition at the Transmitter via Preliminary Training): In the initial stage, the transmitter first emits a sequence of training signals (that consists of only pilot signals) for preliminary channel estimation at LR. Specifically, the training signal sent by the transmitter in this stage can be written as

$$\mathbf{X}_0 = \sqrt{\frac{P_0 T_0}{N_t}} \mathbf{C}_0, \tag{7.5}$$

where $\mathbf{C}_0 \in \mathbb{C}^{T_0 \times N_t}$ is the training data matrix satisfying $\mathrm{Tr}(\mathbf{C}_0^H \mathbf{C}_0) = N_t$ in which $\mathrm{Tr}(\cdot)$ represents the trace of a matrix. The received signal at LR and UR can be written as

$$\text{LR}: \mathbf{Y}_0 = \mathbf{X}_0 \mathbf{H} + \mathbf{W}_0, \tag{7.6}$$
$$\text{UR}: \mathbf{Z}_0 = \mathbf{X}_0 \mathbf{G} + \mathbf{V}_0, \tag{7.7}$$

where $\mathbf{H} \in \mathbb{C}^{N_t \times N_\ell}$, $\mathbf{G} \in \mathbb{C}^{N_t \times N_u}$ are defined as in Section 7.2 and $\mathbf{W}_0 \in \mathbb{C}^{T_k \times N_\ell}$, $\mathbf{V}_0 \in \mathbb{C}^{T_k \times N_u}$ are the AWGN matrices at LR and UR, respectively, in stage 0.

Based on the received signal $\mathbf{Y}_0$ and the known pilot matrix $\mathbf{C}_0$, LR computes a preliminary estimate of $\mathbf{H}$, denoted by $\hat{\mathbf{H}}_0$, and sends this estimate back to the transmitter. By employing the linear minimum mean square error (LMMSE) estimator [18] at LR, the preliminary channel estimate of $\mathbf{H}$ at LR in the initial stage (i.e., stage 0) is given by

$$\hat{\mathbf{H}}_0 = \sigma_h^2 \mathbf{X}_0^H \left(\sigma_h^2 \mathbf{X}_0 \mathbf{X}_0^H + \sigma_w^2 \mathbf{I}_{T_0} \right)^{-1} \mathbf{Y}_0 \triangleq \mathbf{H} + \Delta \mathbf{H}_0, \tag{7.8}$$

where $\Delta \mathbf{H}_0 \in \mathbb{C}^{N_t \times N_\ell}$ stands for the estimation error matrix. By considering orthogonal pilot matrices, i.e., by choosing $\mathbf{C}_0$ such that $\mathbf{C}_0^H \mathbf{C}_0 = \mathbf{I}_{N_t}$, the correlation matrix of $\Delta \mathbf{H}_0$ can be written as

$$\mathrm{E}\left[\Delta \mathbf{H}_0 (\Delta \mathbf{H}_0)^H\right] = N_\ell \left(\frac{1}{\sigma_h^2} \mathbf{I}_{N_t} + \frac{P_0 T_0}{N_t \sigma_w^2} \mathbf{C}_0^H \mathbf{C}_0 \right)^{-1} = N_\ell \left(\frac{1}{\sigma_h^2} + \frac{P_0 T_0}{N_t \sigma_w^2} \right)^{-1} \mathbf{I}_{N_t}. \tag{7.9}$$

The performance of the channel estimate can be measured by the normalized mean square error (NMSE), which is the total MSE normalized by the number of estimated parameters. In this case, the NMSE of $\hat{\mathbf{H}}_0$ is defined as

$$\mathrm{NMSE}_{\mathrm{L}}^{(0)} \triangleq \frac{\mathrm{Tr}\left(\mathrm{E}\left[\Delta \mathbf{H}_0 (\Delta \mathbf{H}_0)^H\right]\right)}{N_t N_\ell} = \left(\frac{1}{\sigma_h^2} + \frac{P_0 T_0}{N_t \sigma_w^2} \right)^{-1}, \tag{7.10}$$

where $N_t N_\ell$ is the number of entries in the channel matrix $\mathbf{H}$ (and, thus, the number of parameters to be estimated).

Stage 1 (Feedback-and-Retraining for DCE): In this stage, LR sends back $\hat{\mathbf{H}}_0$ to the transmitter, who utilizes this information to determine the AN placement in the training signal. Please note that UR is also allowed to intercept the feedback sent by LR but, as can be seen later on, this information does not help UR improve the LMMSE estimate of its own channel.

Specifically, with knowledge of $\hat{\mathbf{H}}_0$, the transmitter then transmits another training signal with AN in the left null space of $\hat{\mathbf{H}}_0$ [7] to degrade UR's channel estimation performance. By assuming that $N_t > N_\ell$, the training signal in stage 1 can be written as

$$\mathbf{X}_1 = \sqrt{\frac{P_1 T_1}{N_t}} \mathbf{C}_1 + \mathbf{A}_1 \mathbf{K}_{\hat{\mathbf{H}}_0}^H, \tag{7.11}$$

where $\mathbf{C}_1 \in \mathbb{C}^{T_1 \times N_t}$ is the pilot matrix satisfying $\mathrm{Tr}(\mathbf{C}_1^H \mathbf{C}_1) = N_t$, $\mathbf{K}_{\hat{\mathbf{H}}_0} \in \mathbb{C}^{N_t \times (N_t - N_\ell)}$ is a matrix with $\mathbf{K}_{\hat{\mathbf{H}}_0}^H \mathbf{K}_{\hat{\mathbf{H}}_0} = \mathbf{I}_{N_t - N_\ell}$ and $\mathbf{K}_{\hat{\mathbf{H}}_0}^H \hat{\mathbf{H}}_0 = \mathbf{0}$, and $\mathbf{A}_1 \in \mathbb{C}^{T_1 \times (N_t - N_\ell)}$ is an AN matrix with entries being i.i.d. $\mathcal{CN}(0, \sigma_{a,1}^2)$. Notice that LR may also suffer from the AN added in (7.11) since the estimate $\hat{\mathbf{H}}_0$ is in general not perfect [15]. Therefore, the pilot signal power P_1 and AN power $\sigma_{a,1}^2$ must be carefully determined. Similarly, the signals received at LR and UR in stage 1 are given by

$$\mathrm{LR} : \mathbf{Y}_1 = \mathbf{X}_1 \mathbf{H} + \mathbf{W}_1, \tag{7.12}$$

$$\mathrm{UR} : \mathbf{Z}_1 = \mathbf{X}_1 \mathbf{G} + \mathbf{V}_1, \tag{7.13}$$

where $\mathbf{W}_1$ and $\mathbf{V}_1$ are the AWGN matrices in stage 1 with entries being i.i.d. with distribution $\mathcal{CN}(0, \sigma_w^2)$ and $\mathcal{CN}(0, \sigma_v^2)$, respectively. The channel estimate obtained at both LR and UR can be refined by making use of the signals received in both phases.

Let us stack the signals in the two stages into the matrix

$$\mathbf{Y} \triangleq \begin{bmatrix} \mathbf{Y}_0 \\ \mathbf{Y}_1 \end{bmatrix} = \begin{bmatrix} \sqrt{\frac{P_0 T_0}{N_t}} \mathbf{C}_0 \\ \sqrt{\frac{P_1 T_1}{N_t}} \mathbf{C}_1 \end{bmatrix} \mathbf{H} + \begin{bmatrix} \mathbf{W}_0 \\ \mathbf{A}_1 \mathbf{K}_{\hat{\mathbf{H}}_0}^H \mathbf{H} + \mathbf{W}_1 \end{bmatrix} \tag{7.14}$$

$$= \begin{bmatrix} \sqrt{\frac{P_0 T_0}{N_t}} \mathbf{C}_0 \\ \sqrt{\frac{P_1 T_1}{N_t}} \mathbf{C}_1 \end{bmatrix} \mathbf{H} + \begin{bmatrix} \mathbf{W}_0 \\ -\mathbf{A}_1 \mathbf{K}_{\hat{\mathbf{H}}_0}^H \Delta \mathbf{H}_0 + \mathbf{W}_1 \end{bmatrix} \tag{7.15}$$

$$\triangleq \bar{\mathbf{C}} \mathbf{H} + \bar{\mathbf{W}}, \tag{7.16}$$

where the second equality follows from (7.8) and the fact that $\mathbf{K}_{\hat{\mathbf{H}}_0}^H \hat{\mathbf{H}}_0 = \mathbf{0}$. In the above, we defined $\bar{\mathbf{C}}$ and $\bar{\mathbf{W}}$ as $\bar{\mathbf{C}} \triangleq \left[\sqrt{\frac{P_0 T_0}{N_t}} \mathbf{C}_0^T, \sqrt{\frac{P_1 T_1}{N_t}} \mathbf{C}_1^T \right]^T$ and $\bar{\mathbf{W}} \triangleq \left[\mathbf{W}_0^T, (-\mathbf{A}_1 \mathbf{K}_{\hat{\mathbf{H}}_0}^H \Delta \mathbf{H}_0 + \mathbf{W}_1)^T \right]^T$. Due to the independence between $\mathbf{W}_0$, $\mathbf{W}_1$, and $\mathbf{A}_1$, the correlation matrix of $\bar{\mathbf{W}}$ can be written as

$$\mathbf{R}_{\bar{\mathbf{W}}} \triangleq \mathrm{E}[\bar{\mathbf{W}} \bar{\mathbf{W}}^H] = \begin{bmatrix} N_\ell \sigma_w^2 \mathbf{I}_{T_0} & \mathbf{0} \\ \mathbf{0} & \left(\mathrm{E}\left[\|\mathbf{K}_{\hat{\mathbf{H}}_0}^H \Delta \mathbf{H}_0 \|^2 \right] \sigma_{a,1}^2 + N_\ell \sigma_w^2 \right) \mathbf{I}_{T_1} \end{bmatrix}. \tag{7.17}$$

From (7.9), (7.10) and the orthogonality principle [18]. We have $\mathrm{E}\left[\|\mathbf{K}_{\hat{\mathbf{H}}_0}^H \Delta \mathbf{H}_0 \|^2 \right] = N_\ell(N_t -$

$N_\ell) \cdot \text{NMSE}_{\text{L}}^{(0)}$. Hence, the corresponding NMSE of $\hat{\mathbf{H}}_1$ can be computed as [18]

$$\text{NMSE}_{\text{L}}^{(1)} = \frac{\text{Tr}\left(\left(\frac{1}{N_\ell \sigma_h^2}\mathbf{I}_{N_t} + \bar{\mathbf{C}}^H \mathbf{R}_{\bar{\mathbf{W}}}^{-1} \bar{\mathbf{C}}\right)^{-1}\right)}{N_t N_\ell} \tag{7.18}$$

$$= \left(\frac{1}{\text{NMSE}_{\text{L}}^{(0)}} + \frac{P_1 T_1 / N_t}{\text{NMSE}_{\text{L}}^{(0)} \cdot (N_t - N_\ell)\sigma_{a,1}^2 + \sigma_w^2}\right)^{-1}, \tag{7.19}$$

where the second equality follows from (7.10) and the premises $\mathbf{C}_0^H \mathbf{C}_0 = \mathbf{I}_{N_t}$ and $\mathbf{C}_1^H \mathbf{C}_1 = \mathbf{I}_{N_t}$.

Similarly, UR can also perform its channel estimate based on the signals received in both stages, i.e.,

$$\mathbf{Z} \triangleq \begin{bmatrix} \mathbf{Z}_0 \\ \mathbf{Z}_1 \end{bmatrix} = \begin{bmatrix} \sqrt{\frac{P_0 T_0}{N_t}}\mathbf{C}_0 \\ \sqrt{\frac{P_1 T_1}{N_t}}\mathbf{C}_1 \end{bmatrix} \mathbf{G} + \begin{bmatrix} \mathbf{V}_0 \\ \mathbf{A}_1 \mathbf{K}_{\hat{\mathbf{H}}_0}^H \mathbf{G} + \mathbf{V}_1 \end{bmatrix}. \tag{7.20}$$

Assuming that UR also employs the LMMSE criterion for estimating $\mathbf{G}$, the NMSE of UR in stage 1 can be shown as

$$\text{NMSE}_{\text{U}}^{(1)} = \left(\frac{1}{\text{NMSE}_{\text{U}}^{(0)}} + \frac{P_1 T_1 / N_t}{(N_t - N_\ell)\sigma_{a,1}^2 \sigma_g^2 + \sigma_v^2}\right)^{-1}, \tag{7.21}$$

where

$$\text{NMSE}_{\text{U}}^{(0)} = \left(\frac{1}{\sigma_g^2} + \frac{P_0 T_0}{N_t \sigma_v^2}\right)^{-1} \tag{7.22}$$

is the NMSE of UR when using only the training signal in stage 0 for channel estimation. Notice from (7.21) and (7.22) that

$$\text{NMSE}_{\text{U}}^{(1)} < \text{NMSE}_{\text{U}}^{(0)} \tag{7.23}$$

whenever $P_1 T_1 > 0$. This is reasonable since UR should be able to obtain a better channel estimate by using both $\mathbf{Z}_0$ and $\mathbf{Z}_1$ instead of only $\mathbf{Z}_0$. In addition, note from (7.21) that NMSE performance of UR does not depend on $\hat{\mathbf{H}}_0$ and, thus, its estimation performance cannot be improved even if UR is able to intercept $\hat{\mathbf{H}}_0$.

Optimal Power Allocation between Training and AN

With (7.19) and (7.21), the set of pilot signal powers P_0, P_1, and AN power $\sigma_{a,1}^2$ can be designed by minimizing the NMSE at LR subject to a constraint on the NMSE at UR and an average energy constraint [1]. The optimization problem is given as follows:

$$\min_{P_0, P_1, \sigma_{a,1}^2 \geq 0} \text{NMSE}_{\text{L}}^{(1)} \tag{7.24a}$$

$$\text{subject to } \text{NMSE}_{\text{U}}^{(1)} \geq \gamma, \tag{7.24b}$$

$$\frac{\text{E}[\|\mathbf{X}_0\|_F^2] + \text{E}[\|\mathbf{X}_1\|_F^2]}{T_0 + T_1} \leq \bar{P}_{\text{ave}}, \tag{7.24c}$$

where γ is the constraint on the NMSE achievable by UR and $\bar{P}_{\text{ave}}$ is the maximum average transmission power. By (7.5) and (7.11), the constraint in (7.24c) can be written as

$$\text{E}[\|\mathbf{X}_0\|_F^2] + \text{E}[\|\mathbf{X}_1\|_F^2] = P_0 T_0 + \left(P_1 + (N_t - N_\ell)\sigma_{a,1}^2\right)T_1 \leq \bar{P}_{\text{ave}}(T_0 + T_1) \triangleq \bar{\mathcal{E}}_{\text{tot}}, \tag{7.25}$$

which represents an average energy constraint on the training signals over both stages and where $\bar{\mathcal{E}}_{\mathrm{tot}}$ can be viewed as the total energy constraint.

Notice that the above problem is interesting only when

$$\left(\frac{1}{\sigma_g^2} + \frac{\bar{\mathcal{E}}_{\mathrm{tot}}}{N_t \sigma_v^2}\right)^{-1} \leq \gamma \leq \sigma_g^2, \tag{7.26}$$

where the value on the left side of the inequality is the minimum NMSE achievable at UR when all power is allocated to pilot signals (i.e., no AN is used to interfere with UR), and the value on the right side of the inequality is the NMSE that UR can obtain by simply taking the mean as the estimate (i.e., no pilot signal is needed). If $\gamma < \left(\frac{1}{\sigma_g^2} + \frac{\bar{\mathcal{E}}_{\mathrm{tot}}}{N_t \sigma_v^2}\right)^{-1}$, the constraint on UR would be redundant, whereas, if $\gamma > \sigma_g^2$, the problem would be infeasible.

Let $\mathcal{E}_0 \triangleq P_0 T_0$ and $\mathcal{E}_1 \triangleq P_1 T_1$ be the energy of the pilot signals in stages 0 and 1, and let $\tilde{\gamma} \triangleq (\frac{1}{\gamma} - \frac{1}{\sigma_g^2})N_t \sigma_v^2$. In this case, the condition in (7.26) reduces to $0 \leq \tilde{\gamma} \leq \bar{\mathcal{E}}_{\mathrm{tot}}$. Then we can reformulate problem (7.24) into the following maximization problem:

$$\max_{\mathcal{E}_0, \mathcal{E}_1, \sigma_{a,1}^2 \geq 0} \quad \mathcal{E}_0 + \frac{(N_t \sigma_w^2 + \sigma_h^2 \mathcal{E}_0)\mathcal{E}_1}{N_t \sigma_w^2 + \sigma_h^2 \mathcal{E}_0 + N_t(N_t - N_\ell)\sigma_h^2 \sigma_{a,1}^2} \tag{7.27a}$$

$$\text{subject to } \mathcal{E}_0 + \frac{\sigma_v^2 \mathcal{E}_1}{(N_t - N_\ell)\sigma_g^2 \sigma_{a,1}^2 + \sigma_v^2} \leq \tilde{\gamma}, \tag{7.27b}$$

$$\mathcal{E}_0 + \mathcal{E}_1 + T_1(N_t - N_\ell)\sigma_{a,1}^2 \leq \bar{\mathcal{E}}_{\mathrm{tot}}. \tag{7.27c}$$

Notice that this problem is non-convex in general. However, this three-variable optimization problem can be converted into an one-dimensional optimization problem that can be solved by a simple one-dimensional line search over a finite interval [1]. The result is stated in the following proposition and the proof can be found in [1].

Proposition 7.3.1 ([1]). *Let $\{\mathcal{E}_0^\star, \mathcal{E}_1^\star, (\sigma_{a,1}^2)^\star\}$ be the optimal solution to the nonconvex optimization problem in (7.27) with $0 \leq \tilde{\gamma} \leq \bar{\mathcal{E}}_{\mathrm{tot}}$.*
(i) For $\eta \triangleq N_t(\frac{\sigma_v^2}{\sigma_g^2} - \frac{\sigma_w^2}{\sigma_h^2}) > \tilde{\gamma}$, the optimal solution is given by $(\sigma_{a,1}^2)^\star = 0$ (i.e., no AN is needed) and any $\mathcal{E}_0^\star, \mathcal{E}_1^\star \geq 0$ such that $\mathcal{E}_0^\star + \mathcal{E}_1^\star = \tilde{\gamma}$ (e.g., $\mathcal{E}_0^\star = \mathcal{E}_1^\star = \tilde{\gamma}/2$).
(ii) For $\eta \leq \tilde{\gamma}$, the optimal value of $\mathcal{E}_0$ (i.e., $\mathcal{E}_0^\star$) can be obtained by solving the following one-variable optimization problem

$$\max_{\mathcal{E}_0} \quad \mathcal{E}_0 + \frac{(N_t \sigma_w^2 + \sigma_h^2 \mathcal{E}_0) \cdot \mathcal{E}_1(\mathcal{E}_0)}{N_t \sigma_w^2 + N_t(N_t - N_\ell)\sigma_h^2 \cdot \sigma_{a,1}^2(\mathcal{E}_0) + \sigma_h^2 \mathcal{E}_0} \tag{7.28a}$$

$$\text{s.t. } \max\{\eta, 0\} \leq \mathcal{E}_0 \leq \tilde{\gamma}, \tag{7.28b}$$

where $\sigma_{a,1}^2(\mathcal{E}_0) = \frac{\bar{\mathcal{E}}_{\mathrm{tot}} - \tilde{\gamma}}{(N_t - N_\ell)[T_1 + \sigma_g^2(\tilde{\gamma} - \mathcal{E}_0)/\sigma_v^2]}$ and $\mathcal{E}_1(\mathcal{E}_0) = \sigma_g^2 \left(\frac{\tilde{\gamma} - \mathcal{E}_0}{\sigma_v^2}\right)(N_t - N_\ell)\sigma_{a,1}^2(\mathcal{E}_0) + \tilde{\gamma} - \mathcal{E}_0$.
The corresponding values of $\mathcal{E}_1^\star$ and $(\sigma_{a,2}^2)^\star$ are given by $\mathcal{E}_1(\mathcal{E}_0^\star)$ and $\sigma_{a,2}^2(\mathcal{E}_0^\star)$, respectively.

Proposition 1 shows that, when $\eta \leq \tilde{\gamma}$, an optimal solution can be obtained via a simple line search with respect to $\mathcal{E}_0$ over the interval $[\max\{\eta, 0\}, \tilde{\gamma}]$. On the other hand, when $\eta > \tilde{\gamma}$, a closed-form solution can be obtained with $(\sigma_{a,1}^2)^\star = 0$ and, e.g., $\mathcal{E}_0^\star = \mathcal{E}_1^\star = \tilde{\gamma}/2$. Notice that η can be viewed as a measure of channel quality difference between the UR and the LR. Therefore, when $\eta > \tilde{\gamma}$ (i.e., when UR experiences a much worse channel condition than LR), there is no need for the transmitter to purposely interfere with UR using AN. However, when $\eta \leq \tilde{\gamma}$, UR's channel condition is comparable or even better than that of LR and, therefore, additional AN should be used to degrade UR's channel estimation performance.

7.3.2 Multiple-stage Feedback and Retraining

From the results presented in the previous section, we can see that the effectiveness of the two-stage feedback-and-retraining scheme relies strongly on the quality of the channel estimate obtained in stage 0 since this is what determines the AN placement in stage 1. However, if the performance constraint at UR is stringent or if the channel toward UR is more favorable, the power that can be allocated for preliminary training would be small and the quality of the channel estimate obtained by LR (and, thus, the transmitter) in stage 0 will be limited. Yet, this can be improved upon by simply increasing the number of feedback-and-retraining stages. Through multiple stages, LR and the transmitter can gradually refine its knowledge of the channel between them and improve the AN placement.

Specifically, suppose that the feedback-and-retraining process is performed K times such that there is a total of $K+1$ stages in the training process (including the preliminary training stage). Moreover, assume that the channels from the transmitter to LR and to UR remain static throughout the entire training process. In stage k, for $k > 0$, the transmitter will obtain, through feedback, knowledge of the channel estimate computed in stage $k - 1$, i.e., $\hat{\mathbf{H}}_{k-1}$, and utilize it to design the training signal to be emitted in the current stage. The training signal in stage k is given by

$$\mathbf{X}_k = \sqrt{\frac{P_k T_k}{N_t}} \mathbf{C}_k + \mathbf{A}_k \mathbf{K}_{\hat{\mathbf{H}}_{k-1}}^H, \tag{7.29}$$

where P_k is the pilot signal power in stage k, $\mathbf{C}_k \in \mathbb{C}^{T_k \times N_t}$ is the pilot matrix satisfying $\mathrm{Tr}(\mathbf{C}_k^H \mathbf{C}_k) = N_t$, $\mathbf{K}_{\hat{\mathbf{H}}_{k-1}} \in \mathbb{C}^{N_t \times (N_t - N_\ell)}$ is a matrix whose columns form an orthonormal basis for the left null space of $\hat{\mathbf{H}}_{k-1}$ (i.e., $\mathbf{K}_{\hat{\mathbf{H}}_{k-1}}^H \mathbf{K}_{\hat{\mathbf{H}}_{k-1}} = \mathbf{I}_{N_t - N_\ell}$ and $\mathbf{K}_{\hat{\mathbf{H}}_{k-1}}^H \hat{\mathbf{H}}_{k-1} = \mathbf{0}$), and $\mathbf{A}_k \in \mathbb{C}^{T_k \times (N_t - N_\ell)}$ is an AN matrix with each entry being i.i.d. $\mathcal{CN}(0, \sigma_{a,k}^2)$.

The signals received by LR and UR in stage k are respectively given by

$$\text{LR} : \mathbf{Y}_k = \mathbf{X}_k \mathbf{H} + \mathbf{W}_k, \tag{7.30}$$

$$\text{UR} : \mathbf{Z}_k = \mathbf{X}_k \mathbf{G} + \mathbf{V}_k, \tag{7.31}$$

where $\mathbf{W}_k$ and $\mathbf{V}_k$ are the AWGN matrices at LR and UR, respectively. Following the derivations given for the two-stage scenario, one can show by induction that the NMSE of the estimate obtained by LR through K stages can be expressed in the recursive form

$$\text{NMSE}_{\text{L}}^{(K)} = \left(\frac{1}{\text{NMSE}_{\text{L}}^{(K-1)}} + \frac{P_K T_K / N_t}{\text{NMSE}_{\text{L}}^{(K-1)} \cdot (N_t - N_\ell)\sigma_{a,K}^2 + \sigma_w^2} \right)^{-1}, \tag{7.32}$$

with the initial $\text{NMSE}_{\text{L}}^{(0)}$ given in (7.10). Similarly, the NMSE of the UR at stage K can be written as

$$\text{NMSE}_{\text{U}}^{(K)} = \left(\frac{1}{\sigma_g^2} + \frac{P_0 T_0}{N_t \sigma_v^2} + \sum_{k=1}^{K} \frac{P_k T_k / N_t}{(N_t - N_\ell)\sigma_{a,k}^2 \sigma_g^2 + \sigma_v^2} \right)^{-1}. \tag{7.33}$$

Given (7.32) and (7.33), the power values $\{P_k\}_{k=0}^{K}$ and $\{\sigma_{a,k}^2\}_{k=1}^{K}$ can be designed similarly by minimizing the NMSE at LR subject to the NMSE at the UR. Similar to (7.24), the problem can be formulated as

$$\min_{P_0, P_k, \sigma^2_{a,k} \geq 0, \ k=1,\ldots,K} \ \text{NMSE}_{\text{L}}^{(K)} \tag{7.34a}$$

$$\text{subject to} \ \ \text{NMSE}_{\text{U}}^{(K)} \geq \gamma, \tag{7.34b}$$

$$\sum_{k=0}^{K} \text{E}[\|\mathbf{X}_k\|_F^2] \leq \bar{P}_{\text{ave}} T_{\text{tot}}, \tag{7.34c}$$

where $T_{\text{tot}} = \sum_{k=0}^{K} T_k$, and (7.34c) is the average energy constraint.

The optimization problem in (7.34) is also non-convex and is, in fact, much more involved compared to (7.24) because of the recursive structure in (7.32). However, as shown in [1], the problem in (7.34) can be handled efficiently by using the monomial approximation and the condensation method. Here, the optimization is performed by solving a sequence of geometric programming (GP) problems with successive convex approximations on UR's NMSE constraint. Notice that GP problems can be efficiently solved by off-the-shelf solvers such as CVX [19] and, thus, an approximate solution can be obtained efficiently and reliably. Details of the condensation method used for solving problem (7.34) can be found in [1].

7.3.3 Simulation Results and Discussions

In this section, we present simulation results to demonstrate the efficacy of the DCE scheme. We consider the wireless system as shown in Figure 7.1 with $N_t = 4$, and $N_\ell = N_u = 2$. The channel matrices $\mathbf{H}$ and $\mathbf{G}$ are assumed to have entries that are i.i.d. complex Gaussian random variables with zero mean and unit variance ($\sigma_h^2 = \sigma_g^2 = 1$). The pilot matrices $\mathbf{C}_k$, $k = 0, \ldots, K$, are randomly drawn from semi-unitary $T_k \times N_t$ matrices. The average power is $P_{\text{ave}} = 30$ dBm ($P_{\text{ave}} = 1$) and the training signal lengths are

$$T_0 = T_1 = \cdots = T_K = \left\lfloor \frac{300}{K+1} \right\rfloor, \tag{7.35}$$

i.e., the total training length is 300 and is partitioned into $(K + 1)$ equal-length segments. In the simulation, we also show an "NMSE lower bound"

$$\text{NMSE lower bound} = \left(\frac{1}{\sigma_g^2} + \frac{P_{\text{ave}} \sum_{k=0}^{K} T_k}{N_t \sigma_w^2} \right)^{-1}, \tag{7.36}$$

which represents the best NMSE performance that can be achieved by LR without using any AN (i.e., $\sigma_{a,k}^2 = 0$ for all k), and corresponds to the NMSE performance of the conventional (non-discriminatory) MIMO channel estimation scheme [16]. With $\sigma_h^2 = \sigma_g^2 = 1$ and $P_{\text{ave}} = 1$, the SNRs at LR and UR are defined as

$$\text{SNR}_{\text{L}} = \frac{\sum_{k=0}^{K} \text{E}[\|\mathbf{X}_k \mathbf{H}\|_F^2]}{\sum_{k=0}^{K} \text{E}[\|\mathbf{W}_k\|_F^2]} = \frac{1}{\sigma_w^2}, \tag{7.37}$$

$$\text{SNR}_{\text{U}} = \frac{\sum_{k=0}^{K} \text{E}[\|\mathbf{X}_k \mathbf{G}\|_F^2]}{\sum_{k=0}^{K} \text{E}[\|\mathbf{V}_k\|_F^2]} = \frac{1}{\sigma_v^2}, \tag{7.38}$$

respectively. It is simply set $\text{SNR}_{\text{L}} = \text{SNR}_{\text{U}}$ throughout the simulations. Each simulation result was obtained by averaging over $1,000$ channel realizations.

In Figure 7.3(a), the NMSE performances of LR and UR are shown for $\gamma = 0.1$ and $\gamma = 0.03$ with only one stage of feedback and retraining (i.e., $K = 1$). The optimal power

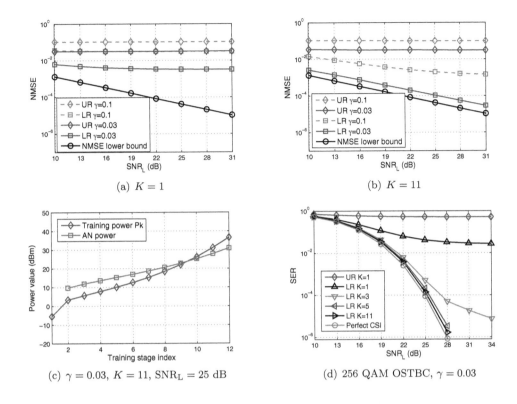

(a) $K = 1$

(b) $K = 11$

(c) $\gamma = 0.03$, $K = 11$, $\text{SNR}_{\text{L}} = 25$ dB

(d) 256 QAM OSTBC, $\gamma = 0.03$

Figure 7.3: Simulation results of NMSE performances and OSTBC detection performance of the proposed DCE scheme for $N_t = 4$, $N_\ell = 2$, $N_u = 2$, and $\text{SNR}_{\text{L}} = \text{SNR}_{\text{U}}$.

values P_0, P_1 and $\sigma^2_{a,1}$ are obtained via (7.24) and Proposition 7.3.1. We can see from Figure 7.3(a) that UR's NMSE is indeed constrained above 0.1 and 0.03 for the two cases, respectively. Moreover, we can see that there is a trade-off between the value of γ and the achievable NMSE at LR. However, a large gap exists between the NMSE of LR and the NMSE lower bound. This loss is due to limitations on the amount of pilot power that can be used in preliminary training, but can be reduced with more feedback-and-retraining stages.

In Figure 7.3(b), the NMSE performances of LR and UR are shown for the case with $K = 11$ feedback-and-retraining stages. The optimal power values $\{P_k\}_{k=0}^{K}$ and $\{\sigma^2_{a,k}\}_{k=1}^{K}$ are obtained by solving (7.34) using the monomial approximation and the condensation method [1]. By comparing Figure 7.3(b) with Figure 7.3(a), we can see that LR's NMSE is greatly reduced compared to the case with $K = 1$ whereas UR's NMSE is still successfully constrained above the respective γ values. Moreover, we can see that the gap between LR's NMSE and the NMSE lower bound is significantly reduced for the case of $\gamma = 0.03$.

In Figure 7.3(c), we show the optimal values of $\{P_k\}_{k=0}^{K}$ and $\{\sigma^2_{a,k}\}_{k=1}^{K}$ for the case with $\gamma = 0.03$ and $\text{SNR}_{\text{L}} = \text{SNR}_{\text{U}} = 25$ dB. We can observe that the optimal P_0 is relatively small due to a stringent NMSE constraint at UR. However, the values of P_k and $\sigma^2_{a,k}$ gradually increase in later stages since LR is able to obtain better knowledge of the forward channel through these stages [see (7.32)].

To demonstrate the effect of DCE on the data transmission, we examine the corresponding data detection performance at the two receivers for the case where a 4-by-4 complex orthogonal space–time block code (OSTBC) is utilized at the transmitter. Each code block contains three QAM source symbols [20]. Both LR and UR will use their channel estimates obtained with DCE to decode the unknown symbols[2]. The average symbol error rates (SERs) of LR and UR are obtained by averaging over 50,000 channel realizations and OSTBCs. In Figure 7.3(d), we show the average SERs for 256-QAM OSTBC and $\gamma = 0.03$. Note that only the SER curve of UR for $K = 1$ is displayed since UR has almost the same average symbol error performances regardless of the values of K. We can see that the SER of LR gradually decreases as K increases whereas that of UR remains around 0.5. Moreover, we can see that the SER performance at LR approaches that in the case with perfect CSI when $K \geq 3$.

7.4 Discriminatory Channel Estimation via Two-way Training

The second DCE scheme that we introduce in this section utilizes two-way training [21–23], i.e., training emitted by both the transmitter and LR, to facilitate channel estimation at both ends. By having LR send the preliminary training signal instead of the transmitter, the need to explicitly feed back the channel to the transmitter is avoided and, more importantly, the constraint set by UR in the preliminary training stage no longer exists. That is, UR is not able to obtain an estimate of the channel between the transmitter and itself from the preliminary training signal since it is sent over the channel between LR and UR. When the channels are reciprocal, the transmitter can directly infer the forward channel (i.e., the channel from itself to LR) with its estimate of the backward channel (i.e., the channel from LR to itself). In this case, the two-way DCE scheme requires only two stages of training, namely, a reverse training stage and a forward training stage. When the channels are nonreciprocal, an additional round-trip training stage, in which a random training signal is broadcast by the transmitter and echoed back by LR using an amplify-and-forward strategy, is needed. The random training signal is known only to the transmitter and is used to estimate the combination of the forward and backward channels. In both cases, AN is superimposed on top of the pilot signal in the final forward training stage to achieve the desired performance discrimination in the channel estimation. The two-way training scheme requires a much smaller training overhead compared to the multistage feedback-and-retraining scheme introduced in the previous section. The two-way DCE schemes for reciprocal and nonreciprocal cases are introduced separately in the following sections.

7.4.1 Two-way DCE Design for Reciprocal Channels

In the two-way DCE scheme, as shown in Figure 7.4, it is necessary to distinguish between the forward and backward channels. Hence, let $\mathbf{H}_f \in \mathbb{C}^{N_t \times N_\ell}$ and $\mathbf{H}_b \in \mathbb{C}^{N_\ell \times N_t}$ be the forward and backward channel matrices, respectively, between the transmitter and LR, with entries being i.i.d. $\mathcal{CN}(0, \sigma_{h_f}^2)$ and $\mathcal{CN}(0, \sigma_{h_b}^2)$. Also, let $\mathbf{G} \in \mathbb{C}^{N_t \times N_u}$ and $\mathbf{F} \in \mathbb{C}^{N_\ell \times N_u}$ be the channel matrices from the transmitter to UR and from LR to UR, respectively, with entries being i.i.d. $\mathcal{CN}(0, \sigma_g^2)$ and $\mathcal{CN}(0, \sigma_f^2)$, respectively. When the channel is reciprocal, both forward and backward channels can be represented by the channel matrix $\mathbf{H} \triangleq \mathbf{H}_f = \mathbf{H}_b{}^T$ (with entries being i.i.d. $\mathcal{CN}(0, \sigma_h^2)$). In this case, the transmitter can obtain an estimate of the forward channel by taking the transpose of the estimated channel matrix obtained through reverse training, i.e., training based on the pilot signals sent from LR to the transmitter. The channel matrices $\mathbf{H}$, $\mathbf{G}$, and $\mathbf{F}$ are assumed to be independent of each other. When the main channel is reciprocal, the two-way DCE scheme can be performed in two

---

[2]It is worthwhile to note that square OSTBCs (i.e., with $N_t = T$) in general cannot be properly decoded without CSI at the receiver. See more discussions in the conclusion section.

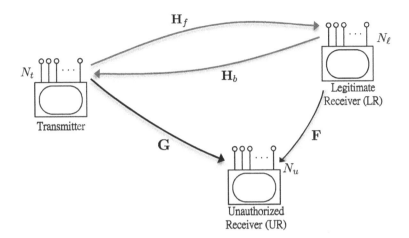

Figure 7.4: A wireless MIMO system consisting of a transmitter, a legitimate receiver (LR), and an unauthorized receiver (UR).

stages, namely, a channel acquisition stage and an AN-assisted training stage, as described below.

Stage 0 (Channel Acquisition at Transmitter via Reverse Training): In the channel acquisition stage, the goal is to allow the transmitter to obtain a reliable estimate of its channel to LR without benefiting the channel estimation process at UR. In the reciprocal case, this can be done by having LR send a training signal in the reverse (or backward) channel to the transmitter. The reverse training signal sent by LR is given by

$$\mathbf{X}_\ell = \sqrt{\frac{P_\ell T_\ell}{N_\ell}} \mathbf{C}_\ell, \tag{7.39}$$

where $\mathbf{C}_\ell \in \mathbb{C}^{T_\ell \times N_\ell}$ is the pilot matrix that satisfies $\mathbf{C}_\ell^H \mathbf{C}_\ell = \mathbf{I}_{N_\ell}$, P_ℓ is LR's transmit power in stage 0, and T_ℓ is the reverse training length. Note that an orthonormal pilot matrix (i.e., $\mathbf{C}_\ell^H \mathbf{C}_\ell = \mathbf{I}_{N_\ell}$) is chosen here because it is optimal in minimizing the channel estimation error in point-to-point channels, as shown in [24]. The received signal at the transmitter is

$$\widetilde{\mathbf{Y}}_t = \mathbf{X}_\ell \mathbf{H}^T + \widetilde{\mathbf{W}}, \tag{7.40}$$

where $\widetilde{\mathbf{W}} \in \mathbb{C}^{T_\ell \times N_t}$ is the AWGN matrix with entries being i.i.d. $\mathcal{CN}(0, \sigma_{\tilde{w}}^2)$. The reverse training signal sent by LR allows the transmitter to obtain an estimate of the forward channel by taking the transpose of its estimate of the backward channel. The LMMSE estimation of $\mathbf{H}$ at the transmitter is denoted by $\widehat{\mathbf{H}}$.

Stage 1 (Forward Training with AN for DCE): With the channel estimate $\widehat{\mathbf{H}}$ obtained in the previous stage, the transmitter then sends a forward training signal with AN inserted in the null space of $\widehat{\mathbf{H}}$ to disrupt the reception at UR. The forward training signal is given by

$$\mathbf{X}_t = \sqrt{\frac{P_t T_t}{N_t}} \mathbf{C}_t + \mathbf{A}\mathbf{K}_{\widehat{\mathbf{H}}}, \tag{7.41}$$

where $\mathbf{C}_t \in \mathbb{C}^{T_t \times N_t}$ is the pilot matrix with $\mathrm{Tr}(\mathbf{C}_t^H \mathbf{C}_t) = N_t$, P_t is the pilot signal power in this stage, and T_t is the training length. Here, $\mathbf{A} \in \mathbb{C}^{T_t \times (N_t - N_\ell)}$ is the AN matrix whose entries are i.i.d. $\mathcal{CN}(0, \sigma_a^2)$ and $\mathbf{K}_{\widehat{\mathbf{H}}} \in \mathbb{C}^{N_t \times (N_t - N_\ell)}$ is a matrix whose column vectors form an orthonormal basis for the left null space of $\widehat{\mathbf{H}}$ (i.e., $\mathbf{K}_{\widehat{\mathbf{H}}}^H \widehat{\mathbf{H}} = \mathbf{0}_{(N_t - N_\ell) \times N_\ell}$ and

$\mathbf{K}_{\widehat{\mathbf{H}}}^H \mathbf{K}_{\widehat{\mathbf{H}}} = \mathbf{I}_{N_t - N_\ell}$). AN is assumed to be statistically independent of the channels and the noise at all terminals. The received signals at LR and UR are respectively given by

$$\mathbf{Y}_\ell = \sqrt{\frac{P_t T_t}{N_t}} \mathbf{C}_t \mathbf{H} + \mathbf{A} \mathbf{K}_{\widehat{\mathbf{H}}}^H \mathbf{H} + \mathbf{W}, \tag{7.42}$$

$$\mathbf{Y}_u = \sqrt{\frac{P_t T_t}{N_t}} \mathbf{C}_t \mathbf{G} + \mathbf{A} \mathbf{K}_{\widehat{\mathbf{H}}}^H \mathbf{G} + \mathbf{V}, \tag{7.43}$$

where $\mathbf{W} \in \mathbb{C}^{T_t \times N_\ell}$ and $\mathbf{V} \in \mathbb{C}^{T_t \times N_u}$ are the AWGN matrices at LR and UR, respectively; their entries are i.i.d. $\mathcal{CN}(0, \sigma_w^2)$ and $\mathcal{CN}(0, \sigma_v^2)$, respectively. By representing $\widehat{\mathbf{H}} = \mathbf{H} + \Delta\mathbf{H}$ and by the fact that $\mathbf{K}_{\widehat{\mathbf{H}}}^H \widehat{\mathbf{H}} = \mathbf{0}$, (7.42) can be rewritten as

$$\mathbf{Y}_\ell = \sqrt{\frac{P_t T_t}{N_t}} \mathbf{C}_t \mathbf{H} - \mathbf{A} \mathbf{K}_{\widehat{\mathbf{H}}}^H \Delta\mathbf{H} + \mathbf{W} \triangleq \bar{\mathbf{C}}\mathbf{H} + \bar{\mathbf{W}}, \tag{7.44}$$

where $\bar{\mathbf{C}} \triangleq \sqrt{\frac{P_t T_t}{N_t}} \mathbf{C}_t$ and $\bar{\mathbf{W}} \triangleq -\mathbf{A} \mathbf{K}_{\widehat{\mathbf{H}}}^H \Delta\mathbf{H} + \mathbf{W}$. Here, $\bar{\mathbf{W}}$ and $\mathbf{H}$ are uncorrelated, i.e., $\mathbb{E}[\bar{\mathbf{W}}^H \mathbf{H}] = \mathbf{0}$, due to the presence of $\mathbf{A}$.

Given $\mathbf{Y}_\ell$, the LMMSE estimate at LR can be computed as

$$\widehat{\mathbf{H}}_\ell = \mathbf{R}_{\mathbf{H}} \bar{\mathbf{C}}^H (\bar{\mathbf{C}} \mathbf{R}_{\mathbf{H}} \bar{\mathbf{C}}^H + \mathbf{R}_{\bar{\mathbf{W}}})^{-1} \mathbf{Y}_\ell, \tag{7.45}$$

where $\mathbf{R}_{\mathbf{H}} = \mathbb{E}[\mathbf{H}\mathbf{H}^H] = N_\ell \sigma_h^2 \mathbf{I}_{N_t}$ and $\mathbf{R}_{\bar{\mathbf{W}}} = \mathbb{E}[\bar{\mathbf{W}}\bar{\mathbf{W}}^H]$. The resulting NMSE of the estimate of $\mathbf{H}$ at LR is given by

$$\mathrm{NMSE_L} = \frac{1}{N_t} \mathrm{Tr}\left(\left(\frac{1}{\sigma_h^2} \mathbf{I}_{N_t} + \frac{P_\ell T_\ell}{N_t} \frac{\mathbf{C}_t^H \mathbf{C}_t}{(N_t - N_\ell)\left(\frac{1}{\sigma_h^2} + \frac{P_\ell T_\ell}{N_\ell \sigma_w^2}\right)^{-1} \sigma_a^2 + \sigma_w^2} \right)^{-1} \right). \tag{7.46}$$

Similarly, the NMSE of the estimate of $\mathbf{G}$ at UR can be computed as

$$\mathrm{NMSE_U} = \frac{1}{N_t} \mathrm{Tr}\left(\left(\frac{1}{\sigma_g^2} \mathbf{I}_{N_t} + \frac{P_\ell T_\ell}{N_t} \frac{\mathbf{C}_t^H \mathbf{C}_t}{(N_t - N_\ell)\sigma_a^2 \sigma_g^2 + \sigma_v^2} \right)^{-1} \right). \tag{7.47}$$

Notice, from (7.46) and (7.47), that increasing the AN power (i.e., σ_a^2) increases the NMSE at both receivers. However, this effect can be reduced at LR by increasing the reverse training power P_ℓ. With limited LR's transmit power, the precision of the channel estimate obtained at the transmitter would be limited and, thus, the power allocated to AN should be determined accordingly so that no excess interference is experienced by LR. Moreover, under a total energy constraint, the more power allocated to reverse training, the less power there is for pilot and AN in the forward. Therefore, the power allocation between reverse training, forward training, and AN must be carefully determined.

Optimal Power Allocation between Pilot Signal and AN

Based on the NMSE expressions given in (7.46) and (7.47), the optimal power allocation among pilot signal and AN in the reverse and forward training stages can be found by minimizing the NMSE at LR subject to a lower constraint on the NMSE at UR. By defining $\mathcal{E}_t \triangleq P_t T_t$ and $\mathcal{E}_\ell \triangleq P_\ell T_\ell$ (which can be viewed as the pilot signal energy at the transmitter and LR, respectively), the proposed optimization problem can be formulated as follows:

$$\min_{\mathcal{E}_\ell, \mathcal{E}_t, \sigma_a^2 \geq 0} \text{NMSE}_L \tag{7.48a}$$

$$\text{subject to} \quad \text{NMSE}_U \geq \gamma, \tag{7.48b}$$

$$\mathcal{E}_\ell + \mathcal{E}_t + (N_t - N_\ell)\sigma_a^2 T_t \leq \bar{\mathcal{E}}_{tot}, \tag{7.48c}$$

$$\mathcal{E}_\ell \leq \bar{\mathcal{E}}_\ell, \tag{7.48d}$$

$$\mathcal{E}_t + (N_t - N_\ell)\sigma_a^2 T_t \leq \bar{\mathcal{E}}_t, \tag{7.48e}$$

where γ is the lower constraint on the NMSE at UR, $\bar{\mathcal{E}}_{tot}$ is the total energy constraint, and $\bar{\mathcal{E}}_\ell$, $\bar{\mathcal{E}}_t$ are the individual energy constraints at LR and the transmitter, respectively. The power allocation problem given above is non-convex and involves an optimization over three variables, i.e., $\mathcal{E}_\ell$, $\mathcal{E}_t$, and σ_a^2. Interestingly, by employing orthogonal forward pilot matrices (i.e., by choosing $\mathbf{C}_t$ such that $\mathbf{C}_t^H \mathbf{C}_t = \mathbf{I}_{N_t}$), the problem can be reformulated as a single-variable optimization problem and can be solved in closed form when either the total energy constraint or the individual energy constraints are redundant. In all other cases, the solution can be obtained via a one-dimensional line search over a finite interval. This is similar to the result obtained in the feedback-and-retraining scheme (c.f. Proposition 7.3.1). Please see [2] for more detailed derivations.

7.4.2   Two-way DCE Design for Nonreciprocal Channels

When the channel between the transmitter and LR is nonreciprocal, the transmitter would not be able to directly infer knowledge of the forward channel through its backward channel estimate. In this case, reverse training is not sufficient for the transmitter to acquire knowledge of the forward channel. In this case, it is necessary to employ an additional round-trip training stage that utilizes an echoed signal (from transmitter to LR and back to the transmitter) to obtain knowledge of the combined forward–backward channel at the transmitter. Then, combined with the backward channel estimate obtained in reverse training, the transmitter can then compute an estimate of the forward channel matrix that can be utilized to design the AN-assisted training in the final stage. The two-way DCE scheme for non-reciprocal channels is described below.

Stage 0 (Channel Acquisition at the Transmitter with Reverse Training): In the reverse training stage, LR emits a training signal

$$\mathbf{X}_{\ell,0} = \sqrt{\frac{P_{\ell,0} T_{\ell,0}}{N_\ell}} \mathbf{C}_{\ell,0} \tag{7.49}$$

similar to that in (7.39), where $P_{\ell,0}$ is the transmission power, $T_{\ell,0}$ is the training length, and $\mathbf{C}_{\ell,0} \in \mathbb{C}^{T_{\ell,0} \times N_t}$ is the pilot signal chosen such that $\mathbf{C}_{\ell,0}^H \mathbf{C}_{\ell,0} = \mathbf{I}_{N_\ell}$. The received signal at the transmitter is given by

$$\widetilde{\mathbf{Y}}_{t,0} = \mathbf{X}_{\ell,0} \mathbf{H}_b + \widetilde{\mathbf{W}}_0 = \sqrt{\frac{P_{\ell,0} T_{\ell,0}}{N_\ell}} \mathbf{C}_{\ell,0} \mathbf{H}_b + \widetilde{\mathbf{W}}_0, \tag{7.50}$$

where $\widetilde{\mathbf{W}}_0 \in \mathbb{C}^{T_{\ell,0} \times N_t}$ is the AWGN matrix with entries being i.i.d. with distribution $\mathcal{CN}(0, \sigma_{\tilde{w}}^2)$. The LMMSE estimate of $\mathbf{H}_b$ at the transmitter is denoted by $\widehat{\mathbf{H}}_b$.

Stage 1 (Channel Acquisition at the Transmitter with Round-Trip Training): In the round-trip training stage, the transmitter first emits a random training signal that is generated upon transmission and known only to itself (and not to LR and UR). Then, this signal is echoed back to the transmitter by LR using an amplify-and-forward strategy. In this case, the effective channel seen at the transmitter will be a combination of the forward and backward channels. Specifically, let $\mathbf{C}_{t,1} \in \mathbb{C}^{T_{t,1} \times N_t}$ be the randomly generated pilot matrix with $\mathrm{Tr}(\mathbf{C}_{t,1}^H \mathbf{C}_{t,1}) = N_t$ and let

$$\mathbf{X}_{t,1} = \sqrt{\frac{P_{t,1} T_{t,1}}{N_t}} \mathbf{C}_{t,1} \tag{7.51}$$

be the signal sent by the transmitter, where $P_{t,1}$ is the pilot signal power and $T_{t,1}$ is the training length in this stage. The received signal at LR is given by

$$\mathbf{Y}_{\ell,1} = \mathbf{X}_{t,1} \mathbf{H}_f + \mathbf{W}_1, \tag{7.52}$$

where $\mathbf{W}_1 \in \mathbb{C}^{T_{t,1} \times N_\ell}$ is the AWGN matrix with entries that are i.i.d. with distribution $\mathcal{CN}(0, \sigma_w^2)$. Upon receiving $\mathbf{Y}_{\ell,1}$, LR amplifies and forwards the received signal back to the transmitter. The echoed signal at the transmitter is given by

$$\begin{aligned}
\widetilde{\mathbf{Y}}_{t,1} &= \alpha \mathbf{Y}_{\ell,1} \mathbf{H}_b + \widetilde{\mathbf{W}}_1 \\
&= \alpha \mathbf{X}_{t,1} \mathbf{H}_f \mathbf{H}_b + \alpha \mathbf{W}_1 \mathbf{H}_b + \widetilde{\mathbf{W}}_1,
\end{aligned} \tag{7.53}$$

where $\widetilde{\mathbf{W}}_1 \in \mathbb{C}^{T_{t,1} \times N_t}$ is the AWGN matrix at the transmitter with entries being i.i.d. with distribution $\mathcal{CN}(0, \sigma_{\tilde{w}}^2)$, and α is the amplifying gain given by

$$\alpha = \sqrt{\frac{P_{\ell,1} T_{t,1}}{P_{t,1} T_{t,1} N_\ell \sigma_{h_f}^2 + T_{t,1} N_\ell \sigma_w^2}} \tag{7.54}$$

where $P_{\ell,1}$ is the power LR spent on echoing the signal. With knowledge of $\mathbf{X}_{t,1}$, the transmitter can obtain an estimate of the combined forward and backward channels, i.e., $\mathbf{H}_f \mathbf{H}_b$. Then, together with the estimate $\mathbf{H}_b$ obtained in the reverse training stage, an estimate of the forward channel matrix $\mathbf{H}_f$, denoted by $\widehat{\mathbf{H}}_{f,t}$, can then be computed.

Stage 2 (Forward Training with AN for DCE): In the forward training stage, the transmitter again sends an AN-assisted training signal to discriminate the channel estimation performance between LR and UR, similar to that described in the previous sections. Given the AN-assisted training signal, the received signals at LR and UR can be expressed as

$$\mathbf{Y}_{\ell,2} = \sqrt{\frac{P_{t,2} T_{t,2}}{N_t}} \mathbf{C}_{t,2} \mathbf{H}_f + \mathbf{A} \mathbf{K}_{\widehat{\mathbf{H}}_{f,t}}^H \mathbf{H}_f + \mathbf{W}_2, \tag{7.55}$$

$$\mathbf{Y}_{u,2} = \sqrt{\frac{P_{t,2} T_{t,2}}{N_t}} \mathbf{C}_{t,2} \mathbf{G} + \mathbf{A} \mathbf{K}_{\widehat{\mathbf{H}}_{f,t}}^H \mathbf{G} + \mathbf{V}_2, \tag{7.56}$$

respectively. The forward training length is denoted by $T_{t,2}$.

In the non-reciprocal case, the NMSE at LR can not be obtained in closed form. However, by assuming that (i) given $\widehat{\mathbf{H}}_b$, the LMMSE estimate of $\mathbf{H}_f$ from (7.55), i.e., $\widehat{\mathbf{H}}_{f,t}$, is statistically independent of the associated error matrix, i.e., $\Delta \mathbf{H}_{f,t}$, and (ii) the transmitter and LR have sufficiently large numbers of antennas, i.e, $N_t, N_\ell \gg 1$, one can obtain an approximation for the NMSE at LR given by [2]

$$\mathrm{NMSE}_L \approx \frac{1}{N_t} \mathrm{Tr} \left(\frac{1}{\sigma_{h_f}^2} \mathbf{I}_{N_t} + \frac{P_{t,2} T_{t,2}}{N_t} \frac{\mathbf{C}_{t,2}^H \mathbf{C}_{t,2}}{(N_t - N_\ell) \sigma_a^2 \left(\sigma_{h_f}^2 - \frac{\sigma_{h_f}^4 P_{t,1} T_{t,1}}{\sigma_{h_f}^2 P_{t,1} T_{t,1} + N_t \sigma_w^2} \frac{N_t \sigma^2}{\beta + N_t \sigma^2} \right) + \sigma_w^2} \right)^{-1} \tag{7.57}$$

where $\sigma^2 \triangleq \frac{\sigma_{h_u}^4 P_{\ell,0} T_{\ell,0}}{\sigma_{h_b}^2 P_{\ell,0} T_{\ell,0} + N_\ell \sigma_{\tilde{w}}^2}$ and

$$\beta = N_\ell \left(\frac{1}{\sigma_{h_b}^2} + \frac{P_{\ell,0} T_{\ell,0}}{N_\ell \sigma_{\tilde{w}}^2} \right)^{-1} + \frac{\sigma_{\tilde{w}}^2}{\alpha^2 \sigma_{h_f}^2 \sigma_w^2} \left(\frac{1}{\sigma_{h_f}^2} + \frac{P_{t,1} T_{t,1}}{N_t \sigma_w^2} \right)^{-1}. \tag{7.58}$$

Please note that both σ^2 and β are also functions of the pilot signal powers, i.e., $P_{\ell,0}$ and $P_{t,1}$. In practice, $\widehat{\mathbf{H}}_{f,t}$ and $\Delta \mathbf{H}_{f,t}$ are only known to be statistically uncorrelated and N_t, N_u should be finite in practice. However, we find via numerical results that the approximations given above are sufficient for us to obtain good power allocation solutions. The NMSE performance at UR can be computed as in the reciprocal case and is given by

$$\text{NMSE}_\text{U} = \frac{1}{N_t} \text{Tr} \left(\left(\frac{1}{\sigma_g^2} \mathbf{I}_{N_t} + \frac{P_{t,2} T_{t,2}}{N_t} \frac{\mathbf{C}_{t,2}^H \mathbf{C}_{t,2}}{(N_t - N_\ell) \sigma_a^2 \sigma_g^2 + \sigma_v^2} \right)^{-1} \right). \tag{7.59}$$

Optimal Power Allocation between Training and AN Signals

Similar to the previous cases, the power allocation between training and AN can also be determined by minimizing the NMSE at LR subject to a lower constraint on the NMSE at UR. By letting $\mathcal{E}_{\ell,0} \triangleq P_{\ell,0} T_{\ell,0}$, $\mathcal{E}_{t,1} \triangleq P_{t,1} T_{t,1}$, $\mathcal{E}_{\ell,1} \triangleq P_{\ell,1} T_{\ell,1}$, and $\mathcal{E}_{t,2} = P_{t,2} T_{t,2}$, the optimization problem can be formulated as:

$$\min_{\mathcal{E}_{\ell,0}, \mathcal{E}_{t,1}, \mathcal{E}_{\ell,1}, \mathcal{E}_{t,2}, \sigma_a^2 \geq 0} \text{NMSE}_\text{L} \tag{7.60a}$$

$$\text{subject to} \quad \text{NMSE}_\text{U} \geq \gamma \tag{7.60b}$$

$$\mathcal{E}_{\ell,0} + \mathcal{E}_{t,1} + \mathcal{E}_{\ell,1} + \mathcal{E}_{t,2} + (N_t - N_\ell) \sigma_a^2 N_t \leq \bar{\mathcal{E}}_{tot} \tag{7.60c}$$

$$\mathcal{E}_{t,1} + \mathcal{E}_{t,2} + (N_t - N_\ell) \sigma_a^2 N_t \leq \bar{\mathcal{E}}_t \tag{7.60d}$$

$$\mathcal{E}_{\ell,0} + \mathcal{E}_{\ell,1} \leq \bar{\mathcal{E}}_\ell. \tag{7.60e}$$

Here, $\bar{\mathcal{E}}_{tot}$ is the total energy constraint and $\bar{\mathcal{E}}_t$ and $\bar{\mathcal{E}}_\ell$ are the individual energy constraints at the transmitter and LR, respectively. To solve this problem, we also resort to the monomial approximation and the condensation method [25, 26] as done in the previous cases.

7.4.3   Simulation Results and Discussions

In this section, we present simulation results to demonstrate the effectiveness of the two-way DCE schemes. Similar to the simulation settings given in the previous section, we consider a MIMO wireless system with $N_t = 4$, $N_\ell = 2$, and $N_u = 2$. The channel matrices, $\mathbf{H}$, $\mathbf{H}_b$, $\mathbf{H}_f$, and $\mathbf{G}$, have entries that are i.i.d. complex Gaussian distributed with zero mean and unit variance, i.e., $\sigma_h^2 = \sigma_{h_b}^2 = \sigma_{h_f}^2 = \sigma_g^2 = 1$. The AWGN matrices, i.e., $\widetilde{\mathbf{W}}$, $\mathbf{W}$, and $\mathbf{V}$, are also assumed to have entries that are i.i.d. complex Gaussian with zero mean and unit variance, i.e., $\sigma_{\tilde{w}}^2 = \sigma_w^2 = \sigma_v^2 = 1$. The forward pilot matrix is assumed to be orthogonal, i.e., $\mathbf{C}_t^H \mathbf{C}_t = \mathbf{C}_{t,2}^H \mathbf{C}_{t,2} = \mathbf{I}_{N_t}$. Moreover, the training lengths are $T_\ell = N_\ell = 2$ and $T_t = N_t = 4$ for the reciprocal case, and $T_{\ell,0} = N_\ell = 2$ and $T_{t,1} = T_{t,2} = N_t = 4$ for the non-reciprocal case. The total length of the training signals sent by the transmitter is denoted by $\bar{T}_t$ and that sent by LR is denoted by $\bar{T}_\ell$. It follows that the total training lengths are $\bar{T}_t = 4 \, (= T_t)$ and $\bar{T}_\ell = 2 \, (= T_\ell)$ in the reciprocal case and are $\bar{T}_t = 8 \, (= T_{t,1} + T_{t,2})$ and $\bar{T}_\ell = 6 \, (= T_{\ell,0} + T_{t,1})$ in the non-reciprocal case. The average transmit power is defined as $\bar{P}_\text{ave} \triangleq \frac{\bar{\mathcal{E}}_{tot}}{\bar{T}_t + \bar{T}_\ell}$ and, thus, a total energy budget can be alternatively expressed in terms of an average power budget. In the simulations, the individual power constraints at the

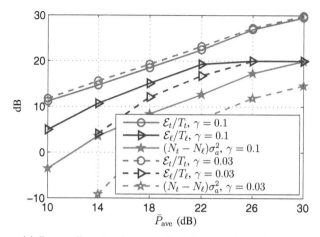

(a) Power allocation between the pilot signal and AN powers

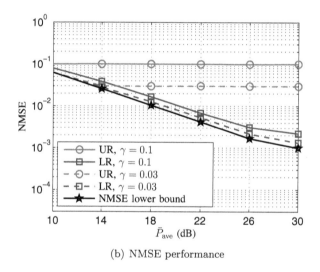

(b) NMSE performance

Figure 7.5: Power allocation and NMSE performance of the two-way DCE scheme for the reciprocal channel case.

transmitter and LR are given by $\bar{P}_t \triangleq \frac{\bar{\mathcal{E}}_t}{T_t} = 30$ dB and $\bar{P}_\ell \triangleq \frac{\bar{\mathcal{E}}_\ell}{T_\ell} = 20$ dB, respectively. Similar to (7.36), we incorporate an NMSE lower bound for comparison:

$$\text{NMSE lower bound} = \left(\frac{1}{\sigma_{h_d}^2} + \frac{\min\{\bar{\mathcal{E}}_t, \ \bar{\mathcal{E}}_{tot}\}}{N_t \sigma_w^2} \right)^{-1}. \tag{7.61}$$

We first focus on the reciprocal channel case. In Figure 7.5(a), we show the optimal power allocation among pilot and AN signals in different stages of the training process. The reverse training power, forward training power, and the AN power are given by $\mathcal{E}_\ell/\bar{T}_\ell$, $\mathcal{E}_t/\bar{T}_t$ and $(N_t - N_\ell)\sigma_a^2$, respectively. We can see from Figure 7.5(a) that the training and AN power values increase approximately at the same rate as the average power $\bar{P}_{\text{ave}}$ increases from 18 dB to 26 dB. This shows that the percentage of power utilized for reverse training, forward training, and AN do not change much under the total energy constraint. However,

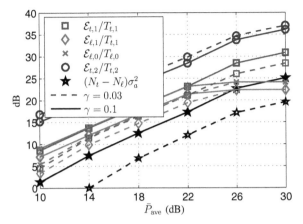

(a) Power allocation between the pilot signal and AN powers

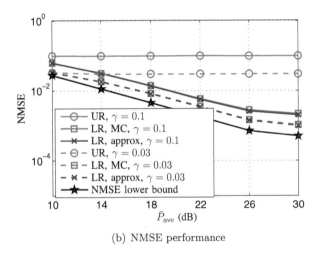

(b) NMSE performance

Figure 7.6: Power allocation and NMSE performance of the two-way DCE scheme for the non-reciprocal channel case.

as $\bar{P}_{ave}$ increases, the power values start to saturate due to the effect of individual energy constraints at the transmitter and LR. Moreover, by comparing between the results for $\gamma = 0.03$ and $\gamma = 0.1$, we can see that it is better to allocate more power to AN and less power to the forward pilot signal as γ increases (i.e., as UR's constraint becomes more stringent). This is because the forward training benefits both LR and UR whereas AN causes interference mainly on UR. However, since the channel estimate at the transmitter is not perfect, an increase in AN power may also degrade the channel estimation performance at LR. The reverse training power also increases with γ since, in this case, the transmitter should be given more accurate knowledge of the channel to maximize the effect of AN usage. In Figure 7.5(b), we show the NMSE at LR and UR versus different values of $\bar{P}_{ave}$. We can see that UR's NMSE is successfully constrained above the value γ while LR's NMSE decreases with $\bar{P}_{ave}$.

In Figures 7.6(a) and 7.6(b), the power allocation and the NMSE performance are shown for the case with non-reciprocal channels. The power allocation behaves similarly as in the reciprocal case shown in Figure 7.5(a). Recall that, in the non-reciprocal case, the power allocation is determined based on the approximate NMSE expresssion in (7.57). In Figure 7.6(b), the approximate NMSE at LR obtained in (7.57) is compared with the exact value obtained through Monte Carlo (MC) simulations. We can see that the approximate NMSE matches closely the results obtained through simulations. For more detailed simulation results, readers are referred to [2].

7.5   Conclusions and Discussions

In this chapter, the DCE scheme was introduced as a novel channel estimation approach to enhance physical layer secrecy. Two DCE schemes were reviewed, namely, the feedback-and-retraining and the two-way DCE schemes. A general DCE scheme involves two stages: the channel acquisition stage, where the transmitter acquires a rough estimate of its channel toward LR, and an AN-assisted forward training stage, where AN is embedded in the null space of the estimated forward channel to degrade the channel estimation performance at UR. In the feedback-and-retraining scheme, channel knowledge at the transmitter is obtained initially by having the transmitter send a preliminary training signal to LR and is further refined through multiple stages of feedback and retraining. In the two-way DCE scheme, a training signal is also sent by LR to facilitate channel estimation directly at the transmitter. When the channel is reciprocal, the transmitter can directly infer its forward channel with knowledge of the backward channel whereas, when the channel is non-reciprocal, an additional round-trip training is required for the transmitter to obtain an estimate of the combined forward and backward channels and, together with the estimate of the backward channel, to infer knowledge of the forward channel. The optimal power values were determined for both schemes by minimizing the NMSE at LR whilst confining the NMSE at UR above some prescribed value. The feedback-and-retraining scheme is less efficient in the early stages since the pilot signals emitted by the transmitter facilitates channel estimation at UR when AN cannot be efficiently utilized. However, the performance can be improved substantially by increasing the number of feedback-and-retraining stages. The two-way DCE scheme avoids this issue by having LR transmit the preliminary training signal instead. However, the channel knowledge at the transmitter is then limited by the available power at LR.

It is important to remark that, with DCE, the effective channel quality between the main and the eavesdropper's channels are effectively discriminated. This provides more room for secrecy coding in the data transmission. In the literature, many signal processing techniques, such as beamforming, precoding, or AN, have been adopted in the data transmission phase to achieve the desired channel quality discrimination. However, we argue that it is more efficient to focus our efforts (e.g., the use of AN) in the channel estimation phase instead of the data transmission phase since the resources need only be applied once during each coherence interval instead of for each symbol period. However, quantitative studies on the trade-off between training and data transmissions in secrecy channels requires further studies. Preliminary studies can be found in [13, 14]. The concept of DCE can also be extended to cases with multiple LRs and URs. However, in these cases, AN must be placed in the null space of all the LR's channels. Optimal power allocation can then be designed to minimize the maximum NMSE among LRs subject to NMSE constraints on all URs. Details can be found in [1]. Moreover, notice that the DCE schemes introduced in this work are based on the LMMSE criterion. However, since UR may not be restricted to the use of the LMMSE channel estimator in practice, it would be interesting to extend the design of the DCE scheme to cases where the Cramér-Rao lower bound (CRLB) is adopted as the performance measure.

Acknowledgment

The works of Y.-W. Peter Hong and Tsung-Hui Chang are supported in part by the National Science Council, Taiwan, under grants NSC 100-2628-E-007-025-MY3 and NSC 101-2218-E-011-043, respectively.

References

[1] T.-H. Chang, W.-C. Chiang, Y.-W. P. Hong, and C.-Y. Chi. Training sequence design for discriminatory channel estimation in wireless MIMO systems. *IEEE Transactions on Signal Processing*, 58(12):6223–6237, December 2010.

[2] C.-W. Huang, T.-H. Chang, X. Zhou, and Y.-W. P. Hong. Two-way training for discriminatory channel estimation in wireless MIMO systems. *IEEE Transactions on Signal Processing*, 61(10):2724–2738, May 2013.

[3] I. Csiszar and J. Korner. Broadcast channels with confidential messages. *IEEE Transactions on Information Theory*, 24(3):339–348, May 1978.

[4] A. Khisti and G. W. Wornell. Secure transmission with multiple antennas. I: The MISOME wiretap channel. *IEEE Transactions on Information Theory*, 56(7):3088–3104, July 2010.

[5] A. Khisti and G. W. Wornell. Secure transmission with multiple antennas. Part II: The MIMOME wiretap channel. *IEEE Transactions on Information Theory*, 56(11):5515–5532, November 2010.

[6] F. Oggier and B. Hassibi. The secrecy capacity of the MIMO wiretap channel. *IEEE Transactions on Information Theory*, 57(8):4961–4972, August 2011.

[7] S. Goel and R. Negi. Guaranteeing secrecy using artificial noise. *IEEE Transactions on Wireless Communications*, 7(6):2180–2189, June 2008.

[8] W.-C. Liao, T.-H. Chang, W.-K. Ma, and C.-Y. Chi. QoS-based transmit beamforming in the presence of eavesdroppers: An optimized artificial-noise-aided approach. *IEEE Transactions on Signal Processing*, 59(3):1202–1216, March 2011.

[9] A. L. Swindlehurst. Fixed SINR solutions for the MIMO wiretap channel. In *Proceedings of IEEE International Conference on Acoustic, Speech and Signal Processing (ICASSP)*, pages 2437–2440, April 2009.

[10] A. Mukherjee and A. L. Swindlehurst. Utility of beamforming strategies for secrecy in multiuser MIMO wiretap channels. In *Proceedings of the 47th Allerton Conference on Communication, Control, and Computing*, pages 1134–1141, September 2009.

[11] H. Jafarkhani S. A. A. Fakoorian and A. L. Swindlehurst. Secure space-time block coding via artificial noise alignment. In *Proceedings of IEEE International Conference on Asilomar Conference on Signals, Systems and Computers (ASILOMAR)*, pages 651–655, November 2011.

[12] T. Yoo and A. Goldsmith. Capacity and power allocation for fading MIMO channels with channel estimation error. *IEEE Transactions on Information Theory*, 52:2203–2214, May 2006.

[13] T.-H. Chang, W.-C. Chiang, Y.-W. P. Hong, and C.-Y. Chi. Joint training and beamforming design for performance discrimination using artificial noise. In *Proceedings of IEEE International Conference on Communications (ICC)*, June 2011.

[14] T.-Y. Liu, S.-C. Lin, T.-H. Chang, and Y.-W. Peter Hong. How much training is enough for secrecy beamforming with artificial noise. In *Proceedings of IEEE International Conference on Communications (ICC)*, June 2012.

[15] S.-C. Lin, T.-H. Chang, Y.-L. Liang, Y.-W. Peter Hong, and C.-Y. Chi. On the impact of quantized channel feedback in guaranteeing secrecy with artificial noise: The noise

leakage problem. *IEEE Transactions on Wireless Communications*, 10(3):901–915, March 2011.

[16] T. F. Wong and B. Park. Training sequence optimization in MIMO systems with colored interference. *IEEE Transactions on Communications*, 52:1939–1947, November 2004.

[17] J. H. Kotecha and A. M. Sayeed. Transmit signal design for optimal estimation of correlated mimo channels. *IEEE Transactions on Signal Processing*, 55(2):546–557, February 2004.

[18] S. M. Kay. *Fundamentals of Statistical Signal Processing: Estimation Theory*. Prentice Hall, 1993.

[19] M. Grant and S. Boyd. CVX: Matlab software for disciplined convex programming, June 2009. http://cvxv.com/cvx

[20] E. G. Larsson and P. Stoica. *Space-Time Block Coding for Wireless Communications*. Cambridge, UK.: Cambridge University Press, 2003.

[21] C. Steger and A. Sabharwal. Single-input two-way SIMO channel: Diversity-multiplexing tradeoff with two-way training. *IEEE Transactions on Wireless Communications*, 7(12):4877–4885, December 2008.

[22] L. P. Withers, R. M. Taylor, and D. M. Warme. Echo-MIMO: A two-way channel training method for matched cooperative beamforming. *IEEE Transactions on Signal Processing*, 56(9):4419–4432, September 2008.

[23] X. Zhou, T. A. Lamahewa, P. Sadeghi, and S. Durrani. Two-way training: Optimal power allocation for pilot and data transmission. *IEEE Transactions on Wireless Communications*, 9(2):564–569, February 2010.

[24] B. Hassibi and B. M. Hochwald. How much training is needed in multiple-antenna wireless links? *IEEE Transactions on Information Theory*, 49(4):951–963, April 2003.

[25] S. Boyd, S.-J. Kim, L. Vandenberghe, and A. Hassibi. A tutorial on geometric programming. *Optimization and Engineering*, 8:67–127, April 2007.

[26] M. Chiang, C. W. Tamd, D. P. Palomar, D. O'Neill, and D. Julian. Power control by geometric programming. *IEEE Transactions on Wireless Communications*, 6(7):2640–2651, July 2007.

Chapter 8

Physical Layer Security in OFDMA Networks

Meixia Tao
Shanghai Jiao Tong University

Jianhua Mo
Shanghai Jiao Tong University

Xiaowei Wang
Shanghai Jiao Tong University

Orthogonal frequency division multiple access (OFDMA) has a major role in the evolution of broadband wireless networks. It enables efficient transmission of a wide variety of data traffic by optimizing power, subcarrier, or bit allocation among different users. Secrecy or private information communications to mobile users are often needed in present and future wireless systems. Hence, it is essential to consider the security demand in radio resource allocation of OFDMA networks. This chapter presents an overview of resource allocation techniques in OFDMA-based wireless networks for providing physical layer security. In particular, a general framework of power and subcarrier allocation in OFDMA downlink networks where users requiring secrecy and nonsecrecy messages coexist is introduced in detail. Through theoretical analysis, several insights into resource allocation with secrecy requirement are provided. At last, possible directions for future research are also given.

8.1 Introduction

Orthogonal frequency division multiplexing (OFDM) has evolved as a leading technique to provide high data rate transmission in broadband wireless networks. The primary advantage of OFDM over single-carrier transmission is its ability to cope with frequency-selective multipath fading without complex equalization filters by dividing a wideband channel into a set of orthogonal narrowband subcarriers. Furthermore, since different subcarriers can be assigned to different users, the OFDM technology naturally provides a multiple access method, known as OFDMA. It offers good system design flexibility by adaptive power, subcarrier, or bit allocation among multiple users according to their channel state information as well as quality-of-service (QoS) requirements. For these advantages, OFDMA has been adopted in many wireless standards, such as 3GPP Long-Term Evolution (LTE) downlink and IEEE 802.16 WiMAX, and it will continue to play an important role in future advanced

137

cellular networks.

In the past decade, tremendous research results have been reported on the resource allocation of OFDMA networks, such as [1–6], where the problem formulation differs mostly in optimization objectives and constraints. The work in [1] appeared as one of the earliest results on margin adaptation for minimizing the total transmit power with individual user rate requirements. Jang and Lee in [2] studied rate adaptation for system sum-rate maximization subject to a total transmit power constraint. In [3], Tao et al. considered a network with heterogenous traffic and studied the power and subcarrier allocation problem for maximizing the sum-rate of nondelay-constrained users while satisfying the basic rate requirement of each delay-constrained user under a total transmit power constraint. Cross-layer optimizations considering utility-function and traffic arrival distribution for OFDMA networks were also studied in several works, e.g., [4–6]. Nevertheless, none of these works take into account the *security* issue, which attracts increasing attention recently in wireless networks.

This chapter aims to provide an overview on resource allocation techniques in OFDMA networks for providing physical layer security. Section 8.2 briefly surveys the related works in the literature. Then the basics of resource allocation for secret communications are reviewed in Section 8.3. Section 8.4 presents in detail a general framework of power and subcarrier allocation in OFDMA downlink networks to maximize the information rate of nonsecrecy users while satisfying minimum secrecy rate for each secrecy user. This is built upon the previous work [7]. Both optimal and low-complexity suboptimal algorithms to solve the problem are introduced. Through theoretical analysis, several insights into resource allocation with secrecy requirement are provided. Practical issues concerning the false channel state information report from malicious users are also discussed. Finally, Section 8.5 summarizes this chapter and outlines some possible directions for future research.

8.2   Related Works on Secure OFDM/OFDM Networks

Securing the OFDM transmission has recently attracted a lot of attention. Upper layer cryptographic approaches have been used to protect the confidentiality of the content, for example, [8–10]. Here, we are only interested in the physical layer security in OFDMA networks.

8.2.1   Secure OFDM Channel

In the OFDM transmission, a guard interval is inserted at the beginning of each OFDM symbol to avoid the intersymbol interference (ISI). In an OFDM-based wiretap channel, under the assumption that both the legitimate receiver and the eavesdropper discard the guard intervals before detection, the channel can be modeled as a set of parallel wiretap channels as shown in Figure 8.1.

The earliest work on secure parallel channels dates back to Yamamoto's works [11–13]. In these works, a so-called secret sharing communication system (SSCS) was studied. In specific, in [11], the SSCS with two or three noiseless channels was studied. The legitimate receiver can access all the channels while the eavesdropper can only access one or two channels. The SSCS with two discrete memoryless broadcast channels was investigated in [12]. In [13], the SSCS with two Gaussian wiretap channels was analyzed. Therein, three cases were considered, (1) two wiretappers cannot cooperate, (2) they can cooperate, and (3) whether or not they can cooperate is not known. The second case in which the two

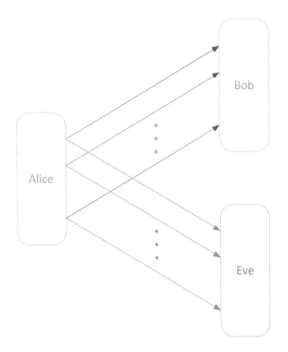

Figure 8.1: The parallel wiretap channels. Legitimate user Alice transmits messages to Bob through parallel channels while Eve eavesdrops on these channels.

wiretappers cooperate can be seen as a special case of the recent works on independent parallel channels such as those in [14] and [15].

Since 2006, there has been a renewed interest for physical layer security in parallel channels. Li et al. [14] studied the composite system consisting of multiple independent parallel channels as shown in Figure 8.1. They showed that the secrecy capacity of the composite system is simply the summation of the secrecy capacities of the individual channels, and using independent codebooks across the subchannels achieves the secrecy capacity. They further derived the optimal power allocation strategy for a system with parallel additive white gaussian noise (AWGN) channels subject to a total power constraint. Liang et al. [15] considered the parallel BCC with independent subchannels. They established the secrecy capacity region of this parallel BCC and showed that independent inputs for each subchannel are optimal to achieve the secrecy capacity region. For the parallel Gaussian BCC, which is an example of parallel BCC with degraded subchannels, they derived the optimal power allocation that achieves the boundary of the secrecy capacity region.

In [16,17], Jorswieck et al. studied the multiuser scenario of parallel channels. In particular, [16] derived the optimal power allocation for single-user and two-user wiretap channels. In [17], the secrecy rate region of downlink OFDM systems with two users was analyzed. An efficient optimal algorithm is derived for the two related problems, namely, weighted sum secrecy rate maximization under transmit power constraints and sum transmit power minimization under secrecy rate constraints. Tsouri and Wulich [18] considered the multiple access channel in which more than one user transmits packets to a common receiver with an external eavesdropper. The authors proposed a method of overloading subcarriers by multiple transmitters to secure OFDM in wireless time-varying channels. This method relies on very tight synchronization between multiple transmitter and single receiver.

In [19] and [20], a friendly jammer was introduced into the secret communication system with multiple orthogonal source–destination links. The power allocation of the jammer was obtained by auction theory. Several different auction schemes were studied and compared.

8.2.2   Secure OFDMA Cellular Networks

The secure OFDMA downlink transmission in a cellular network was considered in [7] and [21]. Wang et al. [7] formulated an analytical framework for resource allocation in a downlink OFDMA-based broadband network with coexistence of secure users (SUs) and normal users (NUs). More details will be given in Section 8.4. In the OFDMA downlink network of [21], there is a base station with N_T antennas, an eavesdropper with $N_E(N_E < N_T)$ antennas, and multiple legitimate receivers equipped with single antennas. The authors formulated the resource allocation problem as a nonconvex optimization problem that takes into account the dynamic circuit power consumption, artificial noise generation, and different QoS requirements. In addition, a trade-off between energy efficiency and secrecy was observed in the simulations.

8.2.3   Secure OFDMA Relay Networks

Recently, the secure OFDM communication has also been combined with relay transmission. Jeong and Kim in [22] studied the multicarrier decode-and-forward (DF) relay system with an eavesdropper. On each subcarrier, there are three possible transmission modes, (1) no communication, (2) direct communication, and (3) relay communication. The optimal mode selection and power allocation of the source and relay over all subcarriers have been determined under the total system power constraint. The resource allocation and scheduling for secure OFDMA DF relay networks was considered by Ng et al. [23]. Similar to [21], the authors in [23] formulated a mixed combinatorial and convex optimization problem taking into account the artificial noise generation and the effect of imperfect channel state information. An efficient iterative and distributed resource allocation algorithm at the relay was proposed.

Authors in [24] and [25] studied the resource allocation (including power allocation, subcarrier pairing, and subcarrier assignment) in OFDMA multipair two-way relay channels. In [24], without knowing the channel state information (CSI) of the eavesdropper, the authors proposed a cooperative jamming scheme to maximize the system sum-rate with a constraint on the information leakage to the eavesdropper. On the other hand, in [25], they assumed that the eavesdropper CSI is known. A near-optimal algorithm using Lagrangian dual decomposition method was proposed for maximizing the secrecy sum-rate.

8.2.4   Secure OFDM with Implementation Issues

Instead of modeling the OFDM transmission as parallel channels, one can consider the practical realization of OFDM transmission and relax the conventional assumption that the eavesdropper is also forced to implement OFDM demodulation. This scenario was studied in [26] and [27]. Therein, the frequency-selective fading channel is modeled as an equivalent MIMO channel, which has been extensively studied in the literature [28–31]. More specifically, Kobayashi et al. [26] studied the frequency-selective broadcast channel with confidential messages where the transmitter sends a common message to receivers 1 and 2 and a confidential message to receiver 1. The receiver 2 is not restricted to use OFDM demodulation, although the receiver 2 is still supposed to drop the guard interval symbols. In the case of a block transmission of N symbols followed by a guard interval of L symbols discarded at both receivers, the frequency-selective channel can be modeled as an $N \times (N + L)$ multiple-input multiple-output (MIMO) Toeplitz matrix. Thus, the equivalent channel can be seen as a MIMO wiretap channel. A Vandermonde precoding that transmits the confidential message in the null space of the channel toward the receiver 2 was proposed and the secrecy degree of freedom was characterized. However, if receiver 2 does not adhere to the decoding rule and is able to access the guard interval symbols, it can extract the confidential message and there is no secrecy rate. Renna et al. [27] considered the more conservative case in which the eavesdropper is free to implement the demodulation while the transmitter and the legitimate receiver adopt OFDM modulation and demodulation. The performance of this scenario was compared with that of the parallel channel model. They also derived the secrecy capacity of the frequency-selective channel, regardless of the transceiver implementations of all the parties. The performance loss caused by the use of a more sophisticated receiver at the eavesdropper was qualified.

8.3 Basics of Resource Allocation for Secret Communications

Before we present the resource allocation in detail for secure OFDMA networks in the next section, some basics of resource allocation for secret communications need to be reviewed.

8.3.1 Power Allocation Law for Secrecy

The definition of secrecy capacity and the fading nature of wireless channels provides opportunities for power allocation in secure transmissions. Recall that the nonzero secrecy rate exists only when the channel gain of the legitimate channel is larger than that of the eavesdrop channel. In particular, the achievable secrecy rate of channels with Gaussian noise is given by [32].

$$C_s = [\log(1 + Ph_M) - \log(1 + Ph_E)]^+ \tag{8.1}$$

where P is the transmit power, h_M and h_E are the channel-to-noise ratios (CNR) of the main channel and the eavesdropper channel, respectively, and $[x]^+ = \max\{0, x\}$. If nonzero secrecy capacity exists, it is easy to show that larger transmit power yields high secrecy capacity just by taking the derivative of the power in the expression of secrecy capacity. Since a nonzero instantaneous secrecy rate cannot be guaranteed all the time due to channel fading, the average secrecy rate over an ergodic fading channel is often used as a performance metric.

In [33], the authors obtained the optimal power allocation for average secrecy rate maximization under average total power constraint, which is expressed as

$$p(h_M, h_E) = \frac{1}{2}\left[\sqrt{\left(\frac{1}{h_E} - \frac{1}{h_M}\right)^2 + \frac{4}{\lambda}\left(\frac{1}{h_E} - \frac{1}{h_M}\right)} - \left(\frac{1}{h_M} + \frac{1}{h_E}\right)\right]^+ \tag{8.2}$$

where λ is a parameter to be chosen such that $\mathbb{E}[p(h_M, h_E]$ satisfies the average total power constraint. Here, notation $\mathbb{E}$ represents statistical average over the joint distribution of channel conditiona h_M and h_E. It is observed that the legitimate user must satisfy $h_M - h_E \geq \frac{1}{\lambda}$ in order to be allocated nonzero power. Intuitively, power should not be wasted on the cases where the secrecy rate is zero or too low. This result is the universal law of power allocation with respect to secrecy rate.

8.3.2 Multiple Eavesdroppers

The traditional wiretap channel model [34] is comprised of three nodes, namely Alice, Bob, and Eve, as in Figure 8.1. However, in practical scenarios, users are networked and share radio resources. The roles of users are not fixed and may change in each transmission. For a certain user, each one of the other users in the same network is a potential eavesdropper.

The presence of multiple eavesdroppers is one of the key features in a network environment with security requirements. In a noncolluding eavesdropper scenario, all eavesdroppers are assumed to be mutually independent and the information is deemed secure if it cannot be eavesdropped by any one of the eavesdroppers. Hence, the equivalent wiretap channel is the strongest among all eavesdroppers. In other words, nonzero secrecy rate is obtained only when the information rate on the main channel is greater than that on any eavesdropper channel. As we will see in the next section, this conclusion will have a significant impact on subcarrier allocation for secret communication in OFDMA networks.

If the eavesdroppers can collude and exchange outputs to decode the message, the network can be regarded as a single-input multiple-output system, where there is a single virtual eavesdropper with multiple receive antennas. In this case, a nonzero secrecy rate can only be obtained when the gain of the main channel is greater than the gain of the virtual eavesdropper channel.

8.4 Resource Allocation for Physical Layer Security in OFDMA Networks

This section introduces a general framework of power and subcarrier allocation for providing physical layer security in OFDMA networks. There are two types of users coexisting in the network. The first type of users have confidential messages to be communicated with the base station (BS) and demand nonzero secrecy rates. These users are referred to as *secure users* (SUs). The other type of users just have normal data traffic and do not worry about security issues at the physical layer. These users are regarded as *normal users* (NUs). All SUs and NUs are legitimate users in the network and they are completely honest to the base station.

The resource allocation problem in such an OFDMA network with the coexistence of SUs and NUs possesses major differences compared with those without secrecy constraint. First, the legitimate subcarriers to be assigned to an SU can only come from the subcarrier set on which this SU has the best channel condition among all the users as discussed in Section 8.3.2. This is because for each SU, any other user in the same network is a potential eavesdropper. As a result, a nonzero secrecy rate on a subcarrier is possible to achieve only if the channel gain of the SU on this subcarrier is the largest among all the users. Note that this observation is very different from that in conventional OFDMA networks where a user is still able to occupy some subcarriers and transmit signals over them even if its channel gains on these subcarriers are not the largest. Secondly, even if an SU has the best channel condition on a given subcarrier, assigning this subcarrier to the SU may not be the optimal solution from the system perspective when the quality of service of NUs is taken into account. This is due to the fact that the achievable secrecy rate is the subtraction of two logarithmic functions as shown in (8.1). In the case of large P, if the gap between h_M and h_E is not large enough, the achievable secrecy rate can be rather low. In this case it may be more beneficial for the system to assign the subcarrier to an NU for transmitting nonconfidential data traffic.

The goal is to find an optimal power and subcarrier allocation policy to maximize the long-term aggregate information rate of all NUs while maintaining a target average secrecy rate of each individual SU under a total power constraint. We solve this problem in dual domain using the decomposition method in an asymptotically optimal manner. Based on the insight derived from the optimal policy, a low-complexity suboptimal algorithm is also introduced.

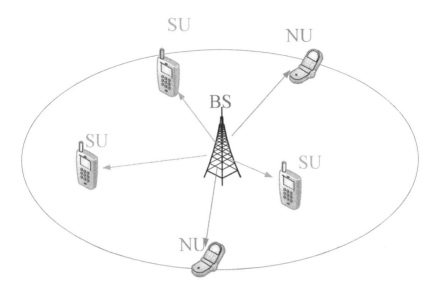

Figure 8.2: The OFDMA downlink network with two types of users: secure users (SUs) and normal users (NUs).

8.4.1 Problem Formulation

We consider the downlink of an OFDMA broadband network with one BS and K mobile users as shown in Figure 8.2. The first K_1 users, indexed as $k = 1, \ldots, K_1$, are SUs. Each of them has confidential messages to communicate with the BS and, therefore, demands a secrecy rate no lower than C_k. The other $K - K_1$ users, indexed as $k = K_1 + 1, \ldots, K$, are NUs, and demand services of best-effort traffic. The total bandwidth is logically divided into N orthogonal subcarriers by using OFDMA with each experiencing slow fading. As a central controller, the BS knows the channel information of all users, finds the allocation policy, and then assigns power and subcarriers to mobile users at each transmission frame according to the instantaneous CSI of all users. We assume full statistical knowledge and instantaneous knowledge of CSI at the BS and that each subcarrier is occupied only by one user at each time frame to avoid multiuser interference. Table 1.1 lists the definitions of the remaining notations used in this section.

The problem is formulated as maximizing the aggregate average information rate of

$p_{k,n}$	Power allocated to user k on subcarrier n
Ω_k	Subcarrier set assigned to user k
$r_{k,n}$	Information rate of NU k on subcarrier n
$r_{k,n}^s$	Secrecy rate of SU k on subcarrier n
$\alpha_{k,n}$	CNR of user k on subcarrier n
$\beta_{k,n}$	Largest CNR of all users on subcarrier n except user k
C_k	Minimum secrecy rate requirement of SU k
P	Power budget
ω_k	Weight parameter
λ, μ_k	Dual variables

Table 8.1: Notations

NUs while satisfying the basic secrecy rate requirement of each SU. Both peak and average power constraints from the BS are considered. Mathematically, the optimization problem is expressed as

$$\max_{\{\mathbf{m}\Omega(\boldsymbol{\alpha}),\mathbf{m}p(\boldsymbol{\alpha})\}} \quad \mathbb{E}\left(\sum_{k=K_1+1}^{K} \omega_k \sum_{n\in\Omega_k} r_{k,n}\right) \tag{8.3}$$

$$\text{subject to} \quad \mathbb{E}\left(\sum_{n\in\Omega_k} r_{k,n}^s\right) \geq C_k, \ 1 \leq k \leq K_1 \tag{8.4}$$

$$\sum_{k=1}^{K}\sum_{n\in\Omega_k} p_{k,n} \leq P \tag{8.5}$$

$$or \quad \mathbb{E}\left(\sum_{k=1}^{K}\sum_{n\in\Omega_k} p_{k,n}\right) \leq P \tag{8.6}$$

$$p_{k,n} \geq 0, \forall k, n \tag{8.7}$$

$$\Omega_1 \cup ... \cup \Omega_K \subseteq \{1, 2, ..., N\}$$

$$\Omega_1, ..., \Omega_K \text{ are disjoint} \tag{8.8}$$

Note that in the above formulation, the average secrecy rate constraint, instead of instantaneous secrecy rate constraints, is imposed. In [23], the authors consider another way to formulate the problem. They put constraint on the secrecy outage probability. Due to the channel fading, zero secrecy rate (or called "secrecy outage") cannot be avoided. These two methods are both practical and reasonable when treated as objective functions or constraints. In addition, when the channel fading process is ergodic, we can change statistical average to time average. Also note the secrecy rate requirement C_k, for $k = 1, \ldots, K_1$ should be set properly in order to maintain the feasibility of the optimization problem. If C_k is set too high, this constraint can never be satisfied even if all the resources are allocated to this SU.

In the example problem, the total transmit power of the BS is subject to either a long-term average or peak constraint, both denoted as power constraint P. Under peak power constraint, the total power should not be higher than the budget P at any transmission. However, under the average power constraint, the instantaneous total power can be higher than P at limited transmissions as long as the long-term average total power is lower than P. We only discuss the optimal policy under average power constraint in the remainder of this chapter.

8.4.2 Optimal Policy

The optimization problem (8.3) is a mixed-integer problem. However, it satisfies the time-sharing condition introduced in [35]. That is, the objective function is concave and the secrecy rate constraint is convex given that $r_{k,n}^s$ is concave in $p_{k,n}$ and that the integral preserves concavity. Therefore, similar to the OFDMA networks without secrecy constraint, we can use a dual approach for resource allocation, and the solution is asymptotically optimal for a large enough number of subcarriers.

Define $\mathcal{P}(\boldsymbol{\alpha})$ as a set of all possible nonnegative power parameters $\{p_{k,n}\}$ at any given system channel condition $\boldsymbol{\alpha}$ satisfying that for each subcarrier n only one $p_{k,n}$ is positive.

The Lagrange dual function is thus given by

$$
\begin{aligned}
g(\mathbf{m}\mu, \lambda) &= \max_{\{p_{k,n}\}\in\mathcal{P}(\mathbf{m}\alpha)} \left\{ \mathbb{E}\left[\sum_{k=K_1+1}^{K} \omega_k \sum_{n=1}^{N} r_{k,n}\left(p_{k,n}(\mathbf{m}\alpha), \mathbf{m}\alpha\right) \right] \right. \\
&\quad + \sum_{k=1}^{K_1} \mu_k \left(\mathbb{E}\left[\sum_{n=1}^{N} r_{k,n}^{s}\left(p_{k,n}(\mathbf{m}\alpha), \mathbf{m}\alpha\right) \right] - C_k \right) \\
&\quad \left. + \lambda \left(P - \mathbb{E}\left[\sum_{k=1}^{K}\sum_{n=1}^{N} p_{k,n}(\mathbf{m}\alpha) \right] \right) \right\},
\end{aligned}
\tag{8.9}
$$

Then the dual problem of the original problem (8.3) is given by

$$
\begin{aligned}
&\min \ g(\mathbf{m}\mu, \lambda) \\
&s.t. \ \mathbf{m}\mu \succeq 0, \lambda \geq 0.
\end{aligned}
\tag{8.10}
$$

Since a dual problem is always convex by definition and can be solved in various gradient-based algorithms, we next focus on finding the expression of the Lagrange dual function, through which the characteristics of the optimal resource allocation policy are found.

By observing (8.9) it is found that the maximization in the Lagrange dual function can be decomposed into N independent subfunctions as:

$$
g(\mathbf{m}\mu, \lambda) = \sum_{n=1}^{N} g_n(\mathbf{m}\mu, \lambda) - \sum_{k=1}^{K_1} \mu_k C_k + \lambda P,
\tag{8.11}
$$

where

$$
g_n(\mathbf{m}\mu, \lambda) = \max_{\{p_{k,n}\}\in\mathcal{P}(\mathbf{m}\alpha)} \mathbb{E}\left[J_n\left(\mathbf{m}\mu, \lambda, \mathbf{m}\alpha, \{p_{k,n}\}_k\right) \right],
\tag{8.12}
$$

with

$$
J_n(\mathbf{m}\mu, \lambda, \mathbf{m}\alpha, \{p_{k,n}\}_k) = \sum_{k=K_1+1}^{K} \omega_k r_{k,n} + \sum_{k=1}^{K_1} \mu_k r_{k,n}^{s} - \lambda \sum_{k=1}^{K} p_{k,n}.
\tag{8.13}
$$

Based on the above decomposition, we now present the optimality conditions of power allocation and subcarrier assignment, respectively.

8.4.2.1 Optimality Condition for Power Allocation

For fixed dual variables $\mathbf{m}\mu$ and λ, the maximization problem in (8.12) is a single-carrier multiple-user power allocation problem. Using Karush-Kuhn-Tucker (KKT) conditions, we obtain the following optimality condition of power allocation:

$$
\begin{aligned}
p_{k,n}^{*} &= \frac{1}{2}\left[\sqrt{\left(\frac{1}{\alpha_{k,n}} - \frac{1}{\beta_{k,n}}\right)^2 + \frac{4\mu_k}{\lambda}\left(\frac{1}{\beta_{k,n}} - \frac{1}{\alpha_{k,n}}\right)} \right. \\
&\quad \left. - \left(\frac{1}{\alpha_{k,n}} + \frac{1}{\beta_{k,n}}\right) \right]^{+}
\end{aligned}
\tag{8.14}
$$

for $k = 1, \ldots, K_1$, and

$$
p_{k,n}^{*} = \left[\frac{\omega_k}{\lambda} - \frac{1}{\alpha_{k,n}} \right]^{+}
\tag{8.15}
$$

for $k = K_1 + 1, \ldots, K$.

We can conclude from (8.15) that the optimal power allocation for NUs follows the conventional waterfilling principle, and the water level is determined by both the weight of the NU and the average power constraint. On the other hand, it is seen from (8.14) that the optimal power allocation for SUs has the same form as the power allocation law (8.2) in conventional fading wiretap channels, as expected. Similar to the observation in (8.2), it is also seen from (8.14) that the SU must satisfy $\alpha_{k,n} - \beta_{k,n} \geq \frac{\lambda}{\mu_k}$ in order to be allocated nonzero power. This means that the power allocation for SU depends on both the channel gain of the SU and the largest channel gain among all the other users. Moreover, it is nonzero only if the former exceeds the latter by the threshold $\frac{\lambda}{\mu_k}$.

8.4.2.2 Optimality Condition for Subcarrier Assignment

Substituting the optimal power allocation (8.14) and (8.15) into (8.12) and comparing all the K possible user assignments for each subcarrier n, we obtain

$$g_n(\mathbf{m}\mu, \lambda) = \mathbb{E}\left[\max_{1 \leq k \leq K} H_{k,n}(\mathbf{m}\mu, \lambda, \mathbf{m}\alpha)\right], \qquad (8.16)$$

where the function $H_{k,n}(\cdot)$ is defined as

$$H_{k,n}(\mathbf{m}\mu, \lambda, \mathbf{m}\alpha) = \mu_k \log\left(\frac{1 + p_{k,n}^* \alpha_{k,n}}{1 + p_{k,n}^* \beta_{k,n}}\right) - \lambda p_{k,n}^* \qquad (8.17)$$

for $1 \leq k \leq K_1$ with $p_{k,n}^*$ given in (8.14), and

$$H_{k,n}(\mathbf{m}\mu, \lambda, \mathbf{m}\alpha) = \omega_k \left[\log \frac{\omega_k \alpha_{k,n}}{\lambda}\right]^+ - \left[\omega_k - \frac{\lambda}{\alpha_{k,n}}\right]^+ \qquad (8.18)$$

for $K_1 < k \leq K$.

It is now clear that the function $H_{k,n}$ plays an important role in determining the optimal subcarrier assignment. In specific, for any given dual variables $\mathbf{m}\mu$ and λ, the subcarrier n will be assigned to the user with the maximum value of $H_{k,n}$. That is, the optimality condition for subcarrier assignment is given by

$$k_n^* = \arg\min_k \max_k H_{k,n}, \quad \text{for } n = 1, \ldots, N. \qquad (8.19)$$

Note that for $k = K_1 + 1, \ldots, K$, $H_{k,n}$ is monotonically increasing in $\alpha_{k,n}$. Therefore, the NU with larger $\alpha_{k,n}$ is more likely to be assigned subcarrier n. We also notice that for $k = 1, \ldots, K_1$, $H_{k,n} > 0$ only when SU k has the largest $\alpha_{k,n}$ among all the K users and satisfies $\alpha_{k,n} > \beta_{k,n} + \lambda/\mu_k$. Otherwise, $H_{k,n} = 0$. In other words, an SU becomes a candidate for subcarrier n only if its CNR is the largest and is λ/μ_k larger than the second largest.

8.4.3 Suboptimal Algorithm

The complexity of the optimal power and subcarrier allocation policy mainly lies in the joint optimization of Lagrange multipliers $\mathbf{m}\mu = (\mu_1, \ldots, \mu_{K_1})$ and λ. The idea of this suboptimal scheme is to first assign the resources to only SUs as if all the NUs were pure eavesdroppers without data transmission. After that, the residual subcarriers and power, if any, are distributed among NUs. By doing this, the joint update of the Lagrange multipliers will be decoupled as detailed below.

In this scheme, the power allocation adopts the expressions in (8.14) and (8.15) except that the parameter $\lambda/\mu_k, k = 1, \ldots, K_1$ in (8.14) is replaced by a new variable ν_k. Also, in (8.15) we define $L_k = \omega_k L_0$, for $k = K_1 + 1, \ldots, K$, where $L_0 = \frac{1}{\lambda}$. Therefore, ν_k and L_0 can be found through two separate binary searches.

8.4.4 Complexity

In the optimal algorithm, if we choose the ellipsoid method to update the dual variables, it converges in $\mathcal{O}((K_1 + 1)^2 \log \frac{1}{\epsilon})$ iterations where ϵ is the accuracy [36]. In the suboptimal algorithm, since $\nu_k (k = 1, \ldots, K_1)$ and L_0 are obtained individually by binary search, the algorithm converges in $\mathcal{O}(K_1 \log \frac{1}{\epsilon} + \log \frac{1}{\epsilon}) = \mathcal{O}((K_1 + 1) \log \frac{1}{\epsilon})$ iterations. In addition, the computational loads in each iteration of both the optimal and suboptimal schemes are linear in $KN|\mathbf{m}\alpha|$ where $|\mathbf{m}\alpha|$ is the number of the training channel realizations. So the suboptimal scheme reduces the complexity by about $\frac{1}{K_1+1}$.

8.4.5 Numerical Examples

In this section, we provide some numerical examples to demonstrate the performance of the optimal and suboptimal resource allocation algorithms. In the simulation setup, we consider an OFDMA network with $N = 64$ subcarriers and $K = 8$ mobile users, among which $K_1 = 4$ are SUs and $K - K_1 = 4$ are NUs. For simplicity, all the weighting parameters ω_k's for NUs are set to one and the secrecy rate requirements C_k's for SUs are set to be identical, denoted as $C_k = R_{SU}$. Let R_{NU} denote the average total information rate of the NUs. The channel on each subcarrier for each user is assumed to be independent and identically distributed Rayleigh fading with unit mean-square value for illustration purposes only.

To evaluate the optimal and suboptimal power and subcarrier adaptation schemes, we introduce two nonadaptive schemes as benchmarks. In these two nonadaptive schemes, subcarrier assignment is fixed beforehand while power allocated to the predetermined subcarrier sets conforms to (8.14) and (8.15). In the first fixed subcarrier assignment scheme, denoted as FSA-1, the 64 subcarriers are equally assigned to the 8 users and thus each user obtains 8 subcarriers. In the second scheme, denoted as FSA-2, the SUs are given higher priority and each is assigned 12 subcarriers, whereas each NU is assigned 4 subcarriers.

We first demonstrate the pair (R_{NU}, R_{SU}) at fixed total transmit SNR = 30 dB in Figure 8.3. First it is observed that using both optimal and suboptimal algorithms, R_{NU} decreases with the increase of R_{SU} and falls sharply to zero at around $R_{SU} = 3.5$ nat/OFDM symbol. It is then observed from Figure 8.3 that the suboptimal algorithm only incurs less than 20% loss in R_{NU} when achieving the same R_{SU} compared with the optimal algorithm. Now comparing the optimal and suboptimal schemes with the nonadaptive ones, both of them earn great advantage over the two benchmarks FSA-1 and FSA-2.

Figure 8.4 and Figure 8.5 show, respectively, the average power consumption and the average number of subcarriers assigned to all SUs with respect to different R_{SU} when the total average transmit power is fixed a 30 dB. From Figure 8.4, we notice that the optimal scheme spends more power on SUs than the suboptimal one. It is observed from Figure 8.5 that the number of occupied subcarriers by SUs increases with the growing of R_{SU} and reaches 32 at the feasible point of $R_{SU} = 3.5$ nat/OFDM symbol for both the optimal and suboptimal schemes. Additionally, the optimal scheme assigns less subcarriers to SUs than the suboptimal one. For FSA-1 and FSA-2, the number of occupied subcarriers by SUs is also increasing with the feasible R_{SU} and is smaller than the number of preassigned subcarriers to SUs for FSA-1 and FSA-2. This indicates that fixed subcarrier assignments waste subcarriers compared with adaptive ones.

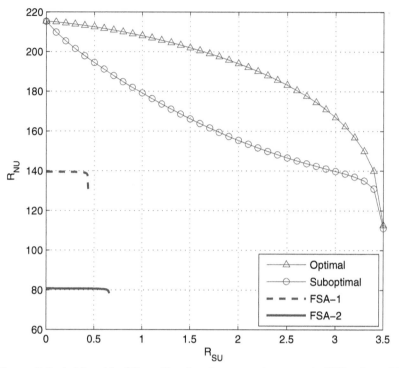

Figure 8.3: Achievable (R_{SU}, R_{NU}) pair at total transmit SNR of 30 dB.

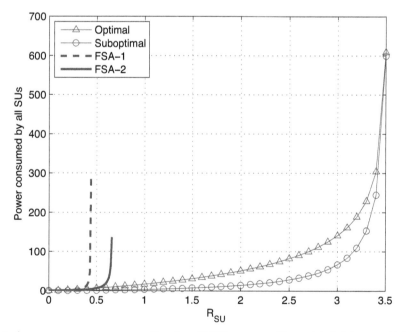

Figure 8.4: Average power consumption by all SUs versus R_{SU} at total transmit SNR of 30 dB.

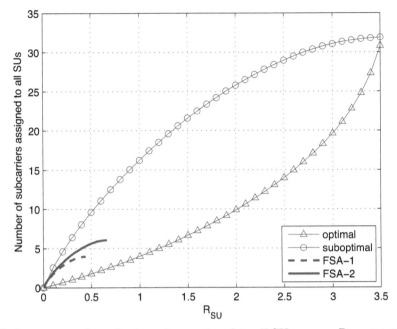

Figure 8.5: Average number of subcarriers assigned to all SUs versus R_{SU} at total transmit SNR of 30 dB.

8.4.6 Discussion on False CSI Feedback

Throughout this section, we assumed that the channel state information of each user obtained at the BS is accurate. If a user deliberately lies and reports a lower channel-to-noise ratio on certain subcarriers, it would get a higher chance of eavesdropping an SUs private message and therefore cause secrecy rate loss. However, on the other side, its own average information rate or secrecy rate would also be reduced due to the less-assigned radio resources. Hence, we argue that there is no incentive for the users to lie about their channel condition.

If, however, there exists a malicious eavesdropper in the network that does not care about its own transmission, the security in the network can be circumvented by sending the BS false channel measurements. We call this *false CSI attack*. To combat the attack from the malicious user, the BS can compensate the CSI received from the malicious user. However, over-compensation may lead to SU's throughput loss because the BS has no knowledge of the real CSI of the malicious user. A more detailed analysis on eavesdrop probability and secrecy loss in the presence of a malicious eavesdropper in OFDMA networks can be found in [37].

8.5 Conclusions and Open Issues

In this chapter, we take OFDMA broadband networks with both SUs and NUs as an example to investigate the power and subcarrier allocation with security requirements. The problem is formulated as maximizing the average aggregate information rate of NUs while satisfying the basic average secrecy rate requirements of SUs. Decomposition method in dual domain can also be used in this optimization problem. Results show that the optimal power allocation for an SU depends on both its channel gain and the largest channel gain among others. We also observe that an SU becomes a valid candidate competing for a subcarrier only if its CNR on this subcarrier is the largest among all and larger enough than the second largest CNR. Numerical results show that the optimal power and subcarrier allocation algorithm effectively boosts the average total information rate of NUs while meeting the basic secrecy rate requirements of SUs.

The current works about physical layer security in OFDMA networks all focus on the downlink transmission, such as [7,21,23]. The resource allocation in the uplink transmission is not explored. Actually, the uplink transmission is much more challenging. For the mobile users, it is hard to obtain the CSI between them and the potential eavesdroppers. Moreover, if there are many eavesdroppers in the network, we need the CSI between each mobile user and eavesdropper to allocate the power and subcarriers, while in downlink transmission, only CSI between the base station and the eavesdropper is needed. A possible solution is to encrypt the uplink transmission by the keys obtained by secured downlink transmission. Thus, it seems to be an interesting problem to find a method to appropriately combine the secret downlink and uplink transmissions.

Another possible research direction is to consider the problem of securing multicast transmission in the OFDMA networks. At present, multicast is a common service in broadband wireless networks, such as MBMS (Multimedia Broadcast Multicast Service) in LTE. Within the framework of physical layer security in OFDMA networks, how to group the mobile users into different multicast groups and how to allocate power and subcarriers to each multicast group are still open.

References

[1] C. Y. Wong, R. S. Cheng, K. B. Letaief, and R. D. Murch, "Multi-user OFDM with adaptive sub-carrier, bit, and power allocation," *IEEE J. Sel. Areas Commu.*, vol. 17, no. 10, pp. 1747–1758, Oct. 1999.

[2] J. Jang and K. B. Lee, "Transmit power adaptation for multiuser OFDM systems," *IEEE J. Sel. Areas Commu.*, vol. 21, no. 2, pp. 171–178, Feb. 2003.

[3] M. Tao, Y.-C. Liang, and F. Zhang, "Resource allocation for delay differentiated traffic in multiuser OFDM systems," *IEEE Trans. Wireless Comm.*, vol. 7, no. 6, pp. 2190–2201, June 2008.

[4] D. S. W. Hui, V. K. N. Lau, and W. H. Lam, "Cross-layer design for OFDMA wireless systems with heterogeneous delay requirements," *IEEE Transactions on Wireless Communications*, vol. 6, no. 8, pp. 2872–2880, Aug. 2007.

[5] N. Mokari, M. R. Javan, and K. Navaie, "Cross-layer resource allocation in OFDMA systems for heterogeneous traffic with imperfect CSI," *IEEE Transactions on Vehicular Technology*, vol. 59, no. 2, pp. 1011–1017, Feb. 2010.

[6] G. Song and Y. Li, "Cross-layer optimization for OFDM wireless networks: Part II: Algorithm development," *IEEE Trans. Wireless Comm.*, vol. 4, no. 2, pp. 625–634, Mar. 2005.

[7] X. Wang, M. Tao, J. Mo, and Y. Xu, "Power and subcarrier allocation for physical-layer security in OFDMA-based broadband wireless networks," *IEEE Transactions on Information Forensics and Security*, vol. 6, no. 3, pp. 693–702, Sept. 2011.

[8] R. S. Owor, K. Dajani, Z. Okonkwo, and J. Hamilton, "An elliptical cryptographic algorithm for rf wireless devices," in *Proceedings of the 39th conference on Winter simulation*, ser. WSC '07. Piscataway, NJ, USA: IEEE Press, 2007, pp. 1424–1429. [Online]. Available: http://dl.acm.org/citation.cfm?id=1351542.1351793.

[9] W.-J. Lin and J.-C. Yen, "An integrating channel coding and cryptography design for OFDM based WLANs," in *IEEE 13th International Symposium on Consumer Electronics, 2009. ISCE '09*, May 2009, pp. 657–660.

[10] D. Reilly and G. Kanter, "Noise-enhanced encryption for physical layer security in an OFDM radio," in *IEEE Radio and Wireless Symposium, 2009. RWS '09*, Jan. 2009, pp. 344–347.

[11] H. Yamamoto, "On secret sharing communication systems with two or three channels," *IEEE Transactions on Information Theory*, vol. 32, no. 3, pp. 387–393, May 1986.

[12] ——, "Coding theorem for secret sharing communication systems with two noisy channels," *IEEE Transactions on Information Theory*, vol. 35, no. 3, pp. 572–578, May 1989.

[13] ——, "A coding theorem for secret sharing communication systems with two Gaussian wiretap channels," *IEEE Transactions on Information Theory*, vol. 37, no. 3, pp. 634–638, May 1991.

[14] Z. Li, R. Yates, and W. Trappe, "Secrecy capacity of independent parallel channels," in *Securing Wireless Communications at the Physical Layer*, R. Liu and W. Trappe, Eds. Springer US, 2010, pp. 1–18.

[15] Y. Liang, H. Poor, and S. Shamai, "Secure communication over fading channels," *IEEE Transactions on Information Theory*, vol. 54, no. 6, pp. 2470–2492, June 2008.

[16] E. Jorswieck and A. Wolf, "Resource allocation for the wire-tap multi-carrier broadcast channel," in *International Conference on Telecommunications, 2008*, June 2008.

[17] E. Jorswieck and S. Gerbracht, "Secrecy rate region of downlink OFDM systems: Efficient resource allocation," in *Proc. 14th Int. OFDM-Workshop(InOWo)*, Hamburg, Germany, Sept. 2009.

[18] G. R. Tsouri and D. Wulich, "Securing OFDM over wireless time-varying channels using subcarrier overloading with joint signal constellations," *EURASIP J. Wirel. Commun. Netw.*, vol. 2009, pp. 6:1–6:18, Mar. 2009. [Online]. Available: http://dx.doi.org/10.1155/2009/437824

[19] Z. Han, N. Marina, M. Debbah, and A. Hjorungnes, "Improved wireless secrecy rate using distributed auction theory," in *International Conference on Mobile Ad-hoc and Sensor Networks*, Dec. 2009, pp. 442–447.

[20] R. Zhang, L. Song, Z. Han, and B. Jiao, "Improve physical layer security in cooperative wireless network using distributed auction games," in *Computer Communications Workshops (INFOCOM WKSHPS), IEEE Conference on*, April 2011, pp. 18–23.

[21] D. Ng, E. Lo, and R. Schober, "Energy-efficient resource allocation for secure OFDM systems," *IEEE Transactions on Vehicular Technology*, vol. 61, no. 6, pp. 2572–2585, July 2012.

[22] C. Jeong and I.-M. Kim, "Optimal power allocation for secure multicarrier relay systems," *IEEE Transactions on Signal Processing*, vol. 59, no. 11, pp. 5428–5442, Nov. 2011.

[23] D. Ng, E. Lo, and R. Schober, "Secure resource allocation and scheduling for OFDM decode-and-forward relay networks," *IEEE Trans. Wireless Comm.*, vol. 10, no. 10, pp. 3528–3540, Oct. 2011.

[24] H. Xing, H. Zhang, Z. Ding, X. Chu, and A. Nallanathan, "Secure resource allocation for OFDMA two-way relay networks via cooperative jamming," in *International ICST Conference on Communications and Networking in China (CHINACOM)*, Aug. 2011.

[25] H. Zhang, H. Xing, X. Chu, A. Nallanathan, W. Zheng, and X. Wen, "Secure resource allocation for OFDMA two-way relay networks," in *Global Telecommunications Conference (GLOBECOM 2012), 2012 IEEE*, Dec.

[26] M. Kobayashi, M. Debbah, and S. Shamai, "Secured communication over frequency-selective fading channels: A practical vandermonde precoding," *EURASIP J. Wirel. Commun. Netw.*, vol. 2009, pp. 2:1–2:19, Mar. 2009. [Online]. Available: http://dx.doi.org/10.1155/2009/386547.

[27] F. Renna, N. Laurenti, and H. Poor, "Physical-layer secrecy for OFDM transmissions over fading channels," *IEEE Transactions on Information Forensics and Security*, vol. 7, no. 4, pp. 1354–1367, Aug. 2012.

[28] A. Khisti and G. Wornell, "Secure transmission with multiple antennas, Part ii: The mimome wiretap channel," *IEEE Transactions on Information Theory*, vol. 56, no. 11, pp. 5515–5532, Nov. 2010.

[29] F. Oggier and B. Hassibi, "The secrecy capacity of the mimo wiretap channel," *IEEE Transactions on Information Theory*, vol. 57, no. 8, pp. 4961–4972, Aug. 2011.

[30] T. Liu and S. Shamai, "A note on the secrecy capacity of the multiple-antenna wiretap channel," *IEEE Transactions on Information Theory*, vol. 55, no. 6, pp. 2547–2553, June 2009.

[31] R. Bustin, R. Liu, H. V. Poor, and S. Shamai, "An MMSE approach to the secrecy capacity of the MIMO Gaussian wiretap channel," *EURASIP J. Wirel. Commun. Netw.*, vol. 2009, pp. 3:1–3:8, Mar. 2009. [Online]. Available: http://dx.doi.org/10.1155/2009/370970.

[32] S. K. Leung-Yan-Cheong and M. E. Hellman, "The Gaussian wire-tap channel," *IEEE Transactions on Information Theory*, vol. 24, no. 4, pp. 451–456, July 1978.

[33] P. K. Gopala, L. Lai, and H. E. Gamal, "On the secrecy capacity of fading channels," *IEEE Transactions on Information Theory*, vol. 54, no. 10, pp. 4687–4698, Oct. 2008.

[34] A. D. Wyner, "The wire-tap channel," *Bell Syst. Tech. J.*, vol. 54, no. 8, pp. 1355–1367, Oct. 1975.

[35] W. Yu and R. Lui, "Dual methods for nonconvex spectrum optimization of multicarrier systems," *IEEE Transactions on Communications*, vol. 54, no. 7, pp. 1310–1322, July 2006.

[36] S. Boyd and L. Vandenberghe, *Convex Optimization*. Cambridge University Press, 2004.

[37] X. Wang, M. Tao, and Y. Xu, "Analysis of false CSI attack by a malicious user in OFDMA networks," in *International Conference on Wireless Communications and Signal Processing (WCSP 2012)*, Oct. 2012.

Chapter 9

The Application of Cooperative Transmissions to Secrecy Communications

Zhiguo Ding
Newcastle University

Mai Xu
Beihang University

Kanapathippillai Cumanan
Newcastle University

Fei Liu
Jiangnan University

The exploitation of user cooperation has been recognized as an important technique to improve the robustness of secure transmissions, where various studies have shown that a careful use of relaying can improve the receptional reliability of the legitimate receivers and decrease eavesdropping capabilities at the same time. In this chapter, a detailed review will be provided to illustrate how to integrate relay transmissions with secrecy communications. Simpler scenarios with single-antenna nodes will be investigated first, where two secure transmission strategies, cooperative jamming and relay chatting, will be focused particularly. Then scenarios with multiple-antenna nodes will be considered, and the design of precoding becomes of interest, where the cases with and without channel information at the transmitters will be discussed separately.

9.1 Introduction

In this chapter, the application of cooperative transmissions to secrecy communications will be considered, where the use of relaying can be motivated as the following. Recall that for a simple noncooperative secrecy communication scenario with one source, one destination, and one eavesdropper, the maximum achievable rate for the wiretap channel is [1]

$$\mathcal{I} = [\mathcal{I}_M - \mathcal{I}_E]^+, \tag{9.1}$$

where $[x]^+ = \max\{0, x\}$, $\mathcal{I}_M = \log[1 + \rho|h_M|^2]$ is the data rate available at the main channel, ρ is the transmit signal-to-noise ratio (SNR), h_M is the attenuation coefficient of the main channel, and $\mathcal{I}_E$ is defined similar to $\mathcal{I}_M$. There are a few interesting observations which can be obtained from the secrecy capacity shown in (9.1). Specifically, the secrecy rate is smaller than the one for nonsecrecy scenarios, due to the existence of the eavesdropper. Furthermore, the value of the secrecy capacity is not only a function of the main channel coefficient but also affected by the channel between the eavesdropper and the source, which makes the secrecy capacity more dynamic and unpredictable. However, in most communication systems, end users have specific requirements for the quality of service. For example, to use many peer-to-peer communications, such as Skype, it is normally expected that voice-only calls require a communication link with at least 30 kilobits per second (kbps), and video calls require a channel with minimal 128 kbps for uploading and downloading. Therefore it is necessary to design spectrally efficient secure transmission protocols and meet the predefined quality of service, which has received a lot of attention recently.

Specifically the use of relays in the context of secrecy communications has been recognized as a promising method to increase the secrecy rate, particularly for the scenario with single-antenna nodes, where each node has limited capability and degrees of freedom [3]. Various secrecy transmission protocols have been developed to combat the challenge that the achievable secrecy rate could be close to zero if the channel connection between the legitimate transceiver is poor [4,5]. For some communication systems, the eavesdropper is not cooperative and hence the legitimate transceiver does not have access to the eavesdropper's channel state information (CSI), which poses a more demanding challenge to implement secrecy transmission in practice.

In this chapter, we first focus on the secrecy communication scenario with single-antenna nodes, where two popular secrecy transmission protocols, cooperative jamming and relay chatting, will be discussed. Analytical results can be demonstrated that cooperative jamming is ideal to stop an untrusted relay intercepting source messages, but could result in information leaking to an external eavesdropper. On the other hand, relay chatting is a secure transmission protocol for the case with external eavesdroppers. The second part of this chapter is to study the design of secrecy transmission protocols for multiple-input multiple-output (MIMO) secrecy communication scenarios. The key idea is either to hide the source information from the eavesdropper by putting it into the null space of the channel matrix between the source and the eavesdropper, or to generate artificial noise not visible to the legitimate receiver by putting it into the null space of the channel matrix between the legitimate transceiver, dependent on whether the eavesdropper's CSI is available. Simulation results will be also provided to demonstrate the performance of these discussed protocols.

9.2   When All Nodes Are Equipped with a Single Antenna

In this section, we first focus on the scenario where all nodes are equipped with a single antenna, either due to the small size of communication devices or limited computational complexity. Two popular secrecy transmission protocols which exploit the use of relays and avoid the requirement of the eavesdropper's CSI will be considered. Specifically, cooperative jamming will be studied first and then relay chatting will be considered. Different to the secrecy transmission schemes assuming the eavesdropper's CSI, both protocols ask the relay to transmit artificial noise. While such jamming information can significantly reduce the receive capability of unintended receivers, it can also cause performance degradation at the legitimate receiver. The difference between cooperative jamming and relay chatting is how to minimize the impact of jamming information on the legitimate receiver, as discussed in

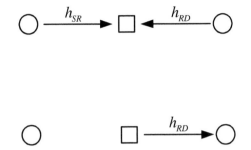

Figure 9.1: The protocol for the cooperative jamming scheme.

the following sections.

9.2.1 Cooperative Jamming

Cooperative jamming was first developed with the motivation that multihop transmissions are preferred in many wireless networks, and it is challenging to keep the source information secure from those intermediate nodes which are untrusted nodes. For example, consider the two-hop transmission scenario with one source, one destination, and one relay. Consider that such a relay is supposed to be cooperative but not to be trusted, which means that the relay will help the source to forward the source message to the destination but also try to intercept the message. To simplify the description, we assume that the relay will use the amplify-forward strategy to relay the source message to the destination. Throughout this chapter, wireless channels are assumed to be complex Gaussian distributed with zero mean and unit variance $(C \sim \mathcal{N}(0,1))$, and time division duplexing is used for simplicity and to exploit the channel reciprocity. Furthermore, the half-duplexing constraint is applied to all nodes. For both cooperative protocols discussed in this chapter, the direct link between the legitimate transmitter and receiver will not be used. Such a two-hop transmission strategy is due to the practical constraints, such as shadowing effects and limited transmission power. In addition, the use of the relay in the context of secrecy transmissions is also helpful to improve secrecy rates. The transmission strategy of cooperative jamming can be obtained as an extension of the schemes proposed in [6–8]. Specifically, during the first time slot, both the source and destination transmit messages at the same time as shown in Figure 9.1, and therefore the relay will observe the mixture of two messages. Conventionally such a situation at the relay is not desirable since it makes it difficult for the relay to separate the mixture, which is perfect for the addressed secrecy communication scenario. At the end of the first time slot, the relay receives

$$y_R = h_{SR}s + h_{RD}x + n_R,$$

where s denotes the source message, x is the artificial noise transmitted by the destination to confuse the eavesdropper, n_R represents the additive Gaussian noise with power $\frac{1}{\rho}$, and h_{RD} denotes the channel between the relay and the destination.

During the second time slot, the relay will forward the mixture $s_R = \frac{y_{R_i}}{\beta}$ where β is the power normalization coefficient, i.e., $\beta = \sqrt{|h_{SR}|^2 + |h_{RD}|^2 + \frac{1}{\rho}}$. We will first focus on the detection at the relay. Due to the half-duplexing constraint, the relay cannot obtain any observation during the second time slot, and therefore the mutual information about s at the relay can be expressed as

$$\mathcal{I}_R = \frac{1}{2}\log\left(1 + \frac{|h_{SR}|^2}{|h_{RD}|^2 + \frac{1}{\rho}}\right),$$

where we assume that the transmission power at the source and the destination is the same. On the other hand, during the second time slot, the destination receives

$$y_D = \frac{1}{\beta}h_{RD}h_{SR}s + \frac{1}{\beta}h_{RD}^2 x + \frac{1}{\beta}h_{RD}n_R + n_D,$$

where n_D is the noise at the destination. Since x was generated by the destination, such artificial noise will not have any negative impact on the detection at the destination and the mutual information between the source message and destination can be expressed as

$$\begin{aligned} \mathcal{I}_D &= \frac{1}{2}\log\left(1 + \frac{\frac{1}{\beta^2}|h_{SR}|^2|h_{RD}|^2}{\frac{1}{\beta^2}|h_{RD}|^2\frac{1}{\rho} + \frac{1}{\rho}}\right) \\ &= \frac{1}{2}\log\left(1 + \frac{\rho^2|h_{SR}|^2|h_{RD}|^2}{\rho|h_{SR}|^2 + 2\rho|h_{RD}|^2 + 1}\right). \end{aligned} \tag{9.2}$$

Now the achievable secrecy rate for the addressed secrecy scenario with an untrusted relay can be expressed as

$$\begin{aligned} \mathcal{I}_1 &= [\mathcal{I}_D - \mathcal{I}_R]^+ \\ &\approx \frac{1}{2}\log\left(\frac{\rho|h_{SR}|^2|h_{RD}|^2}{|h_{SR}|^2 + 2|h_{RD}|^2}\right), \end{aligned} \tag{9.3}$$

where the approximation is obtained with the high SNR approximation due to the fact that the factor $\frac{|h_{SR}|^2}{|h_{RD}|^2 + \frac{1}{\rho}}$ in the expression of $\mathcal{I}_R$ is a constant at high SNRs. As can be observed from the above equation, the high SNR approximation of the secrecy rate is the same as the one for the nonsecrecy case. Therefore by carefully exploiting relaying and designing secrecy transmission protocols, it is possible to avoid performance loss in the context of secrecy communications. However, cooperative jamming can ensure that source messages are not intercepted by the relay, but such a secure performance is not possible if there is an external eavesdropper, as shown in the following.

9.2.1.1   When there Is an External Eavesdropper

Consider that in the addressed relaying scenario, there is an external relay which keeps listening over the two time slots. As a result, the signal model at the eavesdropper can be expressed as

$$\begin{bmatrix} y_{E1} \\ y_{E2} \end{bmatrix} = \begin{bmatrix} h_{SE} & h_{DE} \\ \frac{1}{\beta}h_{RE}h_{SR} & \frac{1}{\beta}h_{RE}h_{RD} \end{bmatrix} \begin{bmatrix} s \\ x \end{bmatrix} + \begin{bmatrix} n_{E1} \\ \hat{n}_{E2} \end{bmatrix}, \tag{9.4}$$

where $\hat{n}_{E2} = \frac{1}{\beta}h_{RE}n_R + n_{E2}$. An intuitive observation from the above equation is that the eavesdropper is able to separate the mixture of s and x. Specifically (9.4) is analog to a set of linear equations. There are two equations to solve two unknown variables, s and x, which are solvable.

To be more rigorous, the closed-form expression of SNR at the eavesdropper will be obtained as follows. Without loss of generality, we only focus on linear detection methods, which result in slight performance loss compared to maximum likelihood detection but are tractable. Specifically zero forcing approaches will be used due to simplicity, but it is noteworthy that zero forcing yields the same performance as minimum mean square error (MMSE) at high SNRs. Therefore applying the manipulation to get $\hat{y}_E = \frac{1}{\beta} h_{RE} h_{RD} y_{E1} - h_{DE} y_{E2}$ yields a signal model without the interference x as

$$\hat{y}_E = \frac{h_{RE}}{\beta}(h_{RD} h_{SE} - h_{DE} h_{SR})s + \frac{1}{\beta} h_{RE} h_{RD} n_{E1} - \frac{1}{\beta} h_{DE} h_{RE} n_R - h_{DE} n_{E2}.$$

After x is removed from the signal model, the mutual information between the source message and the eavesdropper can be obtained as

$$\mathcal{I}_E = \frac{1}{2} \log \left(1 + \frac{\rho^2 |h_{RE}|^2 |(h_{RD} h_{SE} - h_{DE} h_{SR})|^2}{\rho |h_{RE}|^2 (|h_{RD}|^2 + |h_{DE}|^2) + (\rho \eta + 1)|h_{DE}|^2} \right),$$

where $\eta = |h_{SR}|^2 + \rho |h_{RD}|^2$. Now when there is an external eavesdropper, the achievable secrecy rate for cooperative jamming is $\mathcal{I} = \mathcal{I}_D - \mathcal{I}_E$. At high SNR, such an achievable rate can be approximated as

$$\mathcal{I}_2 \approx \frac{1}{2} \log \left(1 + \frac{\frac{|h_{SR}|^2 |h_{RD}|^2}{|h_{SR}|^2 + 2|h_{RD}|^2}}{\frac{|h_{RE}|^2 |(h_{RD} h_{SE} - h_{DE} h_{SR})|^2}{|h_{RE}|^2 (|h_{RD}|^2 + |h_{DE}|^2) + \eta |h_{DE}|^2}} \right), \qquad (9.5)$$

which is a constant and not a function of ρ. To get more insightful understanding about the secrecy rates shown in (9.3) and (9.5), the use of outage probability will be used as follows. When there is no external eavesdropper, the outage probability can be expressed as

$$P(\mathcal{I}_1 < R) \approx P \left(\frac{|h_{SR}|^2 |h_{RD}|^2}{|h_{SR}|^2 + 2|h_{RD}|^2} < \frac{2^{2R}}{\rho} \right), \qquad (9.6)$$

and with an external eavesdropper the outage probability can be shown as

$$P(\mathcal{I}_2 < R) \approx P \left(\frac{\frac{|h_{SR}|^2 |h_{RD}|^2}{|h_{SR}|^2 + 2|h_{RD}|^2}}{\frac{|h_{RE}|^2 |(h_{RD} h_{SE} - h_{DE} h_{SR})|^2}{|h_{RE}|^2 (|h_{RD}|^2 + |h_{DE}|^2) + \eta |h_{DE}|^2}} < 2^{2R} \right), \qquad (9.7)$$

where R is defined as a targeted data rate. Comparing the outage probabilities in the above equations, we can find that the outage probability $P(\mathcal{I}_1 < R)$ can be reduced to zero for any R given $\rho \to \infty$. On the other hand, $P(\mathcal{I}_2 < R)$ will be a constant and not a function of SNR, which means that for any R larger than $\frac{\frac{|h_{SR}|^2 |h_{RD}|^2}{|h_{SR}|^2 + 2|h_{RD}|^2}}{\frac{|h_{RE}|^2 |(h_{RD} h_{SE} - h_{DE} h_{SR})|^2}{|h_{RE}|^2 (|h_{RD}|^2 + |h_{DE}|^2) + \eta |h_{DE}|^2}}$, the probability of the event that the eavesdropper can intercept the source message is one, no matter how large SNR is.

9.2.2 Relay Chatting

Relay chatting was first developed in [9] for the scenario with one source–destination pair and later extended to two-way relaying scenarios in [10]. Similar to cooperative jamming, the key idea of relay chatting is also to generate artificial noise in order to confuse the eavesdropper. However, different to cooperative jamming, the success of relay chatting relies on the selection of relaying and jamming nodes. Specifically a node is selected to transmit jamming information only if its connection to the legitimate receivers is poor, which is important to suppress the negative effect of artificial noise at the legitimate receivers. In

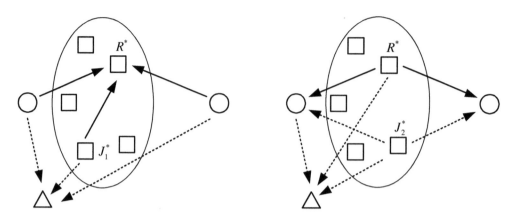

Figure 9.2: The protocol for the relay chatting scheme.

this section, the two-way relaying communication scenario will be considered since the design of relay/jammer selection for such a scenario is quite representative, whereas relay chatting for unidirectional cases can be obtained in a straightforward manner. Consider a bidirectional communication scenario where two sources want to exchange information with each other as shown in Figure 9.2. In addition to two sources, there are K relays which are to help the information exchange between the sources, where there is also an external eavesdropper which tries to intercept the source information. The direct communication link between the two sources will not be used, either due to the shadowing effect or path loss. Similar to the previous section, it is assumed that all nodes are equipped with a single antenna and constrained by the half-duplexing assumption. Furthermore, it is assumed that the eavesdropper has access to the global CSI, but the others only have access to all CSI information except the ones related to the eavesdropper.

The key idea of relay chatting is that the relays are not only acting as a conventional relay by forwarding the source information but also by acting as jamming nodes by generating artificial interference. The secrecy transmission consists of two phases, where there are at least two transmitters during each phase. During each time slot, relays perform jamming, where the beamformer at the relays is designed in a way that the legitimate receivers cannot hear such jamming signals. Particularly, during the first time slot, both sources broadcast their own messages, similar to the idea of physical layer network coding [11]. One of the K relays is chosen to perform as the receiver during the first time slot and listens to the source transmissions. There are a few choices for the criteria of relay selection established in the context of nonsecrecy communication scenarios. Without loss of generality, the max-min criterion is used here [12, 22] and the best relay, denoted by R^*, is chosen based on the following:

$$R^* = \operatorname{argmin} \max \left\{ \min\{|h_1|^2, |g_1|^2\}, \cdots, \min\{|h_K|^2, |g_K|^2\} \right\}, \qquad (9.8)$$

where h_k and g_k denote the channels between the k-th relay and the sources, and $|\cdot|^2$ denotes the absolute square of a complex number. This means that the best relay R^* has a good connection to both source nodes. Another choice of the relay selection criterion is to use the harmonic mean of h_k and g_k, i.e.,

$$R^* = \operatorname{argmin} \max \left\{ \frac{|h_1|^2 |g_1|^2}{|h_1|^2 + |g_1|^2}, \cdots, \frac{|h_K|^2 |g_K|^2}{|h_K|^2 + |g_K|^2} \right\}. \qquad (9.9)$$

An interesting observation is that the use of the above two criteria will yield the same solution.

Due to the existence of the eavesdropper, a relay node is also invited to perform as a jammer and transmit artificial noise in order to confuse the eavesdropper. Apparently such jamming information will also degrade the receiver capability of the best relay R^*. An important step of relay chatting is to select a jamming node during the first phase, denoted by J_1^*, based on the following selection criterion,

$$J_1^* = \operatorname{argmin}\min \left\{ |g_{R^*,1}|^2, \cdots, |g_{R^*,K-1}|^2 \right\}, \tag{9.10}$$

where $g_{R^*,k}$ denotes the channel between the k-th relay and R^*. With such criteria, the legitimate receiver during the first phase, R^*, has the best connection to both sources, and the artificial noise generated by the node J_1^* has the minimum effect at the node R^*. Another way to illustrate is to consider some extreme cases. For example, when there are sufficient relays, it is possible to choose such a jammer whose connection to the legitimate receiver R^* is close to zero, i.e., $|g_{R^*,J_1^*}|^2 \to 0$, which can ensure that the detection of the source information at R^* is free from the artificial noise.

Given the chosen relay and jammer, during the first time slot, the information-forwarding relay receives the following:

$$y_{R^*} = h_{R^*}s_1 + g_{R^*}s_2 + g_{R^*,J_1^*}x_1 + w_{R^*}, \tag{9.11}$$

where s_i is the information-bearing message from the i-th source, x_1 is the artificial noise, and w_{R^*} is the additive white Gaussian noise. Again take $K \to \infty$ and $|g_{R^*,J_1^*}|^2 \to 0$ as an example. For such an extreme example, the above signal model can be simplified as

$$y_{R^*} = h_{R^*}s_1 + g_{R^*}s_2 + w_{R^*}. \tag{9.12}$$

At the end of the first time slot, the relay R^* will generate the following based on its observation:

$$x_{R^*} = \frac{h_{R^*}s_1 + g_{R^*}s_2 + g_{R^*,J_1^*}x_1 + w_{R^*}}{\sqrt{|h_{R^*}|^2 + |g_{R^*}|^2 + |g_{R^*,J_1^*}|^2 + \frac{1}{\rho}}}. \tag{9.13}$$

Similar to physical layer network coding, during the second time slot, the relay, R^*, broadcasts the amplified version of its previous observation, x_{R^*}. At the same time, in order to confuse the eavesdropper, another relay is chosen to transmit jamming signals. Specifically the new jamming relay, denoted as J_2^*, is chosen by using the following criterion:

$$J_2^* = \operatorname{argmin}\min \left\{ \max\{|h_1|^2, |g_1|^2\}, \cdots, \max\{|h_K|^2, |g_K|^2\} \right\}. \tag{9.14}$$

It is interesting to observe that the selection criteria of R^* and J_2^* are quite similar. Note that the two jamming relays, J_1^* and J_2^*, are selected in different ways due to the following reason. The jamming node during the first phase is selected to ensure that the jamming information from such a node will not cause strong interference to R^*, the intended receiver during the first time slot. On the other hand, J_2^* is selected to minimize the effects of artificial interference at both source nodes, the intended receivers during the second time slot. It might be possible that the same relay is selected as the jamming node during the two time slots.

In the following the closed form expression of SNR at the legitimate destinations will be obtained and the impact of relay selection on the system performance will be analyzed.

Without loss of generality, we focus on the first source. After removing its own messages, the first source receives the following observation,

$$y_1 = h_{R^*} \frac{g_{R^*} s_2 + g_{R^*, J_1^*} x_1 + w_{R^*}}{\sqrt{|h_{R^*}|^2 + |g_{R^*}|^2 + |g_{R^*, J_1^*}|^2 + \frac{1}{\rho}}} + h_{J_2^*} x_2 + w_1, \tag{9.15}$$

where x_2 is the jamming signal transmitted during the second time slot and w_1 is the noise at the first source. Different to the secrecy protocols introduced in the previous section, the artificial noise will not be removed completely and therefore the signal-to-interference-and-noise ratio (SINR) is of interest

$$SINR_1 = \frac{|h_{R^*}|^2 |g_{R^*}|^2}{|h_{R^*}|^2 |g_{R^*, J_1^*}|^2 + \frac{1}{\rho}|h_{R^*}|^2 + (|h_{R^*}|^2 + |g_{R^*}|^2 + |g_{R^*, J_1^*}|^2 + \frac{1}{\rho})(|g_{J_2^*}|^2 + \frac{1}{\rho})}. \tag{9.16}$$

And the mutual information for the second source to detection of the first source's information can be written as

$$\mathcal{I}_{2\to1}^M = \log\left(1 + \frac{|h_{R^*}|^2 |g_{R^*}|^2}{|h_{R^*}|^2 |g_{R^*, J_1^*}|^2 + \frac{1}{\rho}|h_{R^*}|^2 + (|h_{R^*}|^2 + |g_{R^*}|^2 + |g_{R^*, J_1^*}|^2 + \frac{1}{\rho})(|g_{J_2^*}|^2 + \frac{1}{\rho})}\right). \tag{9.17}$$

An intuitive observation from the above equation is that provided a sufficient number of relays, the value of the data rate $\mathcal{I}_{2\to1}^M$ can be arbitrarily large. Specifically, when $K \to \infty$, it is possible to have $|g_{R^*}|^2$ and $|h_{R^*}|^2$ approaching to infinity, as well as the channel coefficients related to the jamming nodes close to zero. As a result, the numerator of the fraction in the above equation becomes infinity, and its denominator reduces to zero, which means the overall value of the data rate becomes infinity.

On the other hand, over the two time slots, the eavesdropper receives two observations

$$y_{E1} = h_E s_1 + g_E s_2 + g_{E, J_1^*} x_1 + w_{E1} \tag{9.18}$$

$$y_{E2} = h_{E,R^*} \frac{h_{R^*} s_1 + g_{R^*} s_2 + g_{R^*, J_1^*} x_1 + w_{R^*}}{\sqrt{|h_{R^*}|^2 + |g_{R^*}|^2 + |g_{R^*, J_1^*}|^2 + \frac{1}{\rho}}} + h_{E, J_2^*} x_2 + w_{E2},$$

where h_E and g_E are the channels between the eavesdropper and the sources, g_{E, J_1^*} denotes the channel between the eavesdropper and J_1^*, h_{E,R^*} and h_{E,J_2^*} are defined similarly, and w_{Ei} denotes the noise at the eavesdropper. Define $\mathcal{I}_{2\to1}^E$ as the mutual information for the eavesdropper to detect messages from the second source based on the signal model in (9.18). Therefore the achievable secrecy rate for the message from the second source can be expressed as $\mathcal{I}_{2\to1} = [\mathcal{I}_{2\to1}^M - \mathcal{I}_{2\to1}^E]^+$. As discussed before, when $K \to \infty$, an intuitive observation is that $\mathcal{I}_{2\to1}^M \to \infty$. On the other hand the mutual information at the eavesdropper will become a constant since the relay selection does not have much effect on the channel coefficients related to the eavesdropper, which means that the value of the secrecy rate $\mathcal{I}_{2\to1}$ can be improved by increasing the number of relays. Such a conclusion can be proved by using the following lemma:

Lemma 9.2.1. *For any given targeted data rate $\mathcal{R}$, the outage probability of the achievable secrecy rate for the addressed transmission protocol approaches zero*

$$P\left([\mathcal{I}_{2\to1}^M - \mathcal{I}_{2\to1}^E]^+ < \mathcal{R}\right) \to 0,$$

for $K \to \infty$ and $\rho \to \infty$.

Proof of Lemma 9.2.1. *Due to the space limitation, only the sketch of the proof will be provided here and the reader can find more details from [10]. Ideally the first step of the proof is to find the density function of $\mathcal{I}_{2\to1}^M$ whose expression is quite complicated. So instead the lower bound of the $\mathcal{I}_{2\to1}^M$ can be obtained as follows,*

$$SINR_1 \approx \frac{|h_{R^*}|^2|g_{R^*}|^2}{|h_{R^*}|^2|g_{R^*,J_1^*}|^2 + |g_{J_2^*}|^2 \left(|h_{R^*}|^2+|g_{R^*}|^2+|g_{R^*,J_1^*}|^2\right)} \geq \frac{\frac{|h_{R^*}|^2|g_{R^*}|^2}{(|h_{R^*}|^2+|g_{R^*}|^2)}}{|g_{R^*,J_1^*}|^2 + |g_{J_2^*}|^2}$$

(9.19)

for $K \to \infty$. By using the upper and lower bounds of the harmonic mean, the mutual information $\mathcal{I}_{2\to1}^M$ can be further bounded as

$$\mathcal{I}_{2\to1}^M \geq \log\left(1 + \frac{\frac{1}{2}\min\{|h_{R^*}|^2, |g_{R^*}|^2\}}{|g_{R^*,J_1^*}|^2 + |g_{J_2^*}|^2}\right).$$

(9.20)

Such an expression is of interest since it is quite close to the used criteria and the density functions of its numerator and denominator can be bound. The second step of the proof is to find the upper bound of the mutual information at the eavesdropper as

$$\mathcal{I}_{2\to1}^E \leq \log\left(1 + \frac{|g_E|^2 + |h_{E,R^*}|^2}{\min\{|g_{E,J_1^*}|^2, |h_{E,J_2^*}|^2\}}\right),$$

(9.21)

where we have used the high SNR approximation to remove factors related to noise. Therefore, the outage probability addressed in the lemma can be upper bounded as

$$P_{out} \leq P\left(\frac{\min\{|h_{R^*}|^2, |g_{R^*}|^2\}}{|g_{R^*,J_1^*}|^2 + |g_{J_2^*}|^2} < \tau\right) \leq P\left(\frac{\min\{|h_{R^*}|^2, |g_{R^*}|^2\}}{|g_{R^*,J_1^*}|^2 + \max\{|h_{J_2^*}|^2, |g_{J_2^*}|^2\}} < \tau\right),$$

where $\tau = 2^{\mathcal{R}}\left(1 + \frac{|g_E|^2 + |h_{E,R^*}|^2}{\min\{|g_{E,J_1^*}|^2, |h_{E,J_2^*}|^2\}}\right) - 1$, and the second inequality is to link the addressed probability to the user selection criterion in (9.14). Following some algebra manipulations, it can be shown that the above upper bound reduces to zero when $K \to \infty$. And the proof for the lemma is completed. ∎

9.2.2.1 Remarks

The cooperative jamming scheme cannot work in the context of bidirectional communication scenarios, since it requires the destination to act as a jamming node. For the unidirectional scenarios, cooperative jamming can be viewed as a special case of relay chatting by predefining the destination as the jamming node during the first time slot and not using jamming during the second time slot. Another noteworthy observation is that by setting $\mathcal{R} = 0$, Lemma 9.2.1 implies the following,

$$P\left(\mathcal{I}_{2\to1}^M < \mathcal{I}_{2\to1}^E\right) \to 0,$$

which means that given a sufficient number of relays, the use of the relay chatting protocol can ensure that the secrecy rate is always larger than zero.

The performance of the relay chatting secrecy transmission protocol can be further evaluated by using computer simulation results as shown in Figure 9.3. As can be observed from Figure 9.3(a), the outage probability of the relay chatting scheme can be reduced by increasing the number of relays, which was matched to Lemma 9.2.1. Although a larger targeted data rate causes a worse outage probability, increasing the number of relays to

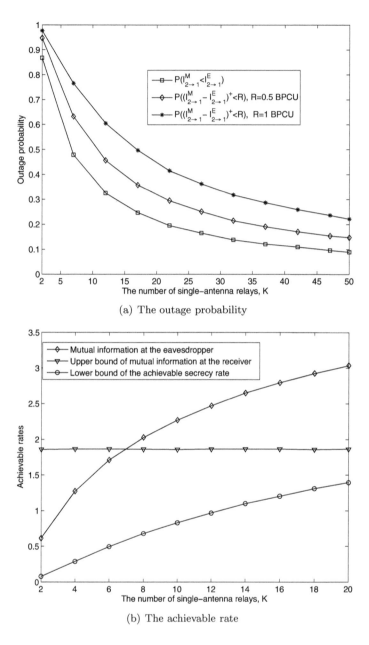

(a) The outage probability

(b) The achievable rate

Figure 9.3: The performance for the secrecy transmission scheme with single-antenna nodes. BPCU stands for bits per channel use. The transmit SNR is $\rho = 30$ dB.

infinity can reduce the outage probability to zero, no matter how large the targeted data is. Therefore, by increasing the number of relays, any predefined quality of service can be met. Figure 9.3(a) provides more details about the impact of the relay selection on the data rate at the legitimate receiver and eavesdropper. Specifically using more relays will not decrease the value of the data rate at the eavesdropper, and the increase of the secrecy rate is due to the improvement of the rate at the legitimate receiver. In order to reduce the mutual information at the eavesdropper to zero, MIMO techniques can be used as discussed in the following section.

9.3 MIMO Relay Secrecy Communication Scenarios

As shown in [2,13], for secrecy scenarios with single-antenna nodes, nonzero secrecy rates can only be achieved when the legitimate transceiver pair have a better channel condition than the eavesdropper. To improve secrecy rates, it is important to combine multiple antenna techniques with secrecy transmissions, where extra degrees of freedom can be obtained by exploiting the spatial diversity [14–16]. For the scenarios with one source, one destination, and one eavesdropper, the authors of [14,15] have considered some asymmetric setups where three nodes have different antennas. For example, for the so-called MISOME case, it is considered that the source and eavesdropper are equipped with multiple antennas, and the destination has a single antenna. The work in [16] considered the scenario where all nodes have multiple antennas, and their study has shown that the following secrecy rate is still achievable and maximum:

$$\mathcal{I} = \left[\log \det(\mathbf{I} + \mathbf{H}_M \mathbf{Q} \mathbf{H}_M^H) - \log \det(\mathbf{I} + \mathbf{H}_E \mathbf{Q} \mathbf{H}_E^H)\right]^+,$$

where $\mathbf{H}_M$ is the channel matrix between the legitimate transceivers, $\mathbf{H}_E$ is the channel matrix between the source and the eavesdropper, and $\mathbf{Q}$ is the covariance matrix of the source messages. An important conclusion made in [16] is that the rank of the covariance matrix $\mathbf{Q}$ will be smaller than the number of transmit and receive antennas, which implies that a loss of degrees of freedom is inevitable in order to avoid the source information being intercepted by the eavesdropper.

Such a loss of degrees of freedom motivates the application of cooperative transmissions to secrecy communications, where the relays can be exploited as an extra dimension to exploit spatial diversity. In practice, the use of relaying is also attractive due to the following reasons. In many communication systems, it is common that the destination is located far away from the source, where multihop transmission has been shown more preferable than direct transmission in terms of power consumption and interference management. In addition, in the context of secrecy communications, the use of cooperative relays is particularly beneficial. For example, the cooperative techniques discussed in the previous sections, such as cooperative jamming and relay chatting, can be applied to significantly improve the achievable secrecy rates and reduce the reception capability of the eavesdropper [6,17].

9.3.1 When CSI of Eavesdroppers Is Known

In the first part of this section we focused on the case that all nodes in the scenario have access to the global CSI, which means that the legitimate transceiver knows the CSI of the eavesdropper [15,16]. Such an assumption can be justified as follows. Consider a digital video broadcasting system with one base station and multiple users, each of which subscribes to a different type of service. Each user could be a potential eavesdropper, and it is important to stop a user from intercepting a service that this user does not subscribe to. But each user in such a system is cooperative in the sense that he/she will inform the base station of his/her CSI, in order to experience a stable service. The design of secrecy transmission protocols for the case without knowing the eavesdropper's CSI will be discussed in the following section. For the addressed cooperative MIMO secrecy communication scenario, the eavesdropper is equipped with M antennas and the other nodes, including the source, the destination, and all L relays, are equipped with N antennas as shown in Figure 9.4. It is assumed that $N > M$, which means that the legitimate transceiver has better capability than the eavesdropper. When $N \leq M$, it is still possible to achieve secrecy transmission by inviting relays to generate artificial noise [17,18], as shown in the following section. Without loss of generality, the decode-forward strategy will be used at the relay [19].

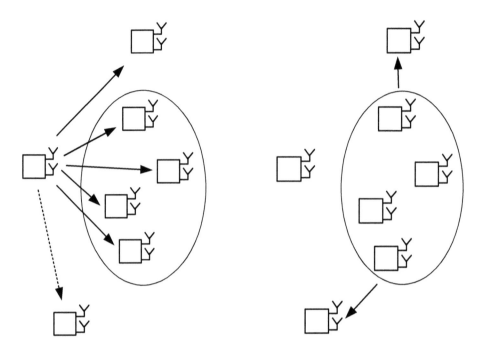

Figure 9.4: The protocol for MIMO relay chatting.

The cooperative secrecy transmission consists of two phases, similar to the ones discussed in the previous section. During the first time slot, the source will broadcast its messages to all other nodes, and assume that K out of the all L relays, named as qualified relays, can decode the source messages correctly. At the second time slot, depending on the predefined quality of service and affordable complexity, there are two options to utilize these K qualified relays. One is to use all the qualified relays by applying distributed beamforming, and the other one is to select a relay with the best connection to the destination. The use of beamforming can result in the best reception reliability, but with extra system overhead to coordinate user cooperation. The use of the best relay is a low-complex solution which can achieve the same diversity gain as beamforming, but with a slight loss of reception reliability. Both the destination and the eavesdropper will try to decode the source messages at the end of the cooperative transmissions.

Prior to transmissions, the source generates the information-bearing vector $\mathbf{s}$, where $\mathbf{s}$ is the $x \times 1$ information-bearing vector and x is a parameter which will be optimized later. During the first time slot, the source transmits the precoded version of the information-bearing message $\tilde{\mathbf{s}} = \mathbf{P}_s \mathbf{s}$, where $\mathbf{P}_s$ is the $N \times x$ precoding matrix, and the design of the precoding matrix is to maximize achievable secrecy data rate with the transmission power constraint $E\{\tilde{\mathbf{s}}^H \tilde{\mathbf{s}}\} = 1$. At the second time slot, we can either invite all qualified relays, or only use the best relay to forward the source messages $\mathbf{s}$. Hence the signal model at the destination and eavesdropper is written respectively as

$$\begin{bmatrix} \mathbf{y}_{M,1} \\ \mathbf{y}_{M,2} \end{bmatrix} = \begin{bmatrix} \mathbf{H}_M \mathbf{P}_s \\ \mathbf{G}_M \tilde{\mathbf{P}}_r \end{bmatrix} \mathbf{s} + \begin{bmatrix} \mathbf{n}_{M,1} \\ \mathbf{n}_{M,2} \end{bmatrix} \quad \& \quad \begin{bmatrix} \mathbf{y}_{E,1} \\ \mathbf{y}_{E,2} \end{bmatrix} = \begin{bmatrix} \mathbf{H}_E \mathbf{P}_s \\ \mathbf{G}_E \tilde{\mathbf{P}}_r \end{bmatrix} \mathbf{s} + \begin{bmatrix} \mathbf{n}_{E,1} \\ \mathbf{n}_{E,2} \end{bmatrix}, \qquad (9.22)$$

where $\mathbf{H}_M$ is the $N \times N$ source–destination channel matrix, $\mathbf{G}_{M,k}$ is the $N \times N$ kth relay–destination channel matrix, $\mathbf{P}_{M,k}$ is the precoding matrix for the k-th relay and the destination, and $\mathbf{G}_M = \begin{bmatrix} \mathbf{G}_{M,1} & \cdots & \mathbf{G}_{M,K} \end{bmatrix}$ and $\tilde{\mathbf{P}}_r = \begin{bmatrix} \mathbf{P}_{r,1} & \cdots & \mathbf{P}_{r,K} \end{bmatrix}$ if all relays have been used for joint beamforming, or $\mathbf{G}_M = \begin{bmatrix} \mathbf{G}_{M,best} \end{bmatrix}$ and $\tilde{\mathbf{P}}_r = \begin{bmatrix} \mathbf{P}_{r,best} \end{bmatrix}$ if only the best relay has been used. The channel matrices $\mathbf{H}_E$ and $\mathbf{G}_E$ have been defined similarly to $\mathbf{H}_M$ and $\mathbf{G}_M$. The design of the precoding matrices at the relay will be discussed later, with the following constraint:

$$\mathcal{E}\{\tilde{\mathbf{P}}_r \mathbf{s}\mathbf{s}^H \tilde{\mathbf{P}}_r^H\} = \begin{cases} K, & \text{all qualified relays are used} \\ 1, & \text{the best relay is used} \end{cases}.$$

Following [20], the achievable secrecy rate for such a cooperative MIMO protocol is written as the difference between the data rates at the destination and the eavesdropper:

$$\mathcal{I}_K = \log \frac{\det \left(\mathbf{I}_x + \rho \mathbf{P}_s^H \mathbf{H}_M^H \mathbf{H}_M \mathbf{P}_s + \rho \tilde{\mathbf{P}}_r^H \mathbf{G}_M^H \mathbf{G}_M \tilde{\mathbf{P}}_r \right)}{\det \left(\mathbf{I}_x + \rho \mathbf{P}_s^H \mathbf{H}_E^H \mathbf{H}_E \mathbf{P}_s + \rho \tilde{\mathbf{P}}_r^H \mathbf{G}_E^H \mathbf{G}_E \tilde{\mathbf{P}}_r \right)}. \tag{9.23}$$

As discussed previously, x is the number of information-bearing symbols, or the rank of the covariance matrix of the transmitted signals, $\mathbf{K} = E\{\tilde{\mathbf{s}}\tilde{\mathbf{s}}^H\}$. Recall that the point-to-point MIMO secrecy rate is written as

$$\mathcal{I}_{p-p} = \log \frac{\det \left(\mathbf{I}_N + \rho \mathbf{H}_M^H \mathbf{K} \mathbf{H}_M \right)}{\det \left(\mathbf{I}_N + \rho \mathbf{H}_E^H \mathbf{K} \mathbf{H}_E \right)}. \tag{9.24}$$

As pointed out in [20], $\mathbf{K}$ is typically rank deficient, which provides us an intuition that the precoding matrix is a tall matrix, e.g., $x < N$. In the following, analytical results will be developed to show how to select x and the precoding matrices $\mathbf{P}_s$ and $\tilde{\mathbf{P}}_r$.

The outage probability for the addressed secrecy scenario can be found by first considering whether there is any qualified relay as shown in the following:

$$P(\mathcal{I} \leq 2R) = \sum_{k=1}^{L} P\left(\mathcal{I}_k < 2R\right) P(K = k) + P\left(\mathcal{I}_{p-p} < 2^{2R} | E\right) P(K = 0) P(E)$$

$$+ P\left(\frac{\det \left(\mathbf{I}_N + \rho \mathbf{H}_M^H \mathbf{K} \mathbf{H}_M \right)}{\det \left(\mathbf{I}_N + \rho \mathbf{H}_E^H \mathbf{K} \mathbf{H}_E \right)} < 1 \right) P(K = 0), \tag{9.25}$$

where E denotes the event of $\frac{\det(\mathbf{I}_N + \rho \mathbf{H}_M^H \mathbf{K} \mathbf{H}_M)}{\det(\mathbf{I}_N + \rho \mathbf{H}_E^H \mathbf{K} \mathbf{H}_E)} > 1$. Note that in the above equation, for notation simplicity, only the event of $\mathcal{I}_k \geq 0$ is considered, which is true for the precoding design shown in the following. The evaluation of the above outage probability will be determined by the choice of the precoding matrices. Due to the space limitation, only the scheme using the best qualified relay is shown in the following, and more details about distributed beamforming and best relay selection can be found in [21].

9.3.1.1 MIMO Secrecy Cooperative Transmission Based on Relay Selection

Recall that the precoding matrix during the second phase will become $\tilde{\mathbf{P}}_r = \mathbf{P}_{r,best}$. In order to find the best relay, first consider that with K qualified relays, the use of n-th relay yields the secrecy rate as

$$\mathcal{I}_{K,n} = \log \frac{\det \left(\mathbf{I}_x + \rho \mathbf{P}_s^H \mathbf{H}_M^H \mathbf{H}_M \mathbf{P}_s + \rho \mathbf{P}_{r,n}^H \mathbf{G}_{M,n}^H \mathbf{G}_{M,n} \mathbf{P}_{r,n} \right)}{\det \left(\mathbf{I}_x + \rho \mathbf{P}_s^H \mathbf{H}_E^H \mathbf{H}_E \mathbf{P}_s + \rho \mathbf{P}_{r,n}^H \mathbf{G}_{E,n}^H \mathbf{G}_{E,n} \mathbf{P}_{r,n} \right)}, \forall n \in \{1, \cdots, K\}. \tag{9.26}$$

The problems of relay selection and precoding design can be jointly considered by formulating the following optimization problem:

$$\underset{n,\mathbf{P}_s,\mathbf{P}_{r,n}}{\text{argmin max}} \quad \mathcal{I}_{K,n} = \log \frac{\det\left(\mathbf{I}_x + \rho \mathbf{P}_s^H \mathbf{H}_M^H \mathbf{H}_M \mathbf{P}_s + \rho \mathbf{P}_{r,n}^H \mathbf{G}_{M,n}^H \mathbf{G}_{M,n} \mathbf{P}_{r,n}\right)}{\det\left(\mathbf{I}_x + \rho \mathbf{P}_s^H \mathbf{H}_E^H \mathbf{H}_E \mathbf{P}_s + \rho \mathbf{P}_{r,n}^H \mathbf{G}_{E,n}^H \mathbf{G}_{E,n} \mathbf{P}_{r,n}\right)}. \quad (9.27)$$

$$s.t. \quad trace\{\mathbf{P}_s^H \mathbf{P}_s\} = 1$$

$$trace\{\mathbf{P}_{r,n}^H \mathbf{P}_{r,n}\} = 1 \quad \forall n \in \{1, \cdots, K\}.$$

It is difficult to find the solution of the addressed maximization as its objective function is too complicated. The following proposition provides an approximation for the optimization problem at high SNR.

Proposition 9.3.1. [21] *At high SNR, the optimization problem in (9.27) is asymptotically equivalent to*

$$\underset{n,\tilde{\mathbf{X}}_{s,2},\tilde{\mathbf{X}}_{n,2}}{\text{argmin max}} \quad \mathcal{I}_{K,n} \approx \log \det\left(\tilde{\mathbf{X}}_{s,2}^H \tilde{\mathbf{H}}_{M,2}^H \tilde{\mathbf{H}}_{M,2} \tilde{\mathbf{X}}_{s,2} + \tilde{\mathbf{X}}_{n,2}^H \tilde{\mathbf{G}}_{n,2}^H \tilde{\mathbf{G}}_{n,2} \tilde{\mathbf{X}}_{n,2}\right) \quad (9.28)$$

$$s.t. \quad trace\{\tilde{\mathbf{X}}_{s,2} \tilde{\mathbf{X}}_{s,2}^H\} = 1 \quad \& \quad trace\{\tilde{\mathbf{X}}_{n,2} \tilde{\mathbf{X}}_{n,2}^H\} = 1 \quad \forall n \in \{1, \cdots, K\}.$$

where $\tilde{\mathbf{H}}_M = \mathbf{H}_M \mathbf{U}_s$, $\tilde{\mathbf{X}}_s = \mathbf{U}_s^H \mathbf{X}_s$, $\tilde{\mathbf{H}}_{M,2}$ *is the* $N \times (N-M)$ *right submatrix of* $\tilde{\mathbf{H}}_M$, $\tilde{\mathbf{X}}_{s,2}$ *is the* $(N-M) \times x$ *lower submatrix of* $\tilde{\mathbf{X}}_s$, *e.g.,* $\tilde{\mathbf{H}}_M = \begin{bmatrix} \tilde{\mathbf{H}}_{M,1} & \tilde{\mathbf{H}}_{M,2} \end{bmatrix}$, $\tilde{\mathbf{X}}_s = \begin{bmatrix} \tilde{\mathbf{X}}_{s,2}^T & \tilde{\mathbf{X}}_{s,2}^T \end{bmatrix}^T$, $\mathbf{X}_s = (\tilde{\mathbf{P}}_s)^{-1} \mathbf{P}_s$, $\tilde{\mathbf{P}}_s = (\mathbf{I}_N - \mathbf{H}_E^H (\mathbf{H}_E \mathbf{H}_E^H)^{-1} \mathbf{H}_E)$, *and* $\mathbf{U}_s$ *is from the eigenvalue decomposition of* $\tilde{\mathbf{P}}_s$, $\tilde{\mathbf{P}}_s = \mathbf{U}_s \Lambda_s \mathbf{U}_s^H$. *The matrices associated with the relays, such as* $\tilde{\mathbf{X}}_{n,2}$ *and* $\tilde{\mathbf{G}}_{n,2}$, *are defined similar to* $\tilde{\mathbf{H}}_{M,2}$ *and* $\tilde{\mathbf{X}}_{s,2}$.

The benefit to using Proposition 9.3.1 is that the denominator of the fraction in the expression of the secrecy rate can be removed, which is important to the application of various convex optimization techniques. Since $\mathbf{U}_s$ is a unitary matrix, the virtual channel matrices, $\tilde{\mathbf{H}}_{M,2}$ and $\tilde{\mathbf{G}}_{n,2}$, are still classical $N \times (N-M)$ random complex Gaussian matrices. Therefore, an interesting observation is that the $N \times N$ MIMO secrecy communication scenario has been degraded to the $N \times (N-M)$ MIMO scenario due to the existence of the eavesdropper, which is consistent with the conclusion in [16] that the rank of the covariance matrix of the information-bearing vector is less than the number of transmit and receive antennas. Such a loss of degrees of freedom is also the motivation to introduce cooperative transmissions into MIMO secrecy communications and compensate the loss of receptional reliability by exploiting the relays as another dimension for exploiting spatial diversity.

While the use of Proposition 9.3.1 enables the use of convex optimization techniques, it is still challenging to obtain the closed-form expressions of the precoding matrices and the resulting SNR, which are important for the analysis of the outage probability. Therefore a suboptimal solution based on block diagonalization is focused in the following, which yields the explicit expression of an achievable diversity-multiplexing trade-off. Perform eigenvalue decomposition as $\tilde{\mathbf{H}}_{M,2}^H \tilde{\mathbf{H}}_{M,2} = \mathbf{U}_M \Lambda_M \mathbf{U}_M^H$ and $\tilde{\mathbf{G}}_{n,2}^H \tilde{\mathbf{G}}_{n,2} = \tilde{\mathbf{U}}_{r,n} \tilde{\Lambda}_{r,n} \tilde{\mathbf{U}}_{r,n}^H$. The use of the diagonalization-based method results in

$$\tilde{\mathbf{X}}_{s,2} = \frac{1}{N-M} \mathbf{U}_M \quad \& \quad \tilde{\mathbf{X}}_{n,2} = \frac{1}{N-M} \tilde{\mathbf{U}}_{r,n}, \quad (9.29)$$

where the constant factor $\frac{1}{N-M}$ is to meet the transmission power constraint. Since the choice of the upper submatrix of $\tilde{\mathbf{X}}_s$ has no impact on the achievable rate, a simple choice of $\tilde{\mathbf{X}}_s$ is to have $\tilde{\mathbf{X}}_s = \begin{bmatrix} \mathbf{0}_{M,N-M}^T & \mathbf{U}_M^T \end{bmatrix}^T$, which means $\mathbf{X}_s = \mathbf{U}_s \tilde{\mathbf{X}}_s$. Similarly we can have $\tilde{\mathbf{X}}_{r,n} = \begin{bmatrix} \mathbf{0}_{M,N-M}^T & \tilde{\mathbf{U}}_{r,n}^T \end{bmatrix}^T$ and $\mathbf{X}_s = \tilde{\mathbf{U}}_s \tilde{\mathbf{X}}_s$. Summarizing all the above steps, the diagonalization-based solutions for the precoding matrices at the source and relays can be written as

$$
\mathbf{P}_s = \frac{1}{N-M}(\mathbf{I}_N - \mathbf{H}_E^H(\mathbf{H}_E\mathbf{H}_E^H)^{-1}\mathbf{H}_E)\mathbf{U}_s \begin{bmatrix} \mathbf{0}_{M\times(N-M)}^T & \mathbf{U}_M^T \end{bmatrix}^T, \tag{9.30}
$$

$$
\mathbf{P}_{r,n} = \frac{1}{N-M}(\mathbf{I}_N - \mathbf{G}_{E,n}^H(\mathbf{G}_{E,n}\mathbf{G}_{E,n}^H)^{-1}\mathbf{G}_{E,n})\mathbf{U}_{r,n} \begin{bmatrix} \mathbf{0}_{M\times(N-M)}^T & \tilde{\mathbf{U}}_{r,n}^T \end{bmatrix}^T.
$$

Among the many choices of precoding, the above diagonalization-based solution is chosen because it only causes small system overhead. For example, each relay precoding matrix $\mathbf{P}_{r,n}$ is only a function of $\mathbf{G}_{E,n}$ and $\mathbf{G}_{M,n}$, the local incoming and outgoing channel information. As a result, there is no need for the relays to communicate with each other for the precoding design, which can significantly reduce the system overhead.

So based on the above choice of the precoding matrices, the reception reliability achieved by the MIMO best relay scheme can be developed and provided in the following theorem.

Theorem 9.3.1. [21] *Based on the orthogonal projection-based precoding, the achievable outage probability for the best relay scheme is asymptotically equivalent to*

$$
P(\mathcal{I} \leq 2R) \;\dot{\leq}\; \rho^{-(L+1)[(N-M-2r)(N-2r)]},
$$

and the achievable diversity-multiplexing trade-off is

$$
d(r) \;\dot{=}\; (L+1)[(N-M-2r)(N-2r)].
$$

- *Remark 1:* Recall that compared to nonsecrecy scenarios, there is a loss of degrees of freedom for secrecy cases in order to avoid the source messages being intercepted by the eavesdropper [16]. Such a loss of degrees of freedom will reduce the reception reliability at the destination, and Theorem 11.4.1 demonstrates that the use of cooperative transmissions can compensate the loss of reception reliability of secrecy communications. Take $N = 2$, $L = 4$, and $M = 1$ as an example. Without using cooperative transmission, the degree of freedom has been reduced, e.g., a 2×2 point-to-point MIMO system degraded to a 2×1 scheme. However, by using cooperative diversity, the achievable diversity gain can be improved from two up to five.

- *Remark 2:* Compared with the scheme without relays, the use of cooperative diversity causes some loss of multiplexing gain, which is due to the fact that two time slots have been used to transmit one same symbol. By using more sophisticated cooperative protocols, such as the nonorthogonal transmission schemes [23, 24], such a loss of multiplexing gain can be avoided. In addition, the path loss effect can further reduce such a loss of multiplexing gain, whereas the direct link between the source and destination becomes insignificant.

- *Remark 3:* The performance of the MIMO secrecy communication protocol can also be evaluated by using computer simulations. First the targeted secrecy data rate is set as $R = 3$ bits per channel use (BPCU). Figure 9.5 demonstrates the outage performance of three schemes, the best relay scheme, the all-relay scheme, and the direct transmission scheme. As can be seen from Figure 9.5, the increase of the number of antennas can increase the achievable secrecy capacity and therefore improve the reception reliability for all schemes. By introducing cooperation into secrecy communications, the outage performance can be increased significantly as shown in Figure 9.5. With the same system setup, Figure 9.5 demonstrates that the two cooperative schemes can achieve the same diversity gain since the curves of the two schemes

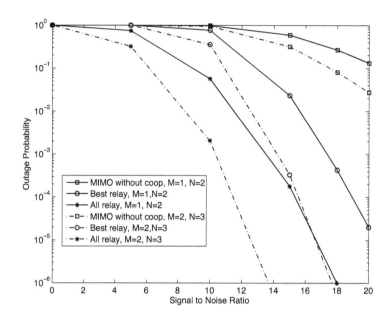

Figure 9.5: The outage probability versus SNR. The targeted secrecy data rate is set as $R = 3$ BPCU.

have the same slope, which is consistent with the analytic results. At intermediate SNR the scheme using all qualified relays can outperform the best relay scheme. The performance difference between the qualified relays and the best relay schemes is analog to the relationship between the maximum ratio combining (MRC) and selection combining, where the former yields the best performance with more computational complexity.

9.3.2   When CSI of Eavesdroppers Is Unknown

As shown in the previous section, an effective method to confuse the eavesdropper is to put the source signals into the null space of the channel matrix between the source and the eavesdropper, where the knowledge of the eavesdropper's CSI is crucial. In many communication systems, particularly for military applications, it is not practical to assume the eavesdropper to be cooperative, and hence the channel information associated with the eavesdropper could be unknown. In such a case, generating artificial noise has been recognized as an efficient method to confuse the eavesdropper. Since the eavesdropper channel is unknown, it is difficult to maximize the effect of artificial noise at the eavesdropper, but it is much less challenging to design a secrecy transmission protocol and minimize the impact of artificial noise on the legitimate receiver. Among various methods to generate artificial noise, we focus on the one that puts the artificial noise into the null space of the matrix for the channels between the legitimate transceiver. Without loss of generality, consider a secrecy communication scenario with one source, one destination, one relay, and one eavesdropper as shown in Fig. 9.6. Differently, it is assumed that the relay is equipped with N antennas and all the other nodes are equipped with M antennas, where the scenario studied in [25] can be viewed as a special case by setting $M = 1$.

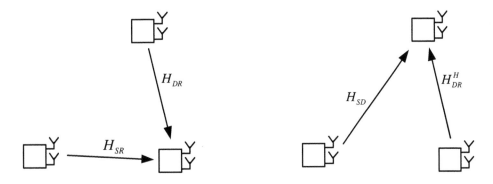

Figure 9.6: The protocol for MIMO cooperative jamming.

9.3.2.1 Description of the Secrecy Transmission Scheme without Knowing the Eavesdropper's CSI

The secrecy transmission consists of two time slots. During the first time slot, the source generates the $x \times 1$ information-bearing vector $\mathbf{s}$, where x is a system parameter to be optimized later and has an impact on the achievable multiplexing gain. According to $\mathbf{s}$, an $(M - x) \times 1$ vector $\mathbf{z}_S$ containing random artificial noise is generated and the $M \times 1$ vector $\tilde{\mathbf{s}}$ is obtained as $\tilde{\mathbf{s}} = \begin{bmatrix} \mathbf{s}^H & \mathbf{z}_S^H \end{bmatrix}^H$. The value of x is constrained by $1 \leq x \leq \min\{M, N\}$, and its choice will be according to a predefined trade-off between the system throughput and the secrecy capability to confuse the eavesdropper. The less x is, the more artificial noise can be generated into the system and used to confuse the eavesdropper, where the disadvantage of a small x results in a loss of data rate. The source transmits the precoded version of $\tilde{\mathbf{s}}$, i.e., $\mathbf{P}_S\tilde{\mathbf{s}}$ and the destination broadcast artificial noise $\mathbf{P}_D\mathbf{z}_D$ at the same time, where the design of the $M \times M$ precoding matrix, $\mathbf{P}_s$, will be discussed at the end of this section, and $\mathbf{z}_D$ is an $M \times 1$ artificial noise vector. A simple choice of the precoding matrix at the destination is to use an identity matrix, and a more sophisticated design will be introduced in the next section. As the end of this phase, the relay receives

$$\mathbf{y}_R = \mathbf{H}_{SR}\mathbf{P}_S\tilde{\mathbf{s}} + \mathbf{H}_{DR}\mathbf{P}_D\mathbf{z}_D + \mathbf{n}_R, \tag{9.31}$$

where $\mathbf{H}_{SR}$ is the $N \times M$ channel matrix between the source and the relay, $\mathbf{H}_{DR}$, $\mathbf{H}_{SE}$, and $\mathbf{H}_{DE}$ are defined similarly, and $\mathbf{n}_R$ denotes the additive complex Gaussian noise.

Similar to cooperative jamming [6], the relay receives the source messages $\mathbf{s}$ and the artificial noise from the destination $\mathbf{z}_D$, but the difference is that the source also generates the artificial noise $\mathbf{z}_S$ which is unknown to the destination. Therefore it is important to avoid having $\mathbf{z}_s$ forwarded to the destination. On the other hand, there is no need to remove the artificial interference $\mathbf{z}_D$, since it was generated by the destination. Therefore a relay detection matrix $\mathbf{W}$ is used as follows:

$$\mathbf{W}\mathbf{H}_{SR}\mathbf{P}_{S2} = \mathbf{0}_{x \times (N-x)}, \tag{9.32}$$

which yields a modified signal model at the relay as

$$\mathbf{W}\mathbf{y}_R = \mathbf{W}\mathbf{H}_{SR}\mathbf{P}_S\tilde{\mathbf{s}} + \mathbf{W}\mathbf{H}_{DR}\mathbf{P}_D\mathbf{z}_D + \mathbf{W}\mathbf{n}_R, \tag{9.33}$$

where $\mathbf{P}_{S2}$ is the $N \times (N - x)$ submatrix of $\mathbf{P}_S$, i.e., $\mathbf{P}_S = \begin{bmatrix} \mathbf{P}_{S1} & \mathbf{P}_{S2} \end{bmatrix}$. To increase system performance, it is preferable to utilize the larger singular values of $\mathbf{H}_{SR}$. Consider that the singular value decomposition (SVD) of $\mathbf{H}_{SR}$ is $\mathbf{H}_{SR} = \mathbf{U}_{SR}\mathbf{\Lambda}_{SR}\mathbf{V}_{SR}^H$, and we let

$$\mathbf{P}_S = \frac{1}{\sqrt{M}}\mathbf{V}_{SR}, \quad \& \quad \mathbf{W} = \tilde{\mathbf{U}}_{SR}^H,$$

where $\tilde{\mathbf{U}}_{SR}$ is an $M \times x$ submatrix of $\mathbf{U}_{SR}$ corresponding to the x largest singular value of $\mathbf{H}_{SR}$, and $\frac{1}{\sqrt{M}}$ is to meet the power constraint at the source, i.e., $\mathcal{E}\{tr\{\mathbf{P}_S\tilde{\mathbf{s}}\tilde{\mathbf{s}}^H\mathbf{P}_S^H\}\} = 1$. By using such a choice of precoding, the signal model at the relay can be written as

$$\tilde{\mathbf{U}}_{SR}^H\mathbf{y}_R = \frac{1}{\sqrt{M}}\tilde{\mathbf{\Lambda}}_{SR}\mathbf{s} + \tilde{\mathbf{U}}_{SR}^H\mathbf{H}_{DR}\mathbf{P}_D\mathbf{z}_D + \tilde{\mathbf{U}}_{SR}^H\mathbf{n}_R, \qquad (9.34)$$

where $\tilde{\mathbf{\Lambda}}_{SR}$ is a diagonal matrix with the x largest singular values of $\mathbf{H}_{SR}$ on its diagonal. As can be observed from the above equation, the artificial noise generated at the source, $\mathbf{z}_S$, has been removed. Note that the dimension of $\tilde{\mathbf{U}}_{SR}^H\mathbf{y}_R$ is x. Given the N relay antennas, the relay can also generate additional noise to confuse the eavesdropper, which means the signals sent by the relay will be

$$\mathbf{s}_R = \frac{1}{\sqrt{N}}\mathbf{P}_R \begin{bmatrix} \mathbf{D}_R\mathbf{W}\mathbf{y}_R \\ \mathbf{z}_R \end{bmatrix}, \qquad (9.35)$$

where $\mathbf{z}_R$ contains $(N - x)$ artificial noise messages generated by the relay, $\mathbf{P}_R$ is an $N \times N$ unitary matrix, and its design will be discussed in the following. Each element of the diagonal matrix $\mathbf{D}_R$ is set as $[\mathbf{D}_R]_{m,m}^{-1} = \sqrt{\frac{1}{M}\lambda_{SR,i}^2 + [\tilde{\mathbf{U}}_{SR}^H\mathbf{H}_{DR}\mathbf{P}_D\mathbf{P}_D^H\mathbf{H}_{DR}^H\tilde{\mathbf{U}}_{SR}]_{m,m} + \frac{1}{\rho}}$, where $[\mathbf{A}]_{i,j}$ denotes the element at the i-th row and j-th column of $\mathbf{A}$. It is straightforward to validate that the use of such a diagonal matrix can ensure the transmission power at the relay is constrained, i.e., $\mathcal{E}\{tr\{\mathbf{s}_R\mathbf{s}_R^H\}\} = 1$.

During the second time slot, the relay transmits $\mathbf{s}_R$ which includes the source messages and artificial noise generated at the destination, similar to cooperative jamming. But to further improve the secure performance, the source is also invited to send additional $(M - x)$ noise signals $\tilde{\mathbf{P}}_S\tilde{\mathbf{z}}_S$, where the choice of the precoding matrix $\tilde{\mathbf{P}}_S$ will be provided later. Again to utilize the large singular values of $\mathbf{H}_{DR}$, first consider the SVD of $\mathbf{H}_{DR}^H = \mathbf{U}_{DR}\mathbf{\Lambda}_{DR}\mathbf{V}_{DR}^H$ and the relay precoding matrix can be set as $\mathbf{P}_R = \mathbf{V}_{DR}$. By using such precoding, the signal model at the destination can be expressed as

$$\mathbf{y}_D = \frac{1}{\sqrt{N}}\mathbf{U}_{DR}\mathbf{\Lambda}_{DR}\begin{bmatrix} \mathbf{D}_R\mathbf{W}\mathbf{y}_R \\ \mathbf{z}_R \end{bmatrix} + \mathbf{H}_{SD}\tilde{\mathbf{P}}_S\tilde{\mathbf{z}}_S + \mathbf{n}_D, \qquad (9.36)$$

where the artificial noise $\mathbf{z}_R$ has been allocated to the $(N - x)$ smallest singular values.

Recall that $\tilde{\mathbf{z}}_S$ was generated at the source and is unknown to the destination, and therefore it is important to ensure that such artificial noise will not degrade the receive capability of the destination. One efficient approach is to design $\tilde{\mathbf{P}}_S$ in a way that $\tilde{\mathbf{z}}_S$ will be occupying the same direction as $\mathbf{z}_R$, which results in the following simplified signal model,

$$\mathbf{y}_D = \frac{1}{\sqrt{N}}\mathbf{U}_{DR}\breve{\mathbf{\Lambda}}_{DR}\begin{bmatrix} \mathbf{D}_R\mathbf{W}\mathbf{y}_R \\ \mathbf{z} \end{bmatrix} + \mathbf{n}_D, \qquad (9.37)$$

where $\mathbf{z}$ is the combination of $\mathbf{z}_R$ and $\mathbf{z}_S$, and $\breve{\mathbf{\Lambda}}_{DR}$ contains the largest x eigenvalues of $\mathbf{\Lambda}_{DR}$. Detailed discussions of $\breve{\mathbf{\Lambda}}_{DR}$ and $\mathbf{z}$ can be found in [27].

9.3.2.2 Detection at the Destination and Eavesdropper

Detection at the destination can be easily accomplished by removing artificial noise, as shown in the following. Denote $\tilde{\mathbf{U}}_{DR}$ as the $M \times x$ submatrix of $\mathbf{U}_{DR}$ corresponding to the

x smallest singulars of $\mathbf{H}_{DR}^H$. Because of the orthogonality among singular vectors, a natural choice of the detection matrix is $\tilde{\mathbf{U}}_{DR}$, and the use of such a detection matrix yields

$$
\begin{aligned}
\tilde{\mathbf{U}}_{DR}^H \mathbf{y}_D &= \frac{1}{\sqrt{N}} \tilde{\mathbf{\Lambda}}_{DR} \mathbf{D}_R \mathbf{W} \mathbf{y}_R + \tilde{\mathbf{U}}_{DR}^H \mathbf{n}_D \qquad (9.38) \\
&= \frac{1}{\sqrt{NM}} \tilde{\mathbf{\Lambda}}_{DR} \mathbf{D}_R \tilde{\mathbf{\Lambda}}_{SR} \mathbf{s} + \frac{1}{\sqrt{N}} \tilde{\mathbf{\Lambda}}_{DR} \mathbf{D}_R \tilde{\mathbf{U}}_{SR}^H \mathbf{n}_R + \tilde{\mathbf{U}}_{DR}^H \mathbf{n}_D,
\end{aligned}
$$

where $\tilde{\mathbf{\Lambda}}_{DR}$ is a diagonal matrix containing the x largest singular values of $\mathbf{H}_{DR}^H$ and the self-interference $\mathbf{z}_D$ has been removed. An important observation from the above signal model is that the source messages have been allocated to the largest singular values of the two channel matrices, which is analog to the idea of eigenvalue alignment [28]. The mutual information between the source and destination will be

$$
\begin{aligned}
\mathcal{I}_M &= \frac{1}{2} \log \det \left(\mathbf{I}_x + \frac{\rho}{M} \tilde{\mathbf{\Lambda}}_{DR}^2 \tilde{\mathbf{\Lambda}}_{SR}^2 \left(\tilde{\mathbf{\Lambda}}_{DR}^2 + N \mathbf{D}_R^{-2} \mathbf{I}_x \right)^{-1} \right), \qquad (9.39) \\
&= \frac{1}{2} \sum_{m=1}^{x} \log \left(1 + \frac{\rho}{M} \frac{\tilde{\lambda}_{DR,m}^2 \tilde{\lambda}_{SR,m}^2}{\tilde{\lambda}_{DR,m}^2 + N [\mathbf{D}_R]_{m,m}^{-2}} \right).
\end{aligned}
$$

At the eavesdropper, the signals observed during the two time slots can be expressed as

$$
\begin{aligned}
\mathbf{y}_{E1} &= \frac{1}{\sqrt{M}} \mathbf{H}_{SE} \mathbf{V}_{SR} \mathbf{s} + \mathbf{H}_{DE} \mathbf{z}_D + \mathbf{n}_{E1}, \\
\mathbf{y}_{E2} &= \frac{1}{\sqrt{N}} \mathbf{H}_{RE} \mathbf{V}_{DR} \begin{bmatrix} \mathbf{D}_R \mathbf{W} \mathbf{y}_R \\ \mathbf{z}_R \end{bmatrix} + \mathbf{H}_{SE} \tilde{\mathbf{P}}_S \tilde{\mathbf{z}}_S + \mathbf{n}_{E2}.
\end{aligned}
$$

The detection at the eavesdropper is analog to a set of linear equations with $2M$ equations and $(3M + N - 2x)$ unknown variables. Apparently, for the special case of $M = N = x$, the eavesdropper has $2M$ observations containing $2M$ messages, and is able to separate the mixture, exactly the same as cooperative jamming. But except for such a special case, the eavesdropper will not be able to intercept the source messages because there are always more unknown variables than the number of observations. Therefore the eavesdropper has to treat all interference as noise and the signal model can be written as

$$
\begin{bmatrix} \mathbf{y}_{E1} \\ \mathbf{y}_{E2} \end{bmatrix} = \begin{bmatrix} \frac{1}{\sqrt{M}} \mathbf{H}_{SE} \tilde{\mathbf{V}}_{SR} \\ \frac{1}{\sqrt{MN}} \mathbf{H}_{RE} \tilde{\mathbf{V}}_{DR} \mathbf{D}_R \tilde{\mathbf{\Lambda}}_{SR} \end{bmatrix} \mathbf{s} + \begin{bmatrix} \tilde{\mathbf{n}}_{E1} \\ \tilde{\mathbf{n}}_{E2} \end{bmatrix} \qquad (9.40)
$$

where $\tilde{\mathbf{n}}_{E1} = \frac{1}{\sqrt{M}} \mathbf{H}_{SE} \bar{\mathbf{V}}_{SR} \mathbf{z}_S + \mathbf{H}_{DE} \mathbf{z}_D + \mathbf{n}_{E1}$, $\tilde{\mathbf{n}}_{E2} = \frac{1}{\sqrt{N}} \mathbf{H}_{RE} \tilde{\mathbf{V}}_{DR} \mathbf{D}_R \tilde{\mathbf{U}}_{SR}^H \mathbf{H}_{DR} \mathbf{P}_D \mathbf{z}_D + \frac{1}{\sqrt{N}} \mathbf{H}_{RE} \tilde{\mathbf{V}}_{DR} \mathbf{D}_R \tilde{\mathbf{U}}_{SR}^H \mathbf{n}_R + \frac{1}{\sqrt{N}} \mathbf{H}_{RE} \tilde{\mathbf{V}}_{DR} \mathbf{z}_R + \mathbf{H}_{SE} \tilde{\mathbf{P}}_S \tilde{\mathbf{z}}_S + \mathbf{n}_{E2}$, and $\mathbf{V}_{DR} = [\tilde{\mathbf{V}}_{DR} \quad \bar{\mathbf{V}}_{DR}]$. And the mutual information at the eavesdropper can be written as

$$
\mathcal{I}_E \approx \frac{1}{2} \log \det \left(\mathbf{I}_x + \begin{bmatrix} \frac{1}{\sqrt{M}} \mathbf{H}_{SE} \tilde{\mathbf{V}}_{SR} \\ \frac{1}{\sqrt{NM}} \mathbf{H}_{RE} \tilde{\mathbf{V}}_{DR} \mathbf{D}_R \tilde{\mathbf{\Lambda}}_{SR} \end{bmatrix} \begin{bmatrix} \frac{1}{\sqrt{M}} \mathbf{H}_{SE} \tilde{\mathbf{V}}_{SR} \\ \frac{1}{\sqrt{NM}} \mathbf{H}_{RE} \tilde{\mathbf{V}}_{DR} \mathbf{D}_R \tilde{\mathbf{\Lambda}}_{SR} \end{bmatrix}^H \mathbf{C}_n^{-1} \right) \quad (9.41)
$$

where the noise covariance matrix $\mathbf{C}_n$ is $\begin{bmatrix} \mathbf{C}_{n1} & \mathbf{C}_{n2} \\ \mathbf{C}_{n2} & \mathbf{C}_{n3} \end{bmatrix}$,

$$\mathbf{C}_{n1} = \frac{1}{M}\mathbf{H}_{SE}\bar{\mathbf{V}}_{SR}\bar{\mathbf{V}}_{SR}^H\mathbf{H}_{SE}^H + \mathbf{H}_{DE}\mathbf{H}_{DE}^H + \frac{1}{\rho}\mathbf{I}_M,$$

$$\mathbf{C}_{n2} = \frac{1}{\sqrt{M}}\mathbf{H}_{DE}\mathbf{H}_{DR}^H\tilde{\mathbf{U}}_{SR}\mathbf{D}_R^H\tilde{\mathbf{V}}_{DR}^H\mathbf{H}_{RE}^H,$$

$$\mathbf{C}_{n3} = \mathbf{H}_{SE}\tilde{\mathbf{P}}_S\tilde{\mathbf{P}}_S^H\mathbf{H}_{SE}^H\frac{1}{N}\mathbf{H}_{RE}\bar{\mathbf{V}}_{DR}\bar{\mathbf{V}}_{DR}^H\mathbf{H}_{RE}^H +$$

$$+ \frac{1}{\rho N}\mathbf{H}_{RE}\tilde{\mathbf{V}}_{DR}\mathbf{D}_R\tilde{\mathbf{U}}_{SR}^H\tilde{\mathbf{U}}_{SR}\mathbf{D}_R^H\tilde{\mathbf{V}}_{DR}^H\mathbf{H}_{RE}^H$$

$$+ \frac{1}{N}\mathbf{H}_{RE}\tilde{\mathbf{V}}_{DR}\mathbf{D}_R\tilde{\mathbf{U}}_{SR}^H\mathbf{H}_{DR}\mathbf{H}_{DR}^H\tilde{\mathbf{U}}_{SR}\mathbf{D}_R^H\tilde{\mathbf{V}}_{DR}^H\mathbf{H}_{RE}^H + \frac{1}{\rho}\mathbf{I}_M.$$

At high SNR, it can be observed that the interference and noise covariance matrix becomes independent to SNR, and therefore $\mathcal{I}_E$ becomes a constant, no matter how large the SNR is.

9.3.2.3 Ergodic Capacity and Outage Probability

The case with $N < M$ and $x < N$ is focused to obtain closed-form expressions for the ergodic capacity and outage probability. Following [2], the achievable secrecy rate for the addressed scenario can be expressed as

$$\mathcal{I} = \frac{1}{2}\left[\mathcal{I}_M - \mathcal{I}_E\right]^+. \tag{9.42}$$

Provided that $1 \le x \le (M-1)$, the expression of the achievable rate can be approximated at high SNR as

$$\mathcal{I} \approx \frac{1}{2}\mathcal{I}_M = \frac{1}{2}\sum_{m=1}^{x}\log\left(1 + \frac{\rho}{M}\frac{\tilde{\lambda}_{DR,m}^2\tilde{\lambda}_{SR,m}^2}{\tilde{\lambda}_{DR,m}^2 + N[\mathbf{D}_R]_{m,m}^{-2}}\right). \tag{9.43}$$

To evaluate the performance achieved by the secrecy protocol, the density function of $[\mathbf{D}_R]_{m,m}$ is required, but such a function is quite difficult to obtain. When $N < M$, we can find out the expression of the mutual information can be significantly simplified by carefully designing the precoding matrix at the destination $\mathbf{P}_D$. Specifically among many choices of the precoding matrix at the destination, and we use the following one which can simplify the analysis

$$\mathbf{P_D} = \frac{1}{\sqrt{\beta}}\mathbf{H}_{DR}^H(\mathbf{H}_{DR}\mathbf{H}_{DR}^H)^{-1}\mathbf{U}_{SR}\mathbf{\Lambda}_{SR}, \tag{9.44}$$

where the factor β is set as $\beta = \mathcal{E}\{tr\{\mathbf{H}_{SR}\mathbf{H}_{SR}^H\}\}\mathcal{E}\{tr\{(\mathbf{H}_{DR}\mathbf{H}_{DR}^H)^{-1}\}\} = \frac{MN^2}{M-N}$. Note that the result about the trace of a Wishart matrix has been used for the calculation of the power constraint factor β [26]. With such a choice of precoding at the destination, we can have

$$[\mathbf{D}_R]_{m,m}^{-2} = \frac{1}{M}\lambda_{SR,m}^2 + [\tilde{\mathbf{U}}_{SR}^H\mathbf{H}_{DR}\mathbf{P}_D\mathbf{P}_D^H\mathbf{H}_{DR}^H\tilde{\mathbf{U}}_{SR}]_{m,m} + \frac{1}{\rho} \approx \left(\frac{1}{M} + \frac{1}{\beta}\right)\lambda_{SR,m}^2. \tag{9.45}$$

As a result, the mutual information can be explicitly related to the eigenvalues as

$$\mathcal{I} \approx \frac{1}{2}\sum_{m=1}^{x}\log\left(1 + \frac{\rho}{M}\frac{\tilde{\lambda}_{DR,m}^2\tilde{\lambda}_{SR,m}^2}{\tilde{\lambda}_{DR,m}^2 + \left(\frac{N}{M} + \frac{N}{\beta}\right)\lambda_{SR,m}^2}\right). \tag{9.46}$$

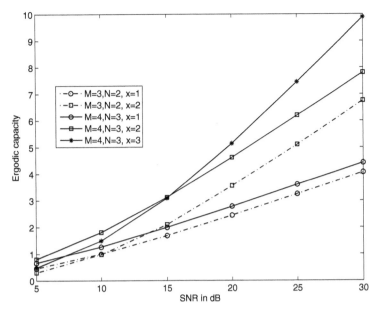

Figure 9.7: Ergodic secrecy capacity versus the SNR, when $N < M$. The precoding matrix $\mathbf{P}_D$ is based on (9.44).

By using the above simplified expression about the secrecy rate, we can obtain an asymptotic behavior of the ergodic capacity.

Lemma 9.3.1. [27] *Provided the S-R and R-D channels are independent and identically Rayleigh distributed. When $M \to \infty$ and N is fixed, the ergodic capacity can be asymptotically expressed as*

$$\mathcal{E}\{\mathcal{I}\} \approx \frac{1}{2}x \cdot \log\left(1 + \frac{N}{1+N}\right). \tag{9.47}$$

When both M and N become infinity, the ergodic capacity can be asymptotically expressed as

$$\mathcal{E}\{\mathcal{I}\} \approx \frac{1}{2}\sum_{m=1}^{x} \log\left(1 + \frac{\rho\mu_{\lambda,m}}{M(1+\alpha)}\right), \tag{9.48}$$

where $\mu_{\lambda,m} = b_{MN}\mu_m + a_{MN}$, $a_{MN} = \left(M^{\frac{1}{2}} + N^{\frac{1}{2}}\right)^2$, $b_{MN} = \left(M^{\frac{1}{2}} + N^{\frac{1}{2}}\right)\left(M^{-\frac{1}{2}} + N^{-\frac{1}{2}}\right)^{\frac{1}{3}}$, and μ_m is a constant.

Furthermore, the outage probability achieved by the above secrecy rate is shown as follows:

Lemma 9.3.2. [27] *Provided the S-R and R-D channels are independent and identically Rayleigh distributed. When the source only sends one data stream, i.e., $x = 1$, the outage probability can be asymptotically expressed as*

$$P(\mathcal{I} < R) \doteq \frac{1}{\rho^{MN}}, \tag{9.49}$$

where $f(\rho)$ is said to be exponentially equal to ρ^d, denoted as $f(\rho) \doteq \rho^d$, when $\lim_{\rho \to \infty} \frac{\log[f(\rho)]}{\log \rho} = d$.

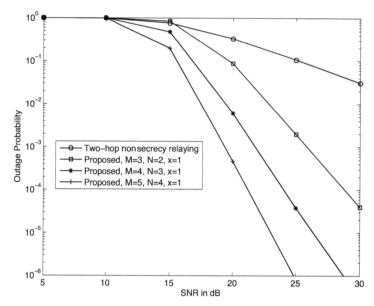

Figure 9.8: Outage probability versus the SNR, when $N < M$. The target data rate is $R = 2$ BPCU. The precoding matrix $\mathbf{P}_D$ is based on (9.44).

9.3.2.4   Numerical Results

The case of $N < M$ is focused on first, where the precoding design shown in (9.44) can be used. In Figure 9.7 the averaged secrecy rate is shown as a function of SNR, where the impact of the antenna number is also demonstrated. As can be observed from the figure, the ergodic capacity can be generally improved by increasing the relay and source/destination

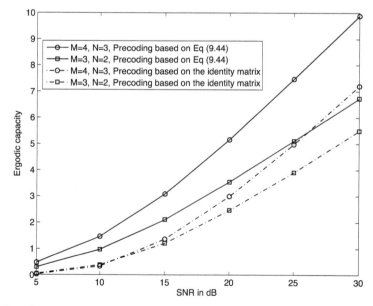

Figure 9.9: Ergodic secrecy capacity versus the SNR with different choices of the precoding matrix $\mathbf{P}_D$. The number of the transmitted data streams is set as $x = \min\{M, N\}$.

antennas. Specifically when the source only transmits a single message, adding one extra antenna at the nodes can increase the ergodic secrecy rate about 2 BPCU, given the SNR is 25 dB. When the parameters about node antennas are fixed, it is interesting to observe that increasing x will improve the averaged secrecy rate, since the achievable multiplexing gain is determined by x, as shown in Lemma 9.3.2. In Figure 9.8, the outage probability achieved by the secrecy protocol has been shown as a function of SNR. The targeted data rate is 2 BPCU. By increasing the number of antennas, the slope of the outage probability curves becomes larger, which implies a larger diversity gain, as can be observed from Figure 9.8. Such an observation is expected, since Lemma 9.3.1 tells us the diversity gain achieved by the addressed protocol is proportional to the number of node antennas. In Figure 9.9, the impact of different choices of $\mathbf{P}_D$ on the ergodic secrecy rate is shown as a function of SNR. In general, the use of the precoding design defined in (9.44) can yield more ergodic capacity than the one based on the identity matrix, but the latter is a more general approach which can be applied to case $N \leq M$.

9.4 Conclusion

In this chapter, the use of cooperative transmissions in secrecy communication systems is focused on. We first considered the secrecy communication scenario with single-antenna nodes, where two popular transmission protocols, cooperative jamming and relay chatting, have been studied. Then we investigated the MIMO secrecy scenario, where the cases with and without the eavesdropper's CSI are considered. We have shown that by manipulating the null space of channel matrices, it is possible to completely stop the eavesdropper intercepting the source information and significantly improve the achievable secrecy rates.

Acknowledgment

The work of Z. Ding and K. Cumanan was supported by the U.K. Engineering and Physical Sciences Research Council under Grant EP/I037423/1.

References

[1] A. D. Wyner, "The wire-tap channel," *Bell Syst. Tech. J.*, vol. 8, pp. 1355–1387, 1975.

[2] M. Bloch, J. Barros, M. R. D. Rodriques, and S. W. McLaughlin, "Wireless information-theoretic security," *IEEE Trans. Inf. Theory*, vol. 54, no. 6, pp. 2515–2534, June 2008.

[3] L. Lai and H. E. Gamal, "The relay-eavesdropper channel: Cooperation for secrecy," *IEEE Trans. Signal Process*, vol. 54, no. 9, pp. 4005–4019, Sep. 2008.

[4] L. Dong, Z. Han, A. Petropulu, and H. Poor, "Improving wireless physical layer security via cooperating relays," *IEEE Trans. Signal Processing*, vol. 58, no. 3, pp. 1875–1888, Mar. 2010.

[5] J. Zhang and M. Gursoy, "Collaborative relay beamforming for secrecy," in *Proc. IEEE International Conference on Communications (ICC'10)*, June 2010, pp. 1– 5.

[6] E. Tekin and A. Yener, "The general Gaussian multiple-access and two-way wiretap channels: Achievable rates and cooperative jamming," *IEEE Trans. Information Theory*, vol. 54, no. 6, pp. 2735–2751, June 2008.

[7] I. Krikidis, J. S. Thompson, and S. McLaughlin, "Relay selection for secure cooperative networks with jamming," *IEEE Trans. Wireless Commun.* vol. 51, no. 10, pp. 5003–5011, Oct. 2009.

[8] R. Zhang, L. Song, Z. Han, and B. Jiao, "Physical layer security for two way relay communications with friendly jammers," in *Proc. IEEE Global Communications Conference*, Dec. 2010.

[9] S. V. S. Adams, D. Goeckel, Z. Ding, D. Towsley, and K. Leung, "Multi-user diversity for secrecy in wireless networks," in *Proc. Information Theory and Applications Workshop (ITA)*, Feb. 2010.

[10] Z. Ding, M. Xu, J. Lu, and F. Liu, "Improving wireless security for bidirectional communication scenarios," *IEEE Trans. Veh. Technol.* vol. 61, no. 6, pp. 2842–2848, July 2012.

[11] S. Zhang, S. Liew, and P. Lam, "Physical layer network coding," in *Proc. 12th Annual International Conference on Mobile Computing and Networking (ACM MobiCom 2006)*, Sep. 2006, pp. 63–68.

[12] Z. Ding, K. K. Leung, D. L. Goeckel, and D. Towsley, "On the study of network coding with diversity," *IEEE Trans. Wireless Commun.* vol. 8, no. 3, pp. 1247–1259, Mar. 2009.

[13] P. K. Gopala, L. Lai, and H. E. Gamal, "On the secrecy capacity of fading channels," *IEEE Trans. Inf. Theory*, vol. 54, no. 10, pp. 4687–4698, Oct. 2008.

[14] P. Parada and R. Blahut, "Secrecy capacity of SIMO and slow fading channels," in *Proc. IEEE International Symposium on Information Theory (ISIT'05)*, pp. 2152–2155, Nov. 2005.

[15] A. Khisti and G. W. Wornell, "Secure transmission with multiple antennas: The MISOME wiretap channel," *IEEE Trans. Inf. Theory*, vol. 56, no. 7, pp. 3088–3104, July 2010.

[16] F. Oggier and B. Hassibi, "The secrecy capacity of the MIMO wiretap channel," *IEEE Trans. Information Theory*, vol. 57, no. 8, pp. 4961–4972, Aug. 2011.

[17] Z. Ding, K. Leung, D. L. Goeckel, and D. Towsley, "Opportunistic relaying for secrecy communications: Cooperative jamming vs relay chatting," *IEEE Trans. on Wireless Commun.* vol. 10, no. 6, pp. 1725–1729, June 2011.

[18] X. He and A. Yener, "Two-hop secure communication using an untrusted relay: A case for cooperative jamming," in *Proc. IEEE Global Telecommunication Conference*, Dec. 2008.

[19] J. N. Laneman, D. N. C. Tse, and G. W. Wornell, "Cooperative diversity in wireless networks: Efficient protocols and outage behavior," *IEEE Trans. Inf. Theory*, vol. 50, no. 12, pp. 3062–3080, Dec. 2004.

[20] F. Oggier and B. Hassibi, "The MIMO wiretap channel," in *Proc. IEEE International Symposium on Information Theory (ISIT'09)*, July 2008.

[21] Z. Ding, K. Leung, D. Goeckel, and D. Towsley, "On the application of cooperative transmission to secrecy communications," *IEEE J. Sel. Areas Commun.*, vol. 30, no. 2, pp. 359 – 368, Feb. 2012.

[22] M. J. Taghiyar, S. Muhaidat, J. Liang, and M. Dianati, "Relay selection with imperfect CSI in bidirectional cooperative networks," *IEEE Commun. Lett.*, vol. 16, no. 1, pp. 57–59, Feb. 2012.

[23] R. U. Nabar, H. Bolcskei, and F. W. Kneubuhler, "Fading relay channels: Performance limits and space-time signal design," *IEEE J. Sel. Areas Commun.*, vol. 22, no. 6, pp. 1099–1109, Aug. 2004.

[24] K. Azarian, H. E. Gamal, and P. Schniter, "On the achievable diversity-multiplexing tradeoff in half-duplex cooperative channels," *IEEE Trans. Inf. Theory*, vol. 51, no. 12, pp. 4152–4172, Dec. 2005.

[25] J. Huang and A. L. Swindlehurst, "Cooperative jamming for secure communications in MIMO relay networks," *IEEE Trans. Signal Process.*, vol. 59, no. 10, pp. 4871–4884, Oct. 2011.

[26] S. V. Antonio and M. Tulino, "Random matrix theory and wireless communications," *Communications and Information Theory*, vol. 1, pp. 1–182, June 2004.

[27] Z. Ding, M. Peng, and H. H. Chen, "A general relaying transmission protocol for MIMO secrecy communications," *IEEE Trans. Comm.*, vol. 60, no, 16, pp. 3461–3471, Nov. 2017.

[28] C. Y. Leow, Z. Ding, and K. Leung, "Joint beamforming and power management for nonregenerative MIMO two-way relaying channels," *IEEE Trans. Vehicular Technology*, vol. 60, no. 9, pp. 4374–4383, Nov. 2011.

Chapter 10

Game Theory for Physical Layer Security on Interference Channels

Eduard Jorswieck
TU Dresden

Rami Mochaourab
Fraunhofer Heinrich Hertz Institute

Ka Ming (Zuleita) Ho
TU Dresden

In this chapter, we propose game-theoretic models for resource allocation in interference channels with physical layer security. Three two-user interference channel scenarios are studied: the peaceful system with public messages, the case with private messages and forward secrecy, and the case with a public feedback channel. These lead to achievable rate, achievable secrecy, and achievable secret-key regions. Next, the outcome of a strategic game in normal form is computed and it is shown that either there exists an equilibrium in dominant strategies or a Nash equilibrium. Finally, three cooperative game models are applied: the Nash bargaining solution, a strategic bargaining algorithm, and the Walrasian market model. They are applicable to all three interference channel scenarios. Numerical illustrations show the utility regions and the outcomes from the different game models are discussed.

10.1   Introduction

The general introduction to information-theoretic or physical layer security is provided in Chapter 1 at the beginning of this book. Therefore, we restrict the introduction to a brief review of the developments of physical layer security for wireless interference networks and the corresponding conflict situation between coexisting links. This motivates us to apply tools from noncooperative and cooperative game theory to understand and resolve the conflict situation.

The notion of perfect secrecy was coined by Shannon in [42]. It took almost twenty years to formulate the degraded wiretap [46] and the nondegraded wiretap channel [11]. In these works it is shown that reliable as well as secure data transmission is possible in general wiretap channels. The extension to continuous input and output alphabets was developed in [24]. It took more than another twenty years before transceiver structures became available to support these wiretap setups. Extensive analysis and designs have been conducted; parts of their results are reported in [6,25,27] and recent tutorial papers [35,43].

In wireless communications scenarios with many active transmitter nodes, *interference* is created at the receivers because of the broadcast nature of the wireless channel. The interference has two negative effects: at first, it lowers the receive signal-to-interference-and-noise ratio (SINR) and thereby disturbs the reliable data transmission. In scenarios with private data, the second negative effect of interference is that it contains private information received at potential eavesdroppers. Then, interference corresponds to *information leakage*.

In multiuser systems, the performance is usually measured in terms of utility regions, i.e., operating points which can be achieved simultaneously for all participating users. The interference creates a conflict situation between the links, and one task of the resource allocation is to find suitable solutions. Game theory provides a systematic approach to resolve these conflict situations under different assumptions on the knowledge and level of cooperation between the links [4, 22].

The impact of the type of the data (public or private) on the achievable utility region can be illustrated by a simple example. Consider the typical cocktail party problem. A set of speakers talk simultaneously to a set of listeners about interesting but public things. Since all are in one room, they all create interference to each other. Each individual listener filters the message interested in by exploiting sophisticated multichannel signal processing (ears plus brain). This situation is illustrated in Figure 10.1 on the left side. If the number of speakers grows large and if there is no cooperation between the individual speakers, the understanding capabilities of each individual pair of speaker and listener will be reduced until almost no novel information can be transferred. It ends in the *tragedy of the commons*. All speakers use up the available resources in the Nash equilibrium.

The situation changes dramatically, if the speakers have private information for a subset of listeners (for example for only one). A common behavior is to speak as muted as possible into the ear of the intended listener to prevent eavesdropping. This is illustrated in Figure 10.1 on the right side. In wireless communication this corresponds to transmitting the private data into specific directions (beamforming) taking into account the information leakage. Therefore, the problem as explained above for public information does not occur and the resources can be better utilized. There exist parallels between the motivating example and secure wireless data transmission, e.g., the more information about the potential eavesdroppers is available at the speaker (or transmitter), the better it can adapt. If perfect channel state information (CSI) to the eavesdroppers is available, the secrecy capacity can be computed and achieved by coding and signal processing.

In this chapter, the observation from the cocktail party problem is analyzed for current wireless communication systems. In particular, we investigate the application of game theory to find reasonable efficient operating points for multiantenna wireless multipoint-to-multipoint systems under secrecy constraints. The utility functions for such secrecy scenarios already "punish" part of the transmit signal which is leaked as interference at the unintended receiver. Thereby, even in noncooperative games, a less selfish and more altruistic operating point is obtained compared to the Nash equilibrium (NE) in the peaceful setting [20]. As a first step in this chapter, we characterize the NE and observe that the links

Figure 10.1: On the left: A typical Manhattan cocktail party with public information. All speakers talk as loud as required and the listener must follow the conversation of interest despite interference. (© copyright: Anton Oparin/shutterstock.com) On the right: A private communication channel by spatially directed whispering not disturbing other receivers and not allowing for eavesdropping (beamforming for secret data transmission). (© copyright: stockyimages/shutterstock.com)

still have an incentive to cooperate and further improve their noncooperative outcome with secrecy utility functions.

There exist many different models for cooperative games. To name a few, games in characteristic form with and without transferable utilities, bargaining games (including axiomatic and evolutionary), and exchange markets. In this chapter, we decided to choose one representative from axiomatic bargaining, distributed bargaining, and an exchange market. Therefore, we apply three approaches to find cooperative operating points:

- the Nash bargaining solution (NBS) [30]

- a distributed implementation of a Pareto optimal bargaining solution [32]

- and the Walras equilibrium [31].

We have chosen the NBS because it is very popular and linked to proportional fair scheduling. The disadvantage of NBS is its implementation. Therefore, we have chosen a distributed bargaining algorithm as a second model and finally the Walrasian market model to include microeconomic aspects. The properties and implementation of the three cooperative game models are studied and their performances are compared to their peaceful counterparts by numerical simulations.

10.2   System Models and Scenarios

The general system model is the two-user multiple-input single-output (MISO) interference channel. It is general enough to capture the effect of spatial precoding at the transmitter side and fixed receive filters (simple receivers) [5]. It is sufficiently simple to obtain analytical results and an in depth understanding for the system design. We compare three different scenarios which are all based on the interference channel:

- the peaceful or standard MISO interference channel with public messages,

- the MISO interference channel with private messages (forward secrecy rates),

- and the MISO interference channel with private messages and a public discussion channel.

The development of these three scenarios corresponds to increasing complexity. The basic interference channel is characterized by its input alphabets $\mathcal{X}_1, \mathcal{X}_2$, its output alphabets $\mathcal{Y}_1, \mathcal{Y}_2$, and its conditional probability distribution $p(\boldsymbol{y}_1, \boldsymbol{y}_2 | \boldsymbol{x}_1, \boldsymbol{x}_2)$ describing the channel.

Note that there exists a huge body of work on multiple-antenna interference channels which we cannot review within the scope of this book chapter. Interested readers are referred to the monograph [5] and references listed therein. Only at certain parts (e.g., for non-cooperative solutions and bargaining algorithms), we provide a brief selection of references.

The practical motivation to consider the MISO interference channel is that it models all scenarios in which two (or in general multiple) links share the same frequency and transmit at the same time. This can occur in nonlicensed transmission modes (e.g., WLAN), in multicell communications with frequency reuse factor equal to one, in wireless sensor networks, and in many more.

10.2.1   Standard MISO Interference Channel

The standard single-antenna interference channel has a long history in information theory [1,7]. The capacity region is still unsolved, even if larger achievable rate regions and tighter outer bounds were derived (see, e.g., [10,15]).

The peaceful MISO interference channel is shown in Figure 10.2. It consists of two transmitters and two receivers. The messages by both links are public, but receiver 1 is only interested in the message from transmitter 1, and receiver 2 only in message from transmitter 2, respectively.

We assume that the receivers treat the additional interference created by the other transmitter simply as noise and apply single-user detection. Denote the quasistatic block flat-fading channels as in Figure 10.2 with $\boldsymbol{h}_{11}$ and $\boldsymbol{h}_{22}$ for the direct links and $\boldsymbol{h}_{ij}$ for the link from transmitter i to receiver j. The beamforming vectors at the transmitters are called $\boldsymbol{w}_1$ and $\boldsymbol{w}_2$. The additive white Gaussian noises at the two receivers are independently and identically distributed with variance σ_n^2. Then, the achievable rate region under individual

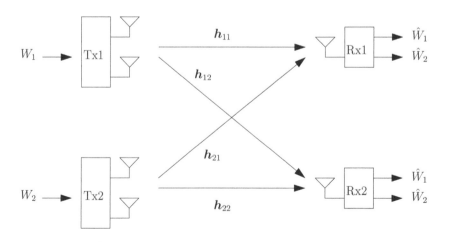

Figure 10.2: The two-user MISO interference channel (illustrated for $n_T = 2$ transmit antennas).

power constraints $||\boldsymbol{w}_1|| \leq 1$ and $||\boldsymbol{w}_2|| \leq 1$ is given by [21],

$$\mathcal{R} = \bigcup_{||\boldsymbol{w}_1|| \leq 1, ||\boldsymbol{w}_2|| \leq 1} \{R_1(\boldsymbol{w}_1, \boldsymbol{w}_2), R_2(\boldsymbol{w}_1, \boldsymbol{w}_2)\}, \tag{10.1}$$

where the achievable rates R_1 and R_2 are both functions of the two beamforming vectors $\boldsymbol{w}_1, \boldsymbol{w}_2$ and they are given by

$$R_1(\boldsymbol{w}_1, \boldsymbol{w}_2) = \log\left(1 + \frac{|\boldsymbol{w}_1^H \boldsymbol{h}_{11}|^2}{\sigma_n^2 + |\boldsymbol{w}_2^H \boldsymbol{h}_{21}|^2}\right) \text{ and } R_2(\boldsymbol{w}_1, \boldsymbol{w}_2) = \log\left(1 + \frac{|\boldsymbol{w}_2^H \boldsymbol{h}_{22}|^2}{\sigma_n^2 + |\boldsymbol{w}_1^H \boldsymbol{h}_{12}|^2}\right). \tag{10.2}$$

The typical signal-to-interference-and-noise (SINR) expression can be detected in the Shannon capacity formula in (10.2).

Definition 10.2.1. *A rate tuple* $(R_1, R_2) \in \mathcal{R}$ *is **Pareto optimal** if there is no other tuple* $(\tilde{R}_1, \tilde{R}_2) \in \mathcal{R}$ *with*

$$(\tilde{R}_1, \tilde{R}_2) \geq (R_1, R_2) \text{ and } (\tilde{R}_1, \tilde{R}_2) \neq (R_1, R_2).$$

The Pareto boundary is the set consisting of all Pareto optimal points. In order to compute the rate regions and to characterize certain operating points on the Pareto boundary, the following characterization for the two-user case is useful. For the proof the reader is referred to [19]. Here, we provide only the intuition behind. Any point on the Pareto boundary of the rate region is achievable with the beamforming strategies

$$\boldsymbol{w}_1(\lambda_1') = \frac{\lambda_1' \boldsymbol{w}_1^{\mathrm{MRT}} + (1 - \lambda_1') \boldsymbol{w}_1^{\mathrm{ZF}}}{||\lambda_1' \boldsymbol{w}_1^{\mathrm{MRT}} + (1 - \lambda_1') \boldsymbol{w}_1^{\mathrm{ZF}}||} \quad \text{and} \quad \boldsymbol{w}_2(\lambda_2') = \frac{\lambda_2' \boldsymbol{w}_2^{\mathrm{MRT}} + (1 - \lambda_2') \boldsymbol{w}_2^{\mathrm{ZF}}}{||\lambda_2' \boldsymbol{w}_2^{\mathrm{MRT}} + (1 - \lambda_2') \boldsymbol{w}_2^{\mathrm{ZF}}||} \tag{10.3}$$

for some $0 \leq \lambda_1', \lambda_2' \leq 1$ and with beamforming vectors $\boldsymbol{w}_k^{\mathrm{MRT}} = \frac{\boldsymbol{h}_{11}}{||\boldsymbol{h}_{11}||}$ and $\boldsymbol{w}_k^{\mathrm{ZF}} = \frac{\Pi_{\boldsymbol{h}_{12}}^{\perp} \boldsymbol{h}_{11}}{||\Pi_{\boldsymbol{h}_{12}}^{\perp} \boldsymbol{h}_{11}||}$,

with $\Pi_{\boldsymbol{h}_{kl}}^{\perp} := \boldsymbol{I} - \frac{\boldsymbol{h}_{kl} \boldsymbol{h}_{kl}^H}{||\boldsymbol{h}_{kl}||^2}$. The intuition behind the parameterization in (10.3) is that power should be either directed into the direction of the direct channel to increase signal power (the term corresponding to MRT) or into the null space of the interference channel (the term corresponding to ZF). All other directions would be a waste of the transmit power. Another useful parameterization of the beamforming vectors in (10.3) is given by

$$\boldsymbol{w}_k(\lambda_k) = \sqrt{\lambda_k} \frac{\Pi_{\boldsymbol{h}_{kl}} \boldsymbol{h}_{kk}}{||\Pi_{\boldsymbol{h}_{kl}} \boldsymbol{h}_{kk}||} + \sqrt{1 - \lambda_k} \frac{\Pi_{\boldsymbol{h}_{kl}}^{\perp} \boldsymbol{h}_{kk}}{||\Pi_{\boldsymbol{h}_{kl}}^{\perp} \boldsymbol{h}_{kk}||}, \tag{10.4}$$

where $\lambda_k \in [0, \lambda_k^{\mathrm{MRT}}]$ with $\lambda_k^{\mathrm{MRT}} = ||\Pi_{\boldsymbol{h}_{kl}} \boldsymbol{h}_{kk}||^2 / ||\boldsymbol{h}_{kk}||^2$ and $\Pi_{\boldsymbol{h}_{kl}} := \frac{\boldsymbol{h}_{kl} \boldsymbol{h}_{kl}^H}{||\boldsymbol{h}_{kl}||^2}$. There is a unique mapping from λ_k to λ_k' and vice versa.

10.2.2 MISO Interference Channel with Private Messages

In this scenario, we slightly modify the model from Section 10.2.1. Now, the two messages W_1 and W_2 are private and intended only for the corresponding receivers. In order to understand the differences, the encoding and decoding processes are described in detail in the following. The scenario is illustrated in Figure 10.3.

In the definitions, we follow closely [26]. At the transmitters we have stochastic encoders mapping from the message sets to codewords, i.e., $f_t(\boldsymbol{x}_t|w_t), \boldsymbol{x}_t \in \mathcal{X}_t^n, w_t \in \mathcal{W}_t$, and $\sum_{\boldsymbol{x}_t \in \mathcal{X}_t^n} f_t(\boldsymbol{x}_t|w_t) = 1$. The decoding functions are mappings $\psi_t : \mathcal{Y}_t^n \mapsto \mathcal{W}_t, t \in \{0, 1\}$. The secrecy levels at the receivers are measured by their equivocation rates $\frac{1}{n}H(W_2|Y_1)$ and $\frac{1}{n}H(W_1|Y_2)$, respectively.

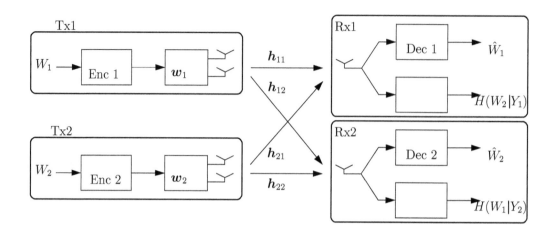

Figure 10.3: The two-user MISO interference channel with private messages (illustrated for $n_T = 2$ transmit antennas).

An $(M_1, M_2, n, P_e^{(n)})$-Code for the interference channel consists of two encoding functions f_1, f_2, two decoding functions ψ_1, ψ_2, and the maximum error probability $P_e^{(n)} = \max\{P_{e,1}^{(n)}, P_{e,2}^{(n)}\}$ with

$$P_{e,t}^{(n)} = \sum_{w_1, w_2} \frac{1}{M_1 M_2} Pr\left[\psi_t(\boldsymbol{y}_t) \neq w_t | (w_1, w_2) \text{ sent}\right]. \tag{10.5}$$

Definition 10.2.2 (Achievable secrecy rate). *A **secrecy rate pair** (sR_1, sR_2) is achievable if for any $\epsilon_0 > 0$ there is a $(M_1, M_2, n, P_e^{(n)})$-Code such that $M_t \geq 2^{n_S R_t}$ and $P_e^{(n)} \leq \epsilon_0$ and security constraints $n_S R_1 - H(W_1|\boldsymbol{y}_2) \leq n\epsilon_0$, $n_S R_2 - H(W_2|\boldsymbol{y}_1) \leq n\epsilon_0$ are satisfied.*

From Lemma 1 in [20], we know that the Gaussian MISO interference channel with private messages in Figure 10.3 has the following achievable secrecy rate pair:

$$sR_1 = \underbrace{\log\left(1 + \frac{\rho|\boldsymbol{w}_1^H \boldsymbol{h}_{11}|^2}{1 + \rho|\boldsymbol{w}_2^H \boldsymbol{h}_{21}|^2}\right)}_{\text{information term}} - \underbrace{\log\left(1 + \rho|\boldsymbol{w}_1^H \boldsymbol{h}_{12}|^2\right)}_{\text{secrecy term}} \tag{10.6}$$

$$sR_2 = \log\left(1 + \frac{\rho|\boldsymbol{w}_2^H \boldsymbol{h}_{22}|^2}{1 + \rho|\boldsymbol{w}_1^H \boldsymbol{h}_{12}|^2}\right) - \log\left(1 + \rho|\boldsymbol{w}_2^H \boldsymbol{h}_{21}|^2\right). \tag{10.7}$$

The difference of the two expressions in (10.6) and (10.7) correspond to the difference of the mutual information expressions (for link one): $I(\boldsymbol{x}_1; \boldsymbol{y}_1) - I(\boldsymbol{x}_1; \boldsymbol{y}_2|\boldsymbol{x}_2)$. It is assumed that the eavesdropper can decode successfully its own message before trying to overhear the private message of the other link.

The inverse noise variance is denoted by ρ. Analog to the standard MISO interference channel, we define the achievable secrecy rate region as

$$sR = \bigcup_{||\boldsymbol{w}_i|| \leq 1} \{sR_1, sR_2\}. \tag{10.8}$$

The Pareto boundary of the achievable secrecy rate region is characterized by the same beamforming parameterization as before, i.e., any point on the Pareto boundary of the secrecy rate region is achievable with the beamforming strategies in (10.3) and (10.4). The proof is provided in [20] but the intuitive argument is the same as before because the same signal power and interference power terms occur in the secrecy rate expressions.

10.2.3 MISO Interference Channel with Public Feedback and Private Messages

In this section, we extend the scenario from the last section further and include the possibility for each link to use a public feedback channel that is overheard by the other receiver. The transmitters $Tx1$ and $Tx2$ intend to send independent messages $W_1 \in \mathcal{W}_1$ and $W_2 \in \mathcal{W}_2$ to the desired receivers $Rx1$ and $Rx2$, respectively. This is done in N channel uses and via public feedback channels while ensuring information-theoretic security (to be defined rigorously below). Receiver k receives the signal $y_k = \boldsymbol{h}_{1k}^H \boldsymbol{w}_1 x_1 + \boldsymbol{h}_{2k}^H \boldsymbol{w}_2 x_2 + n_k$.

The definition of the achievable key rate follows the permissible secret-sharing strategies from [2]. The following protocol is applied: first, both transmitters use the wiretap interference channel N times for their symbols X_1, X_2. The two receivers observe Y_1, Y_2. Then, each receiver uses its public discussion channel to send messages U_1, U_2 back to the transmitters. Finally, the source and destination pairs should generate their secret keys $K_1 = (X_1, U_1)$ and $L_1 = (Y_1, U_1)$ as well as $K_2 = (X_2, U_2)$ and $L_2 = (Y_2, U_2)$.

One important assumption for establishing the achievable secret-key rates is that the public feedback channel induces no delay or errors and offers infinite rate without power constraints. These ideal assumptions are relaxed in [3] and achievable secret-key rates are derived with and without local randomness. In [9], a wiretap channel is considered where

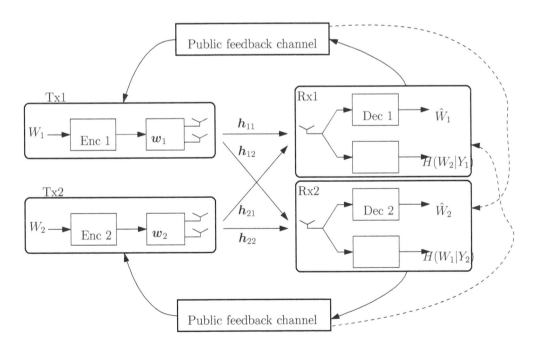

Figure 10.4: The two-user MISO interference channel with private messages and public discussion channels (illustrated for $n_T = 2$ transmit antennas).

the users obtain correlated observations from an input sequence which is controlled by the legitimate transmitter and the secret-key capacity is derived.

The rate R_k is an achievable key rate through the channel (X_k, Y_k, Y_l) with $l \neq k$ if for every $\epsilon > 0$, there exists a permissible secret-sharing strategy as described above such that

$$\Pr[K_k \neq L_k] < \epsilon, \tag{10.9}$$

$$\frac{1}{N} I(K_k; \boldsymbol{y}_k, U_l) < \epsilon, \tag{10.10}$$

$$\frac{1}{N} H(K_k) > R_k - \epsilon, \tag{10.11}$$

$$\frac{1}{N} \log |\mathcal{U}_k| < \frac{1}{N} H(K_k) + \epsilon, \tag{10.12}$$

for sufficiently large N. The first condition in (10.9) corresponds to the requirement that both transmitter and receiver obtain the same secret key. The second condition in (10.10) corresponds to the requirement that the amount of information at the eavesdropper about the secret key K_1 is limited. The third condition in (10.11) corresponds to the requirement that a certain secret-key rate R_k should be achieved. Finally the condition in (10.12) corresponds to the requirement that the generated key is close to uniformly distributed.

In the description of the encoding and decoding setup, we follow closely the approach from [45, Proof Theorem 1]. First, assume that both sources send a sequence of identically and independently distributed symbols $\boldsymbol{x}_1 \in \mathcal{X}_1^N, \boldsymbol{x}_2 \in \mathcal{X}_2^N$ both of length N, each distributed according to $p(X_1), p(X_2)$, over the interference channel. With average power constraints $\mathbb{E}[|X_1|^2] \leq P_1$ and $\mathbb{E}[|X_2|^2] \leq P_2$ and the law of large numbers, the sequences $\boldsymbol{x}_1$ and $\boldsymbol{x}_2$ satisfy the power constraints $\frac{1}{N} \sum_{j=1}^{N} |X_k(j)|^2 \leq P_k$ for $k = \{1, 2\}$. The two observations are $\boldsymbol{y}_1 \in \mathcal{Y}_1^N$ at receiver 1 and $\boldsymbol{y}_2 \in \mathcal{Y}_2^N$ are at receiver 2, respectively.

Next, let us treat the two receivers separately and consider receiver 1: In order to transmit a sequence $\boldsymbol{u}_1$ of N symbols independent of $(\boldsymbol{x}_1, \boldsymbol{y}_1, \boldsymbol{y}_2)$, it sends $\boldsymbol{u}_1 + \boldsymbol{y}_1$ back to the transmitter over its public feedback channel. This creates a conceptual wiretap channel from receiver 1 with input symbol U_1 to the transmitter 1 in the presence of the eavesdropper (receiver two) where the transmitter one observes $(U_1 + Y_1, X_1)$ while the receiver two observes $(U_1 + Y_1, Y_2)$. Applying the continuous alphabet extension of the result in [12, Theorem 3], the achievable secrecy rate is given by

$$R_{S,1} = \max_{U_1} \left[I(U_1; U_1 + Y_1, X_1) - I(U_1; U_1 + Y_1, Y_2) \right]. \tag{10.13}$$

The same argument applies for the second receiver and the achievable secrecy rate for its input symbol U_2, independent of (X_2, Y_1, Y_2), under eavesdropping by receiver one is given by

$$R_{S,2} = \max_{U_2} \left[I(U_2; U_2 + Y_2, X_2) - I(U_2; U_2 + Y_2, Y_1) \right]. \tag{10.14}$$

These two expressions (10.13) and (10.14) are achievable key rates for the original wiretap interference channel with public discussion. In [45, Proof of Theorem 1], it is shown that these two secret rates evaluate to

$$R_{S,k} = \max_{X_k : \mathbb{E}[|X_k|^2] \leq P_k} I(X_k; Y_k) - I(Y_k; Y_l) \text{ for } l \neq k. \tag{10.15}$$

In [36], several-achievable secret key rate regions are introduced and compared. Here, we follow a different approach which is amendable to the game-theoretic model. The expressions for the secret key rates in (10.15) can be approximated as follows,

$$R_{S,k} \leq \max_{X_k : \mathbb{E}[|X_k|^2] \leq P_k} I(X_k; Y_k) - I(Y_k; \tilde{Y}_l) \text{ for } l \neq k,$$

where $\tilde{Y}_l$ is the received signal at receiver l after cancellation of its own signal x_l. Thereby, we obtain from [18, Theorem 1] that the following rates are achievable approximate secret key rates for the MISO wiretap interference channel with public discussion and under the assumption that the eavesdropper first decodes its own message:

$$K_1 = \log \left(1 + \frac{|\boldsymbol{h}_{11}^H \boldsymbol{w}_1|^2}{\left(\sigma_n^2 + |\boldsymbol{h}_{21}^H \boldsymbol{w}_2|^2\right) \left(1 + \rho |\boldsymbol{h}_{12}^H \boldsymbol{w}_1|^2\right)} \right) \tag{10.16}$$

$$K_2 = \log \left(1 + \frac{|\boldsymbol{h}_{22}^H \boldsymbol{w}_2|^2}{\left(\sigma_n^2 + |\boldsymbol{h}_{12}^H \boldsymbol{w}_1|^2\right) \left(1 + \rho |\boldsymbol{h}_{21}^H \boldsymbol{w}_2|^2\right)} \right).$$

Following the definition of the utility regions in the previous two sections, the approximate achievable secret-key rate region is given by

$$\mathcal{K} = \bigcup_{||\boldsymbol{w}_1||^2 \leq P, ||\boldsymbol{w}_2||^2 \leq P} (K_1, K_2). \tag{10.17}$$

The characterization of the beamforming vectors $\boldsymbol{w}_1$ and $\boldsymbol{w}_2$ that can achieve the complete Pareto boundary is the same as in (10.3) and (10.4) (see [18, Corollary 2]).

10.2.4 Discussion and Comparison of Scenarios

The three scenarios and their achievable (secret) transmission rates are summarized below. With the Pareto boundary attaining beamforming vector parameterization, let us define

$$g_k = ||\Pi_{\boldsymbol{h}_{kl}} \boldsymbol{h}_{kk}||^2, \quad \breve{g}_k = ||\Pi_{\boldsymbol{h}_{kl}}^\perp \boldsymbol{h}_{kk}||^2, \quad g_{kl} = ||\boldsymbol{h}_{kl}||^2. \tag{10.18}$$

Then, we have for the direct and cross-channel gains [31, Lemma 1]

$$|\boldsymbol{h}_{kk}^H \boldsymbol{w}_k(\lambda_k)|^2 = \left(\sqrt{\lambda_k g_k} + \sqrt{(1 - \lambda_k)\breve{g}_k} \right)^2 \quad \text{and} \quad |\boldsymbol{h}_{kl}^H \boldsymbol{w}_k|^2 = \lambda_k g_{kl}. \tag{10.19}$$

The achievable rates for the MISO interference channel with public messages (corresponding to the peaceful system) are given by

$$R_1(\lambda_1, \lambda_2) = \log \left(1 + \frac{\left(\sqrt{\lambda_1 g_1} + \sqrt{(1 - \lambda_1)\breve{g}_1} \right)^2}{\sigma^2 + \lambda_2 g_{21}} \right) \tag{10.20}$$

$$R_2(\lambda_1, \lambda_2) = \log \left(1 + \frac{\left(\sqrt{\lambda_2 g_2} + \sqrt{(1 - \lambda_2)\breve{g}_2} \right)^2}{\sigma^2 + \lambda_2 g_{21}} \right).$$

The achievable secrecy rates for the MISO interference channel with private messages and without public discussion channels read

$$_S R_1(\lambda_1, \lambda_2) = \log \left(1 + \frac{\rho \left(\sqrt{\lambda_1 g_1} + \sqrt{(1 - \lambda_1)\breve{g}_1} \right)^2}{1 + \rho \lambda_2 g_{21}} \right) - \log \left(1 + \rho \lambda_1 g_{12}\right) \tag{10.21}$$

$$_S R_2(\lambda_1, \lambda_2) = \log \left(1 + \frac{\rho \left(\sqrt{\lambda_2 g_2} + \sqrt{(1 - \lambda_2)\breve{g}_2} \right)^2}{1 + \rho \lambda_1 g_{12}} \right) - \log \left(1 + \rho \lambda_2 g_{21}\right). \tag{10.22}$$

The achievable approximate secret-key rates for the MISO interference channel with private messages and with public discussion channel are

$$K_1(\lambda_1, \lambda_2) = \log\left(1 + \frac{\left(\sqrt{\lambda_1 g_1} + \sqrt{(1-\lambda_1)\breve{g}_1}\right)^2}{(\sigma^2 + \lambda_2 g_{21})(1 + \rho\lambda_1 g_{12})}\right)$$

$$K_2(\lambda_1, \lambda_2) = \log\left(1 + \frac{\left(\sqrt{\lambda_2 g_2} + \sqrt{(1-\lambda_2)\breve{g}_2}\right)^2}{(\sigma^2 + \lambda_1 g_{12})(1 + \rho\lambda_2 g_{21})}\right). \tag{10.23}$$

We expect that the achievable rate region for the MISO interference channel with public messages is the largest, because only transmit power constraints apply. If a public feedback channel is available from the receivers to their corresponding transmitters, the achievable secret key rate can only be larger than without this public discussion, otherwise the public feedback channel would simply not be used. This intuition is indeed correct: the following inequality chain holds for all $1 \le k \le 2$:

$$R_k \ge K_k \ge {}_sR_k \qquad \Longrightarrow \qquad \mathcal{R} \supseteq \mathcal{K} \supseteq {}_s\mathcal{R}. \tag{10.24}$$

The first inequality in (10.24) follows from $I(Y_k; Y_l) \ge 0$ and the second inequality in (10.24) follows from the Markov chain property $Y_1 \to X_1 \to Y_2$ and the data processing inequality $I(Y_1; Y_2) \le I(X_1; Y_2)$. Please note that the not-approximated secret-key rate does not fulfil the inequality chain in (10.24). In [18], we see that the achievable secret-key region can be smaller or larger than the secrecy region.

The approximate secret-key rate is smaller than the achievable rates of the peaceful system because in the lower part of (10.23), the additional interference term $\rho|\boldsymbol{h}_{12}^H \boldsymbol{w}_1|^2 \left(\sigma_n^2 + |\boldsymbol{h}_{21}^H \boldsymbol{w}_2|^2\right)$ shows up. This term is zero if and only if $|\boldsymbol{h}_{12}^H \boldsymbol{w}_1|^2 = 0$, i.e., ZF transmission is applied or the interference channels are zero. Equality for the complete achievable performance region holds if the interference channels are equal to zero, i.e., $\boldsymbol{h}_{12} = \boldsymbol{h}_{21} = 0$ and two parallel communication channels are available for the two links.

Compared to the signal-to-leakage-and-interference-and-noise ratio (SLINR), which is given by

$$SLINR_1 = \frac{|\boldsymbol{h}_{11}^H \boldsymbol{w}_1|^2}{\sigma_n^2 + |\boldsymbol{h}_{12}^H \boldsymbol{w}_1|^2 + |\boldsymbol{h}_{21}^H \boldsymbol{w}_2|^2}, \tag{10.25}$$

the secret-key rate has an additional advantage for the eavesdropper proportional to the product of interference and leakage. This can be observed if the lower part in (10.23) is expanded and compared to the lower part in (10.25).

10.3   Noncooperative Solutions

In the last section, the achievable performance regions for three different scenarios with and without private messages and with and without public feedback channels are presented. The performance of one link is directly affected by the choice of the beamforming strategy of the other link. This leads to a clear conflict situation which we solve using simple tools from noncooperative game theory. The transmitters are assumed to operate autonomously without central controller for coordination. They also do not cooperate in a sense that they communicate before choosing their strategies. Therefore, we model the situation as a static one-shot game in strategic form. We introduce three solution concepts: dominant strategies, worst-case strategies, and Nash equilibrium strategies and explain their applications. Next, we predict the outcome of the beamforming game for the three scenarios and show the impact of the private messages on the outcome of the game.

10.3.1 Noncooperative Games in Strategic Form

The basic tools from noncooperative game theory (as explained and applied for wireless communications and signal processing, e.g., in [22, 28]) are applied to the three specific MISO interference channel scenarios.

A game in strategic form is described by a triple $\{\mathcal{N}, \mathcal{S}, (u_1, ..., u_K)\}$. In our scenarios, we have two-player games, i.e., the set of players $\mathcal{N} = \{1, 2\}$. The set of strategies $\mathcal{S}$ is decomposed into the strategies of player one and player two $\mathcal{S} = \mathcal{S}_1 \times \mathcal{S}_2$. In our case the strategies are the beamforming vectors $\boldsymbol{w}_1$ and $\boldsymbol{w}_2$ and the optimization space is given by $\mathcal{S}_k = \mathbb{C}^{n_T}$. The utility functions u_1, u_2 for our two-player game are given by the achievable rates R_k, the achievable secrecy rates $_sR_k$, or the achievable secret key rates K_k, depending on the scenario of interest.

One important implication from the characterization of the optimal beamforming vectors in the last section is that we can reduce the strategy space of the two players significantly. Using the parameterization in (10.3) or in (10.4), the strategy space is the interval between zero and one or zero and λ^{MRT}, i.e., $\mathcal{S}_k = [0, 1]$ and $\mathcal{S} = [0, 1]^2$.

The players can choose either a deterministic strategy for the upcoming game, they play a *pure strategy*, or they decide stochastically on the next move, a certain strategy λ_1 with a certain probability $0 \leq p_1(\lambda_1)$ with $\int_0^1 p_1(\lambda_1)d\lambda_1 = 1$, called a *mixed strategy*.

If the best strategy of one player does not depend on the choice of the other player, the best strategy is called a *dominant strategy*. For the dominant strategy λ_1^{DS} it holds that

$$u_1(\lambda_1^{DS}, \lambda_2) \geq u_1(\lambda_1, \lambda_2) \qquad \text{for all} \quad \lambda_2 \in [0, 1]. \tag{10.26}$$

The solution concept of dominant strategies is very strong and easily motivated because it does not make any assumptions on the strategy of the other player and we do not have to assume a certain behavior. If both players have dominant strategies then the strategy tuple $\lambda_1^{DS}, \lambda_2^{DS}$ is called *equilibrium in dominant strategies*. Note that such a dominant strategy equilibrium does not necessarily exist for all games.

If a dominant strategy does not exist, a rather conservative approach is to choose the strategy which maximizes the utility for the worst-case strategy of the other player. The worst-case strategy λ_1^{WC}, also called *maxmin strategy*, fulfills

$$\lambda_1^{WC} = \arg\min_{0 \leq \lambda_1 \leq 1} \max_{0 \leq \lambda_2 \leq 1} \min u_1(\lambda_1, \lambda_2). \tag{10.27}$$

Alternatives for more optimistic or realistic players could be to look at the best strategy or the average strategy, respectively.

Another important solution concept was introduced more than 50 years ago, the *Nash equilibrium* (NE). The idea is to find an equilibrium state such that no player has an incentive to deviate unilaterally from the equilibrium. The pure NE strategies $\lambda_1^{NE}, \lambda_2^{NE}$ satisfy for all $1 \leq k \neq l \leq 2$

$$u_k(\lambda_k^{NE}, \lambda_l^{NE}) \geq u_k(\lambda_k, \lambda_l^{NE}) \qquad \text{for all} \quad \lambda_k \in [0, 1]. \tag{10.28}$$

The definition of the NE can be extended to mixed strategies. The strategy space is then given by a probability distribution (pd) p_1 over $\lambda_1 \in [0, 1]$ and p_2 over $\lambda_2 \in [0, 1]$. In order to evaluate the utility, the average with respect to the two probability measures is computed, i.e., for player one with strategies p_1, p_2,

$$u_1(p_1, p_2) = \mathbb{E}_{\lambda_1, \lambda_2}\left[u_1(\lambda_1, \lambda_2)\right] = \int_0^1 \int_0^1 p_1(\lambda_1) p_2(\lambda_2) u_1(\lambda_1, \lambda_2) \, d\lambda_1 \, d\lambda_2. \qquad (10.29)$$

Then, the definition of the *NE for mixed strategies* can be easily stated for the NE pds p_1^{NE}, p_2^{NE} similar to (10.28) for $1 \le k \ne l \le 2$ as

$$u_k(p_k^{NE}, p_l^{NE}) \ge u_k(p_k, p_l^{NE}) \qquad \text{for all} \quad p_k. \qquad (10.30)$$

Another intuitive motivation for the NE defined above is by the *best response dynamics* (BRD). For the pure strategy case, a best response of one player is a correspondence between the strategy space of the other player and its own strategy space. If it is a singleton, it is called the best response function. It solves the following programming problems (best response) for player one and player two:

$$br_1(\lambda_2) = \operatorname{argmin}\max_{\lambda_1} u_1(\lambda_1, \lambda_2) \quad \text{and} \quad br_2(\lambda_1) = \operatorname{argmin}\max_{\lambda_2} u_2(\lambda_1, \lambda_2). \qquad (10.31)$$

Playing alternately the best response strategy starting with certain λ_2^0 results in

$$br_1(br_2(br_1(br_2(...br_1(\lambda_2^0))))),$$

which will lead (if it converges) to the NE in pure strategies defined above. The reason is because playing the best response in the NE will give the same strategies as before. The NE can be understood as a fixed point of the best responses. If the BRD converges from any starting point, the NE is called *globally stable*.

For the beamforming scenarios in the MISO interference channel, the best response dynamic has an engineering interpretation. Interestingly, the iterative procedure of playing the best responses corresponds to the distributed algorithm illustrated in Figure 10.5.

In Figure 10.5, on the left hand side, a one-dimensional best response is shown for a two-player game. The curve BR1 denotes the best response of player one given the strategy of player two. In the best response dynamics (the dashed step line in the lens), the horizontal

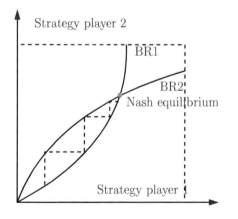
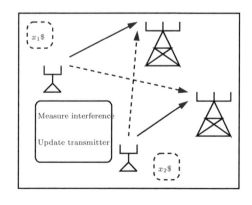

Figure 10.5: Nash equilibrium and best response dynamics and distributed algorithm.

step corresponds to the optimization of the strategy of player one for fixed strategy of player two. In the example, the process converges to the intersection point of both best response curves, the NE. On the right hand side of Figure 10.5, a two-user MIMO interference channel scenario is illustrated. Each iteration of the best responses on the left (one step to the right and one step up) correspond to one iteration of the distributed implementation on the right (measure the noise plus interference covariance matrix and update the transmit strategy).

In general, there are two fundamental questions about the NE for a game in strategic form. The first is its *existence* and the second about *uniqueness*. Note that uniqueness is not sufficient for global stability but necessary. One important result is the existence of the NE in mixed strategies. It follows from a fixed-point theorem that there always exists a NE in mixed strategies for our two-player beamforming game. The existence of a NE in pure strategies needs to be shown. However, the following characterization is useful in showing existence of a pure NE. A game in strategic form with strategy space that consists of nonempty, compact, convex subsets of the Euclidean space and utility functions u_i, *continuous* in the strategy space and *quasiconcave* in the strategy of player i has a NE in pure strategies [14].

If existence is proved, the next question about uniqueness is in general difficult. The specific structures of the utility functions and strategy spaces need to be exploited.

Note that other solution concepts for games in strategic form exist. One prominent example is the correlated equilibrium, which is similar to the NE in mixed strategies as defined above in (10.30). Here, the probability distributions over the strategy spaces of the two players are allowed to be dependent. Instead of the product of the marginal distributions, the joint probability distribution $p(p_1, p_2)$ is used.

The noncooperative game model described above is used for example to find operating points in multicarrier interference channels [41] and in MIMO multiuser systems [40].

10.3.2   Solution for the MISO Interference Channel Scenarios

The solution concepts for the game in strategic form explained in the last section are applied to the following three beamforming games in pure strategies with $\mathcal{N} = \{1, 2\}$ and $\Lambda = [0, 1]^2$:

$$\{\mathcal{N}, \Lambda, (R_1, R_2)\}, \qquad \{\mathcal{N}, \Lambda, (_sR_1, _sR_2)\}, \qquad \{\mathcal{N}, \Lambda, (K_1, K_2)\}. \qquad (10.32)$$

For the MISO interference channel with public messages, we know from [21, Proposition 3], that a NE in dominating strategies exists. The NE strategies are maximum ratio transmission (MRT), i.e., $\lambda_1^{NE} = \lambda_1^{MRT}$ and $\lambda_2^{NE} = \lambda_2^{MRT}$ implying $\boldsymbol{w}_1^{NE} = \frac{\boldsymbol{h}_{11}}{||\boldsymbol{h}_{11}||}$ and $\boldsymbol{w}_2^{NE} = \frac{\boldsymbol{h}_{22}}{||\boldsymbol{h}_{22}||}$. These are dominating strategies, i.e., the strategy choice of the other transmitter does not influence its own strategy. Since this strategy maximizes the achievable SINR regardless of the interference created to the other player, it is called *selfish*.

For the MISO interference channel with private messages without public discussion channel, we know from [20, Theorem 3] that there exists a unique NE in pure strategies for the MISO interference channel secrecy rate game. The beamforming vectors that achieve the NE can be obtained from Algorithm 1.

Finally, for the MISO interference channel with private messages and public discussion channel, we have from [18, Lemma 4] that there exists a Nash equilibrium for the beamforming game in MISO wiretap interference channels with public discussion. Furthermore, it is unique and it has a dominating strategy.

Algorithm 1: Best response dynamics for the-two user MISO interference channel with private messages

---

Result: Beamforming vectors corresponding to NE for game $\{\mathcal{N}, \Lambda, (_SR_1, _SR_2)\}$
Input: Channel realizations $\boldsymbol{h}_{11}, \boldsymbol{h}_{12}, \boldsymbol{h}_{21}, \boldsymbol{h}_{22}$ and noise variance σ_n^2
initialization: $\lambda_1^0 = 0$ and $\lambda_2^0 = 1$, $_SR_1^0 = {_SR_2^0} = 0$, $\ell = 1$;
while $|_SR_1^\ell - {_SR_1^{\ell-1}}| + |_SR_2^\ell - {_SR_2^{\ell-1}}| > \epsilon$ **do**

$\quad \Big|\ \lambda_1^\ell = \text{argmin} \max_{0 \le \lambda_1 \le 1} {_SR_1(\lambda_1, \lambda_2^{\ell-1})}$;

$\quad \Big|\ \lambda_2^\ell = \text{argmin} \max_{0 \le \lambda_2 \le 1} {_SR_2(\lambda_1^\ell, \lambda_2)}$;

$\quad \Big|\_ \ell = \ell + 1$;

Output: Optimal $\lambda_1^{\ell-1}, \lambda_2^{\ell-1}$

Note that the achievable secret key rate depends on the *effective signal-to-noise ratio (SNR)*, i.e., $R_{S,k} = \log(1 + SNR_{S,k})$

$$SNR_{S,k}(\lambda_k, \lambda_l) = \frac{\left(\sqrt{\lambda_k g_k} + \sqrt{(1-\lambda_k)\breve{g}_k}\right)^2}{(\sigma_n^2 + g_{lk}\lambda_l)(1 + \rho g_{kl}\lambda_k)}. \tag{10.33}$$

For fixed strategy of link l, the best response problem for link k is then given by

$$\max_{0 \le \lambda_k \le \lambda_k^{\text{MRT}}} SNR_{S,k}(\lambda_k, \lambda_l). \tag{10.34}$$

The solution of (10.34) is independent of λ_l and therefore the best response strategy of player k is also the dominant strategy Nash equilibrium given in the next theorem.

Theorem 10.3.1. *The beamforming game $\{\mathcal{N}, \Lambda, (K_1, K_2)\}$ has a unique dominant strategy Nash equilibrium given as*

$$\lambda_k^{DS} = \frac{g_k}{\breve{g}_k(1 + \rho g_{kl})^2 + g_k}, \quad k \ne l. \tag{10.35}$$

Proof: The proof is provided in Appendix 10.7.1.

Interestingly, the NE of the peaceful system results in a dominating strategy (namely MRT) and the NE of the approximate secret key rate game results in another dominating strategy, too. The MISO interference channel with private messages has a BRD which converges to a unique NE. Note further, that all NE lead to inefficient operating points, i.e., achievable rates that are not on the Pareto boundary. This observation is confirmed in the numerical illustrations presented in Section 10.5.

From (10.35), the asymptotic behavior of the NE of the beamforming game with secret key rates can be examined at low and high SNR. At low SNR ($\rho \to 0$), λ_k^{DS} converges to $g_k/(\breve{g}_k + g_k) = \lambda_k^{MRT}$. That is, at low SNR the Nash equilibrium corresponds to joint MRT beamforming. At high SNR ($\rho \to \infty$), λ_k^{DS} converges to zero corresponding to ZF beamforming.

10.4 Cooperative Solutions

In contrast to noncooperative games, in this section, we apply models from cooperative game theory to the three MISO interference channel scenarios. The main difference is that the players can negotiate before the play and make binding agreements. This requires the possibility to communicate before the strategies are chosen. Additionally, an external controller should check that the players behave according to their binding agreements, sometimes called exogenous enforcement.

Note that we do not consider games in characteristic form, including coalitional games with and without transferable utilities, because we consider only two-player games in this chapter.

10.4.1 Bargaining Solutions

The broad class of bargaining games, that also includes cooperative games, can be classified into axiomatic, behavioral, and strategic bargaining games. In general, the bargaining game is defined by the triple $\{\mathcal{N}, \mathcal{U}, \boldsymbol{c}\}$: the set of players $\mathcal{N} = \{1, 2\}$, the utility region $\mathcal{U}$ which is either $\mathcal{R}$, $_S\mathcal{R}$, or $\mathcal{K}$ for our three scenarios, and the conflict point $\boldsymbol{c}$ is the outcome which is realized if the players cannot agree. The conflict point $\boldsymbol{c}$ can be fixed and a-priori known or can be part of the bargaining process. The utility region is the set of potential outcomes of the game. If there exists at least one point $\boldsymbol{u} \in \mathcal{U}$ with $u_i > c_i$ for all $i \in \mathcal{N}$, then it is a *bargaining problem*.

One very important axiomatic solution concept is the *Nash bargaining solution* (NBS). A set of reasonable properties of the solution $\boldsymbol{u}^*$ is formulated in terms of axioms: feasibility ($\boldsymbol{u} \geq \boldsymbol{c}$), Pareto-optimality, symmetry, independence of irrelevant alternatives, and invariance to linear transform. Under the requirement that the utility region is convex, the NBS $\boldsymbol{u}^*$ is characterized (see Section 10.4.2 for details) by

$$\boldsymbol{u}^* = \text{argmin}_{\boldsymbol{u} \in \mathcal{U}} \quad (u_1 - c_1) \cdot (u_2 - c_2). \tag{10.36}$$

Note that there are other axiomatic solution concepts including the Kalai-Smorodinsky solution, the proportional solution, and the egalitarian solution.

A convenient way to represent the preferences of the players over the strategy space, is within an *Edgeworth box* [13]. It is applicable for two-player games with scalar strategy spaces. Originally, it was developed for market situations in which the axes correspond to goods and the utility of the players is a function of the amount of goods they possess. For the first time in [31] the idea is proposed to identify the beamforming weights with goods and apply the Edgeworth box approach.

For the representation in Figure 10.6, we identify the good one of player one by $x_1^{(1)} = \lambda_1$ and the good two of player one by $x_2^{(1)} = \lambda_2^{MRT} - \lambda_2$. Similarly, for player two $x_1^{(2)} = \lambda^{MRT} - \lambda_1$ and $x_2^{(2)} = \lambda_2$. The indifference curves are defined for a fixed utility tuple ϕ_1', ϕ_2' as follows:

$$I_1(x_1^{(1)}, \phi_1') = x_2^{(1)} : u_1(x_1^{(1)}, x_2^{(1)}) = \phi_1' \text{ and } I_2(x_2^{(2)}, \phi_2') = x_1^{(2)} : u_2(x_1^{(2)}, x_2^{(2)}) = \phi_2'. \tag{10.37}$$

The utility function can be replaced by the achievable rate R_k, the secrecy rate $_S R_k$, or the approximate achievable secret key rate K_k.

The Edgeworth box [13], [16, Chapter 5], is illustrated in Figure 10.7. The box is constructed in joining Figure 10.6 (left) and Figure 10.6 (right). Thus, the Edgeworth box has two points of origin, O_1 and O_2, corresponding to consumer 1 and consumer 2, respectively. The initial amounts of goods of the consumers define the size of the box (width is λ_1^{MRT}, height is λ_2^{MRT}). The bundles $(x_1^{(1)}, x_2^{(1)})$ and $(x_1^{(2)}, x_2^{(2)})$ make up an *allocation* $\boldsymbol{x} = ((x_1^{(1)}, x_2^{(1)}), (x_1^{(2)}, x_2^{(2)}))$. Every point in the box denotes an allocation, i.e., an assignment of bundles to each consumer. The consumers' preferences in the Edgeworth box can be revealed according to their indifference curves. The dark region in Figure 10.7 is called

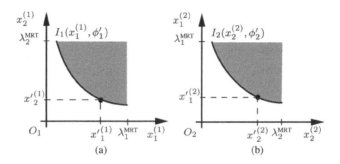

Figure 10.6: An illustration of the indifference curves.

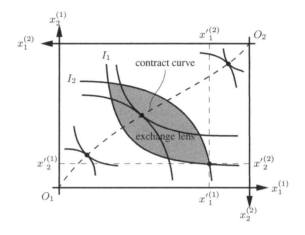

Figure 10.7: An illustration of an Edgeworth box.

the *exchange lens* and contains all allocations that are Pareto improvements to the initial outcome.

The locus of all Pareto-optimal points in the Edgeworth box is called the *contract curve* [13]. On these points, the indifference curves are tangent, and are characterized by the following condition [13, p. 21] for the achievable rates R_k:

$$\frac{\partial R_1(x_1^{(1)}, x_2^{(1)})}{\partial x_1^{(1)}} \frac{\partial R_2(x_1^{(2)}, x_2^{(2)})}{\partial x_2^{(2)}} = \frac{\partial R_2(x_1^{(2)}, x_2^{(2)})}{\partial x_1^{(2)}} \frac{\partial R_1(x_1^{(1)}, x_2^{(1)})}{\partial x_2^{(1)}}. \tag{10.38}$$

Based on the characterization in (10.38), a closed-form expression for the Pareto boundary of the achievable rate region $\mathcal{R}$ is provided in [31, Theorem 2]. For the achievable secrecy region $_S\mathcal{R}$ and the approximate secret-key region $\mathcal{K}$ Equation (10.38) can be solved, too.

Theorem 10.4.1. *The Pareto boundary of the approximate secret key region $\mathcal{K}$ corresponds to the beamforming parameters on the the contract curve ($x_1^{(1)}$ as a function of $x_2^{(2)}$) which is characterized by the root of the following cubic equation,*

$$a[x_1^{(1)}]^3 + b[x_1^{(1)}]^2 + c[x_1^{(1)}] + d = 0 \tag{10.39}$$

in the range $[0, \lambda_1^{MRT}]$ *where*

$$a = -(g_{12} - C)^2 g_1 - (g_{12} - C)^2 \breve{g}_1 \qquad (10.40)$$

$$b = g_1(g_{12} - C)^2 - 2(g_{12} - C)\sigma^2 g_1 - 2\breve{g}_1(g_{12} - C)(\sigma^2 + C) \qquad (10.41)$$

$$c = 2(g_{12} - C)\sigma^2 g_1 - g_1\sigma^4 - \breve{g}_1(\sigma^2 + C)^2 \qquad (10.42)$$

$$d = g_1\sigma^4 \qquad (10.43)$$

with

$$C = \frac{(\sigma^2 + g_{21}x_2^{(2)}) \left(\sqrt{g_2 x_2^{(2)}} - \sqrt{\breve{g}_2(1 - x_2^{(2)})} \right) g_{12}}{(\sigma^2 + g_{21}x_2^{(2)}) \left(\sqrt{\frac{g_2}{x_2^{(2)}}} - \sqrt{\frac{\breve{g}_2}{1-x_2^{(2)}}} \right) - \left(\sqrt{x_2^{(2)}g_2} + \sqrt{\breve{g}_1(1 - x_2^{(2)})} \right) g_{21}}. \qquad (10.44)$$

Proof: The proof is provided in Appendix 10.7.2.

The Pareto boundary of the achievable secrecy region $_S\mathcal{R}$ can be calculated similarly and is left as an exercise for the interested reader.

As a second cooperative solution concept, we apply a bargaining algorithm based on the Edgeworth box representation. In [32], a distributed algorithm is developed which is guaranteed to converge to an operating point on the Pareto boundary of the achievable rate region. The bargaining process is structured in stages. A bargaining stage, indexed with s, can span several bargaining steps. During a bargaining stage s, players 1 and 2 choose $\lambda_1^{(t)}$ and $\lambda_2^{(t)}$ (leading to a new operating point in the Edgeworth box) in the aim of achieving Pareto improvement to the last operating point. Player 1 and 2 adapt $\lambda_1^{(t)}$ and $\lambda_2^{(t)}$ in each bargaining step t based on the signaling between them. In our bargaining process, two bits signaling from each player are sufficient in each bargaining step to reveal information on the position in the Edgeworth box. The choice of $\lambda_k^{(t+1)}$ is determined depending on the position of the initial allocation with respect to the lens. This leads to a case study within the Edgeworth box [32]. The important property of the bargaining algorithm is that it is guaranteed to converge to the Pareto boundary.

10.4.2   Nash Bargaining Solution

Axiomatic bargaining solutions are successfully applied to find operating points for interference channels in a multicarrier setting in [23], for MISO interference channels in [33], and for multiple-input multiple-outpiut (MIMO) interference channels in [8].

For the axiomatic bargaining solution [34], we need to define the conflict point c which is achieved if the two players cannot agree on any outcome. Two popular choices are $c = 0$ or $c = c^{NE}$, where c^{NE} is the outcome for the noncooperative solution $\lambda_1^{NE}, \lambda_2^{NE}$. The Nash bargaining solution (NBS) fulfills the following axioms under the initial requirement that the utility region is convex:

A1 *Feasibility*: The outcome of the bargaining cannot be worse than the conflict point c.

A2 *Pareto-optimality*: The solution should be on the Pareto boundary of the utility region.

A3 *Symmetry*: If the utility region is symmetric, and the conflict point, too, then the bargaining solution should be symmetric, too.

A4 *Independence of irrelevant alternatives:* If the bargaining outcome for a certain utility region results in a point u which lies in a subset of the original region, then the bargaining solution for this smaller region should be also u.

The only bargaining outcome which fulfills all four axioms A1 to A4 can be computed by $\max_{x \in \mathcal{U}} \prod_{k=1}^{K} (x_k - c_k)$. By solving (10.38), it is possible to obtain a closed-form expression for the Pareto boundary of the rate region. Denote the Pareto boundary of the set $\mathcal{R}$ by $\mathcal{P}(\mathcal{R})$. Then, in order to find the NBS solution point in the rate region $\mathcal{R}$ and solve

$$\max_{x}(x_1 - c_1^{NE})(x_2 - c_2^{NE}) \quad \text{s.t.} \quad x \in \mathcal{P}(\mathcal{R}), \tag{10.45}$$

the corresponding one-dimensional line search over λ_1 can be performed.

10.4.3 Bargaining Algorithm in the Edgeworth Box

In order to plot the curves in Figure 10.6 the following closed-form expressions for the indifference curves (here for player one) are helpful:

$$I_1^R(x_1^{(1)}, \phi_1') = \frac{1}{g_{21}} \left(\frac{\left(\sqrt{x_1^{(1)} g_1} + \sqrt{(1 - x_1^{(1)}) \breve{g}_1} \right)^2}{2^{\phi_1'} - 1} - \sigma^2 \right), \tag{10.46}$$

$$I_1^S(x_1^{(1)}, \phi_1') = \frac{1}{g_{21}} \left(\frac{\left(\sqrt{x_1^{(1)} g_1} + \sqrt{(1 - x_1^{(1)}) \breve{g}_1} \right)^2}{2^{\phi_1'} - 1 + \rho x_1^{(1)} g_{12}} - \sigma^2 \right), \tag{10.47}$$

$$I_1^K(x_1^{(1)}, \phi_1') = \frac{1}{g_{21}} \left(\frac{\left(\sqrt{x_1^{(1)} g_1} + \sqrt{(1 - x_1^{(1)}) \breve{g}_1} \right)^2}{(2^{\phi_1'} - 1)(1 + \rho x_1^{(1)} g_{12})} - \sigma^2 \right). \tag{10.48}$$

One important observation is that all three indifference curves are *quasiconcave* functions of $x_1^{(1)}$. This follows from [37] because the denominator is linear in $x_1^{(1)}$ and the numerator is concave in $x_1^{(1)}$.

The three indifference curves from (10.46), (10.47), and (10.48) are illustrated in Figure 10.8. For fixed $x_1^{(1)}$, all points above the line result in higher utility than ϕ_1'. The algorithm outline above is implemented for all three utility functions and illustrated in Section 10.5.3.

10.4.4 Walras Equilibrium Solution

Solution concepts for MIMO interference channels with pricing are derived in [38] and a generalized framework is developed in [39].

As a third cooperative solution concept, we propose a *competitive market model*. In a competitive market, the consumers buy quantities of goods and also sell goods they possess such that they maximize their profit. Each good has a price and every consumer takes the prices as given. The prices of the goods are not determined by consumers, but arbitrated by markets. In our case, the arbitrator determines the prices of the goods. For a detailed description of this solution concept with an application for MISO interference channels with public messages, the interested reader is referred to [17]. The goods correspond to the parameterizations of the efficient beamforming vectors. Let p_k denote the unit price of good k. In order to be able to buy goods, each consumer k is endowed with a budget $\lambda_k^{MRT} p_k$ which is the worth of his initial amounts of goods.

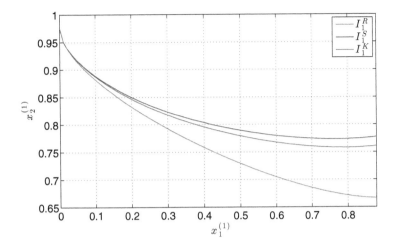

Figure 10.8: Indifference curves for three utility functions $R_k, _S R_k, K_k$.

The demand problem for user $k = 1$ and utility R_1 is given by

$$\max R_1(x_1^{(1)}, x_2^{(1)}) \quad \text{s.t.} \quad p_1 x_1^{(1)} + p_2 x_2^{(1)} \leq \lambda_1^{MRT} p_1. \tag{10.49}$$

Similarly for the second user. In the Walras equilibrium, the prices p_1 and p_2 are chosen such that the demand equals the supply. The supply of good one is λ_1^{MRT} and the supply of good two is λ_2^{MRT}. The demands of user one of good one and good two are $x_1^{(1)}$ and $x_2^{(1)}$, respectively, whereas the demands of user two of good one and good two are $x_1^{(2)}$ and $x_2^{(2)}$, respectively. Denote the solution to the demand problems as a function of the prices p_1, p_2 by $x_1^{*(k)}(p_1, p_2)$ and $x_2^{*(k)}(p_1, p_2)$ for $k = 1, 2$. Then, the Walras equilibrium prices p_1^*, p_2^* satisfy

$$x_1^{*(1)}(p_1^*, p_2^*) + x_1^{*(2)}(p_1^*, p_2^*) = \lambda_1^{MRT} \quad \text{and} \quad x_2^{*(1)}(p_1^*, p_2^*) + x_2^{*(2)}(p_1^*, p_2^*) = \lambda_2^{MRT}. \tag{10.50}$$

For the prices calculated in (10.50) we obtain the allocation on the contract curve where the corresponding indifference curves are tangent. The line passing through this allocation with slope $-p_1/p_2$ defines the budget sets of the consumers. In Figure 10.9, for the MISO interference channel with public messages an Edgeworth box is plotted for a sample channel realization including the Walras equilibrium. A very important property of the Walras equilibrium is that it is always Pareto optimal. This can be observed in Figure 10.9 since it lies on the contract curve.

The difference between the market model for the MISO interference channel with private messages and with public discussion channel and the market model of the setting with public messages is the utility functions of the consumers and the total supply of the goods. In the setting with public discussion, the utility functions of the consumers are the approximate secret-key rates in (10.23) which we write in terms of the goods as $K_1(x_1^{(1)}, x_2^{(1)})$ and $K_2(x_1^{(2)}, x_2^{(2)})$. The supply of goods one and two are λ_1^{DS} and λ_2^{DS}, respectively, which correspond to the Nash equilibrium strategies in Theorem 10.3.1. In this setting, each consumer k is initially endowed with the total supply of his good λ_k^{DS}. Accordingly, the budget of consumer k is $\lambda_k^{DS} p_k$ where p_k is the price of good k.

For given prices (p_1, p_2), the demand problem for user one (similarly for user two) is given by

$$\max K_1(x_1^{(1)}, x_2^{(1)}) \quad \text{s.t.} \quad p_1 x_1^{(1)} + p_2 x_2^{(1)} \leq \lambda_1^{DS} p_1. \tag{10.51}$$

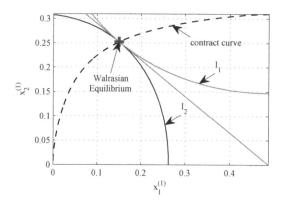

Figure 10.9: Edgeworth box which depicts the allocation for the Walrasian prices.

Let the solution to the consumer demand problems in (10.51) as a function of the prices be $x_1^{*(k)}(p_1, p_2)$ and $x_2^{*(k)}(p_1, p_2)$ for $k = 1, 2$, and define the aggregate excess demand of goods one and two respectively as

$$z_1(p_1, p_2) = x_1^{*(1)}(p_1, p_2) + x_1^{*(2)}(p_1, p_2) - \lambda_1^{DS}, \tag{10.52}$$

$$z_2(p_1, p_2) = x_2^{*(1)}(p_1, p_2) + x_2^{*(2)}(p_1, p_2) - \lambda_2^{DS}. \tag{10.53}$$

The Walrasian equilibrium which equates the demand to the supply of each good corresponds to the prices (p_1^*, p_2^*) such that $z_1(p_1^*, p_2^*) = 0$ and $z_2(p_1^*, p_2^*) = 0$. The existence of a Walras equilibrium in this setting is guaranteed having the following property [16, Theorem 5.5]:

Theorem 10.4.2. *The utility function $K_k(x_1^{(k)}, x_2^{(k)})$ for consumer $k = 1, 2$, is strongly increasing and strictly quasiconcave in $(x_1^{(k)}, x_2^{(k)})$.*

Proof: The proof is provided in Appendix 10.7.3.

An important property of the aggregate excess demand functions which guarantees the uniqueness of the Walras equilibrium is the gross substitute property [29, Proposition 17.F.3]. The gross substitute property is satisfied if whenever the price of one good increases, then the demand of the other good strictly increases [29, Definition 17.F.2].

Theorem 10.4.3. *The aggregate excess demand functions in (10.52) and (10.53) have the gross substitute property.*

Proof: The proof is provided in Appendix 10.7.4.

The unique Walras equilibrium can be calculated iteratively by a price adjustment (tâtonnement) process. In [44], the following discrete time tâtonnement process is proposed:

$$p_i^{(t+1)} = \max\left[p_i^{(t)} + a_i z_i(p_1^{(t)}, p_2^{(t)}), 0\right], \quad i = 1, 2, \tag{10.54}$$

with the step size $a_i > 0$. It is shown in [44] that the process in (10.54) is globally convergent if the aggregate excess demand functions have the gross substitute property. This condition is satisfied in our setting according to Theorem 10.4.3.

The realization of the tâtonnement process is as follows: in each iteration t, given the prices $(p_1^{(t)}, p_2^{(t)})$ the consumers calculate their demands of the goods independently and send them to the arbitrator. The arbitrator calculates the aggregate excess demand functions $z_1(p_1^{(t)}, p_2^{(t)})$ and $z_2(p_1^{(t)}, p_2^{(t)})$ and updates the prices according to (10.54). The arbitrator

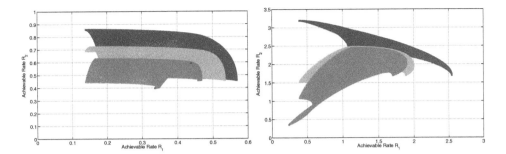

Figure 10.10: The two-user MISO interference channel: achievable rate region with single-user decoding $\mathcal{R}$ (blue), achievable approximate secret key region with public discussion $\mathcal{K}$ (green), achievable secrecy rate region $_S\mathcal{R}$ (red), two transmit antennas, two links, left side –5 dB SNR, right side 5 dB SNR.

then forwards the new prices to the consumers. This process is repeated until convergence to the Walras equilibrium prices.

10.5   Illustrations and Discussions

In this section, we provide numerical assessment of the different solution concepts for the three difference utility functions for the two-user MISO interference channel. First, the utility regions are compared for representative system parameters. Next, the noncooperative operating points NE as well as their cooperative counterparts are discussed. Finally, the iterative bargaining algorithm is explained by numerical simulations.

10.5.1   Comparison of Utility Regions

In Figures 10.10 and 10.11, the achievable rate region of a two-antenna two-user MISO wiretap interference channel is shown with and without public feedback in comparison with the achievable rate for the peaceful system.

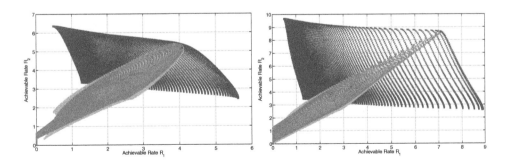

Figure 10.11: The two-user MISO interference channel: achievable rate region with single-user decoding $\mathcal{R}$ (blue), achievable approximate secret key region with public discussion $\mathcal{K}$ (green), achievable secrecy rate region $_S\mathcal{R}$ (red), two transmit antennas, two links, left side 15 dB SNR, right side 25 dB SNR.

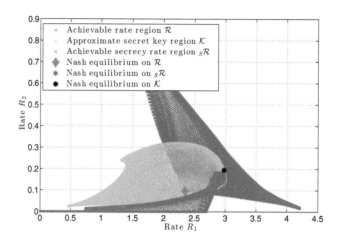

Figure 10.12: Achievable rate regions and noncooperative operating points for a two-antenna sample channel realization at 5 dB SNR.

It can be observed that the peaceful system achieves the largest region in all four cases whereas the gain by the approximate secret key rate with public feedback channel over the secrecy rate region is significant for small SNR but reduces for higher SNR.

In Figure 10.11, the medium and high SNR behavior can be observed. The ZF operating point converges to the Pareto boundary of all three achievable rate regions and is asymptotically sum rate optimal.

10.5.2 Noncooperative and Cooperative Operating Points

First, we compare the noncooperative operating points, namely the three Nash equilibria for the MISO interference channel scenarios. In Figure 10.12, the achievable rate regions for a sample channel realization with two transmit antennas and 5 dB SNR is shown. The three NE operating points are marked as magenta diamond for $\mathcal{R}$, as blue star for $_s\mathcal{R}$, and as black star for $\mathcal{K}$.

One interesting observation in Figure 10.12 is that the NE of the largest region $\mathcal{R}$ is Pareto-dominated by of the both other NES . This implies that for utility functions which do take the information leakage and thereby the interference caused to the other link into account, the outcome of the noncooperative game is improved even though the region itself is smaller. The reason for this is the tragedy of the commons. Note that this is in contrast to the NBS. There by axiom A4—independence of irrelevant alternatives—such a case cannot occur.

From the opposite point of view, this means that the distributed algorithm which corresponds to one implementation of the NE as described in Figure 10.5 results in more efficient operating points if the achievable secrecy rates $_sR_k$ or the approximate achievable secret key rate expressions K_k are used as the utility.

In Figure 10.13, the axiomatic bargaining solutions are illustrated. All NBS operating points are on the Pareto boundary of their respective regions. Therefore, no clear Pareto order can be induced in the example. The NBS in $\mathcal{R}$ provides the largest rate for link one but the smallest rate for link two.

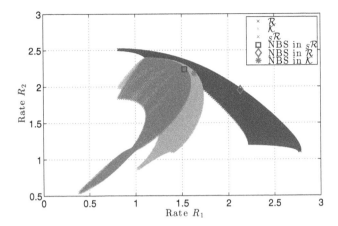

Figure 10.13: Achievable rate regions and NBS operating points for a two-antenna sample channel realization at 5 dB SNR.

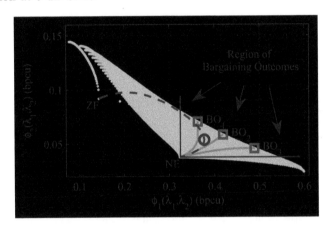

Figure 10.14: SINR region of a two-user MISO IFC with SNR = 0 dB and two antennas at the transmitters. The bargaining outcomes are marked with squares for three different initial step lengths $\delta_1^{(0)}, \delta_2^{(0)}$ with $BO1 : (0.02, 0.01)$, $BO2 : (0.015, 0.01)$, and $BO3 : (0.01, 0.01)$.

10.5.3 Bargaining Algorithm Behavior

The bargaining algorithm described above converges to a point on the Pareto boundary. For the achievable rate region in MISO interference channels, in Figure 10.14 the SINR region is shown including the trajectory of the bargaining algorithm from the NE to the Pareto boundary. The resulting operating point depends on the initial step length $\delta_1^{(0)}$ and $\delta_2^{(0)}$.

The same bargaining algorithm can be applied to the secrecy rate region and the simulation results in Figure 10.15 show one sample trajectory in the achievable secrecy rate region.

Finally, the same bargaining algorithm is applied to the approximate achievable secret key rate region, and the simulation results in Figure 10.16 show one sample trajectory in the achievable approximate secret key region.

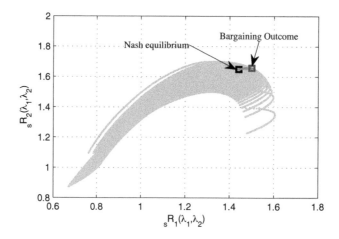

Figure 10.15: Bargaining iterations for the secrecy rate region.

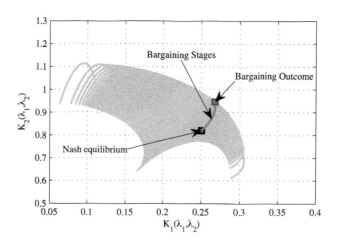

Figure 10.16: Bargaining iterations for the secret key rate region.

Table 10.1: Summary of Scenarios and Results

Scenario	public	private forward	private backward
Utility	achievable rate R_k	achiev. secrecy rate $_sR_k$	approx. secret key rate K_k
Region	$\mathcal{R}$ is largest	$_s\mathcal{R}$ is smallest	$\mathcal{K}$ in between
NE	MRT	best responses unique stable	dominant strategy
NBS Bargain Walras	line-search converges closed form	line search converges tâtonnement	line search converges tâtonnement

10.6   Conclusions

In this chapter, the multiple-antenna interference channel is analyzed in different scenarios: with public messages, with private messages, with and without public feedback channel.

Since transmission of one link creates interference or information leakage on the other link, a resulting conflict situation is described. Solution concepts from game theory are applied to find suitable operating points. Nash equilibrium, Nash bargaining solution, a bargaining algorithm, and Walras equilibrium are described and compared.

The results in this chapter are summarized in Table 10.1.

Finally, we argue that utility functions based on private message transmission which take into account the information leakage are more suitable for noncooperative implementation. From another perspective, this means that selfish utility functions require a certain type of cooperation (bargaining) or regulation (pricing).

Acknowledgment

The work of Eduard Jorswieck and Zuleita Ho has been partly performed in the framework of the European research project DIWINE, which is partly funded by the European Union under its FP7 ICT Objective 1.1—The Network of the Future. The work of Eduard Jorswieck is partly funded by the Federall ministry of Germany and Research of the Federall Republic of Germany (Fördes Kewuzeiclieu 16 KIS 0009, Propleylase.) The authors alone are responsible for athe context of the chapter.

The authors would like to thank Erik Larsson for discussions on game theoretic models and Somayeh Salimi for discussions on secret key generation in multi node networks. Rami Mochaourab would like to thank Thomas Wirth for interesting discussions.

10.7 Appendix: Proofs

For the results which are not available in the literature and which are derived in this chapter, we provide the proofs.

10.7.1 Proof of Theorem 10.3.1

The stationary point of problem (10.34) is given by

$$\frac{\partial SNR_{S,k}}{\partial \lambda_k} = \frac{\left(\sqrt{\lambda_k g_k} + \sqrt{(1-\lambda_k)\breve{g}_k}\right)}{\sigma_n^2 + g_{lk}\lambda_l}$$
$$\times \left(\frac{\left(\sqrt{\frac{g_k}{\lambda_k}} - \sqrt{\frac{\breve{g}_k}{1-\lambda_k}}\right)}{(1+\rho g_{kl}\lambda_k)} - \frac{\rho g_{kl}\left(\sqrt{\lambda_k g_k} + \sqrt{(1-\lambda_k)\breve{g}_k}\right)}{(1+\rho g_{kl}\lambda_k)^2}\right) = 0. \quad (10.55)$$

Since $\left(\sqrt{\lambda_k g_k} + \sqrt{(1-\lambda_k)\breve{g}_k}\right)/\sigma_n^2 + g_{lk}\lambda_l > 0$, the condition in (10.55) reduces to

$$\left(\sqrt{\frac{g_k}{\lambda_k}} - \sqrt{\frac{\breve{g}_k}{1-\lambda_k}}\right)(1+\rho g_{kl}\lambda_k) = \rho g_{kl}\left(\sqrt{\lambda_k g_k} + \sqrt{(1-\lambda_k)\breve{g}_k}\right). \quad (10.56)$$

Multiplying both sides of (10.56) by $\sqrt{\lambda_k(1-\lambda_k)}$ we get

$$\left(\sqrt{g_k(1-\lambda_k)} - \sqrt{\breve{g}_k\lambda_k}\right)(1+\rho g_{kl}\lambda_k) = \rho g_{kl}\left(\lambda_k\sqrt{(1-\lambda_k)g_k} + (1-\lambda_k)\sqrt{\lambda_k\breve{g}_k}\right)$$

(10.57a)

$$\sqrt{g_k(1-\lambda_k)}\,(1+\rho g_{kl}\lambda_k - \rho g_{kl}\lambda_k) = \sqrt{\breve{g}_k\lambda_k}\,(1+\rho g_{kl}\lambda_k + \rho g_{kl}(1-\lambda_k)) \quad (10.57b)$$

$$\sqrt{g_k(1-\lambda_k)} = \sqrt{\breve{g}_k\lambda_k}(1+\rho g_{kl}). \quad (10.57c)$$

Squaring both sides of (10.57c) and solving for λ_k we get the expression in (10.35).

10.7.2   Proof of Theorem 10.4.1

Substituting the following partial derivatives

$$\frac{\partial SNR_{S,1}}{\partial x_1^{(1)}} = \frac{\left(\sqrt{x_1^{(1)}g_1} + \sqrt{(1-x_1^{(1)})\breve{g}_1}\right)}{\sigma_n^2 + g_{21}(\lambda_2^{MRT} - x_2^{(1)})}$$
$$\times \left(\frac{\left(\sqrt{\frac{g_1}{x_1^{(1)}}} - \sqrt{\frac{\breve{g}_1}{1-x_1^{(1)}}}\right)}{(1+\rho g_{12}x_1^{(1)})} - \frac{\rho g_{12}\left(\sqrt{x_1^{(1)}g_1} + \sqrt{(1-x_1^{(1)})\breve{g}_1}\right)}{(1+\rho g_{12}x_1^{(1)})^2}\right) \quad (10.58a)$$

$$\frac{\partial SNR_{S,1}}{\partial x_2^{(1)}} = -\frac{\left(\sqrt{x_1^{(1)}g_1} + \sqrt{(1-x_1^{(1)})\breve{g}_1}\right)^2 g_{21}}{(1+\rho g_{12}x_1^{(1)})\left(\sigma_n^2 + g_{21}(\lambda_2^{MRT} - x_2^{(1)})\right)^2} \quad (10.58b)$$

$$\frac{\partial SNR_{S,2}}{\partial x_2^{(2)}} = \frac{\left(\sqrt{x_2^{(2)}g_2} + \sqrt{(1-x_2^{(2)})\breve{g}_2}\right)}{\sigma_n^2 + g_{12}(\lambda_1^{MRT} - x_1^{(2)})}$$
$$\times \left(\frac{\left(\sqrt{\frac{g_2}{x_2^{(2)}}} - \sqrt{\frac{\breve{g}_2}{1-x_2^{(2)}}}\right)}{(1+\rho g_{21}x_2^{(2)})} - \frac{\rho g_{21}\left(\sqrt{x_2^{(2)}g_2} + \sqrt{(1-x_2^{(2)})\breve{g}_2}\right)}{(1+\rho g_{21}x_2^{(2)})^2}\right) \quad (10.58c)$$

$$\frac{\partial SNR_{S,2}}{\partial x_1^{(2)}} = -\frac{\left(\sqrt{x_2^{(2)}g_2} + \sqrt{(1-x_2^{(2)})\breve{g}_2}\right)^2 g_{12}}{(1+\rho g_{21}x_2^{(2)})\left(\sigma_n^2 + g_{12}(\lambda_1^{MRT} - x_1^{(2)})\right)^2} \quad (10.58d)$$

in Equation (10.38) and arranging the terms we get

$$(\sigma^2 + g_{12}x_1^{(1)})\frac{\left(\sqrt{g_1/x_1^{(1)}} - \sqrt{\breve{g}_1/(1-x_1^{(1)})}\right)}{\left(\sqrt{x_1^{(1)}g_1} + \sqrt{(1-x_1^{(1)})\breve{g}_1}\right)} = C \quad (10.59)$$

where C is a function of $x_2^{(2)}$ given in (10.44). From (10.59) we calculate the following:

$$(\sigma^2 + g_{12}x_1^{(1)})\left(\sqrt{g_1/x_1^{(1)}} - \sqrt{\breve{g}_1/(1-x_1^{(1)})}\right) = C\left(\sqrt{x_1^{(1)}g_1} + \sqrt{(1-x_1^{(1)})\breve{g}_1}\right). \quad (10.60)$$

Multiplying both sides by $\sqrt{x_1^{(1)}(1 - x_1^{(1)})}$ and arranging the terms we get

$$\sqrt{g_1(1 - x_1^{(1)})}\left((\sigma^2 + g_{12}x_1^{(1)}) - Cx_1^{(1)}\right) = \sqrt{\breve{g}_1 x_1^{(1)}}\left(C(1 - x_1^{(1)}) + (\sigma^2 + g_{12}x_1^{(1)})\right).$$

$$(10.61)$$

Squaring both sides of the equation above we have

$$g_1(1 - x_1^{(1)})\left((\sigma^2 + g_{12}x_1^{(1)}) - Cx_1^{(1)}\right)^2 = \breve{g}_1 x_1^{(1)}\left(C(1 - x_1^{(1)}) + (\sigma^2 + g_{12}x_1^{(1)})\right)^2. \quad (10.62)$$

Solving for $x_1^{(1)}$ we get the cubic equation in (10.39).

10.7.3 Proof of Theorem 10.4.2

Consider consumer one. The analysis is analogous for consumer two. Since consumer preference is invariant to positive monotonic transforms [16, Theorem 1.2] we can equivalently analyze the following utility function:

$$f_1(x_1^{(1)}, x_2^{(1)}) = \log\left(\frac{\left(\sqrt{x_1^{(1)}g_1} + \sqrt{(1 - x_1^{(1)})\breve{g}_1}\right)^2}{\left(\sigma^2 + \lambda_2^{DS}g_{21} - x_2^{(1)}g_{21}\right)\left(1 + \rho x_1^{(1)}g_{12}\right)}\right). \quad (10.63)$$

In order to prove that $f_1(x_1^{(1)}, x_2^{(1)})$ is strongly increasing it is shown in [31, Appendix B] that it is sufficient to prove that

$$\frac{\partial f_1(x_1^{(1)}, x_2^{(1)})}{\partial x_1^{(1)}} > 0 \text{ and } \frac{\partial f_1(x_1^{(1)}, x_2^{(1)})}{\partial x_2^{(1)}} > 0. \quad (10.64)$$

The conditions in (10.64) are satisfied since

$$\frac{\partial f_1(x_1^{(1)}, x_2^{(1)})}{\partial x_1^{(1)}} = \frac{\sqrt{\frac{g_1}{x_1^{(1)}}} - \sqrt{\frac{\breve{g}_1}{(1 - x_1^{(1)})}}}{\sqrt{g_1 x_1^{(1)}} - \sqrt{(1 - x_1^{(1)})\breve{g}_1}} - \frac{\rho g_{12}}{1 + \rho x_1^{(1)}g_{12}} > 0 \text{ for } x_1^{(1)} < \lambda_1^{DS}, \quad (10.65)$$

$$\frac{\partial f_1(x_1^{(1)}, x_2^{(1)})}{\partial x_2^{(1)}} = \frac{g_{21}}{\sigma^2 + \lambda_2^{DS}g_{21} - x_2^{(1)}g_{21}} > 0. \quad (10.66)$$

The function $f_1(x_1^{(1)}, x_2^{(1)})$ is strictly quasiconcave since the numerator inside the logarithm in (10.63) is a concave function in $x_1^{(1)}$ and the denominator is linear in $x_1^{(1)}$ and $x_2^{(1)}$ following the results in [37].

10.7.4 Proof of Theorem 10.4.3

Consumer one's demand problem in (10.51) can be equivalently written as

$$\max \quad \log\left(\frac{\left(\sqrt{x_1^{(1)}g_1} + \sqrt{(1 - x_1^{(1)})\breve{g}_1}\right)^2}{\left(\sigma^2 + \lambda_2^{DS}g_{21} - x_2^{(1)}g_{21}\right)\left(1 + \rho x_1^{(1)}g_{12}\right)}\right) \quad (10.67)$$

$$\text{s.t.} \quad p_1 x_1^{(1)} + p_2 x_2^{(1)} \leq \lambda_1^{DS} p_1. \quad (10.68)$$

Since the solution of (10.67) will satisfy the budget constraint with equality, we can reformulate the consumer demand problem as

$$\max_{0\leq x_1^{(1)}\leq \lambda_1^{DS}} \log\left(\frac{\left(\sqrt{x_1^{(1)}g_1}+\sqrt{(1-x_1^{(1)})\breve{g}_1}\right)^2}{\left(\sigma^2+\lambda_2^{DS}g_{21}-\frac{p_1}{p_2}(x_1^{(1)}+\lambda_1^{DS})g_{21}\right)\left(1+\rho x_1^{(1)}g_{12}\right)}\right).\qquad(10.69)$$

We rewrite the above problem as

$$\max_{0\leq x_1^{(1)}\leq \lambda_1^{DS}} \underbrace{\log\left(\frac{\left(\sqrt{x_1^{(1)}g_1}+\sqrt{(1-x_1^{(1)})\breve{g}_1}\right)^2}{\left(\sigma^2+\lambda_2^{DS}g_{21}-\frac{p_1}{p_2}(x_1^{(1)}+\lambda_1^{DS})g_{21}\right)}\right)}_{\phi_1(x_1^{(1)},p_1,p_2)}\underbrace{-\log\left(1+\rho x_1^{(1)}g_{12}\right)}_{\phi_2(x_1^{(1)})}.\qquad(10.70)$$

Since $\phi_1(x_1^{(1)},p_1,p_2)$ is monotonically increasing with $x_1^{(1)}\in[0,\lambda_1^{DS}]$ and $\phi_2(x_1^{(1)})$ is monotonically decreasing with $x_1^{(1)}$, in the optimum of (10.70) we must have

$$\frac{\partial\phi_1(x_1^{(1)},p_1,p_2)}{\partial x_1^{(1)}}=-\frac{\partial\phi_2(x_1^{(1)})}{\partial x_1^{(1)}}.\qquad(10.71)$$

Increasing p_1 increases $\frac{\partial\phi_1(x_1^{(1)},p_1,p_2)}{\partial x_1^{(1)}}$. In order for

$$-\frac{\partial\phi_2(x_1^{(1)})}{\partial x_1^{(1)}}=-\frac{\rho g_{12}}{1+\rho x_1^{(1)}g_{12}}\qquad(10.72)$$

to increase, $x_1^{(1)}$ must decrease. Thus increasing the price of good one decreases the demand of good one by consumer one. The behavior of the demand of good two can be analyzed using the budget constraint. Having $x_2^{(1)}=\frac{p_1}{p_2}(\lambda_1^{DS}-x_1^{(1)})$ then increasing p_1 increases $x_2^{(1)}$ since $x_1^{(1)}$ decreases according to the analysis before. Similarly for the second good. If p_2 increases, then $x_2^{(2)}$ decreases and $x_1^{(2)}$ increases. Accordingly, the aggregate excess demand functions in (10.52) and (10.53) satisfy the gross substitute property.

References

[1] R. Ahlswede. The capacity region of a channel with two senders and two receivers. *Ann. Prob.*, 2:805–814, 1974.
[2] R. Ahlswede and I. Csiszàr. Common Randomness in Information Theory and Cryptography—Part I: Secret Sharing. *IEEE Transactions on Information Theory*, 39(4):1121–1132, 1993.
[3] H. Ahmadi and R. Safavi-Naini. Secret key establishment over noisy channels. In Joaquin Garcia-Alfaro and Pascal Lafourcade, editors, *Foundations and Practice of Security*, volume 6888 of *Lecture Notes in Computer Science*, pages 132–147. Springer, Berlin, 2012.
[4] T. Alpcan and T. Basar. *Network Security: A Decision and Game Theoretic Approach*. Cambridge University Press, 2010.

[5] E. Björnson and E. A. Jorswieck. *Optimal Resource Allocation in Coordinated Multi-Cell Systems.* Foundations and Trends in Communications and Information Theory, Vol. 9, no. 2-3, Jan. 2013, pp, 113–381. Now Publishers, Delpt., 2013.

[6] M. Bloch and J. Barros. *Physical-Layer Security: From Information Theory to Security Engineering.* Cambridge University Press, 2011.

[7] A. B. Carleial. Interference channels. *IEEE Trans. on Inf. Theory*, 24:60–70, 1978.

[8] Z. Chen, S. A. Vorobyov, C.-X. Wang, and J. Thompson. Nash bargaining over MIMO interference systems. In *Communications, 2009. ICC '09. IEEE International Conference on*, pages 1–5, June 2009.

[9] T.-H. Chou, V. Y. F. Tan, and S. C. Draper. The sender-excited secret key agreement model: Capacity theorems. In *Communication, Control, and Computing (Allerton), 2011 49th Annual Allerton Conference on*, pages 928–935, Sept. 2011.

[10] M. H. M. Costa. On the Gaussian interference channel. *IEEE Trans. on Inf. Theory*, 31:607–615, 1985.

[11] I. Csiszár and J. Körner. Broadcast Channels with Confidential Messages. *IEEE Transactions on Information Theory*, Vol. 24(No. 3):339–348, May 1978.

[12] I. Csiszár and J. Körner. Broadcast channels with confidential messages. *IEEE Trans. on Information Theory*, 24:339–348, 1978.

[13] F. Y. Edgeworth. *Mathematical Psychics: An Essay on the Application of Mathematics to the Moral Sciences.* London, U.K.: C. K. Paul, 1881.

[14] D. Fudenberg and J. Tirole. *Game Theory.* MIT Press, 1993.

[15] T. Han and K. Kobayashi. A new achievable rate region for the interference channel. *IEEE Trans. on Information Theory*, 27:49–60, 1981.

[16] G. A. Jehle and P. J. Reny. *Advanced Microeconomic Theory.* Addison-Wesley. Pearson Education, 2nd edition, 2003.

[17] E. Jorswieck and R. Mochaourab. *Mechanisms and Games for Dynamic Spectrum Allocation*, chapter Walrasian Model for Resource Allocation and Transceiver Design in Interference Networks. Cambridge University Press, to appear 2013.

[18] E. A. Jorswieck. Secret key region in multiple antenna wiretap interference channels with public discussion. In *subm. to ITG Conf. on Source and Channel Coding*, 2013.

[19] E. A. Jorswieck, E. G. Larsson, and D. Danev. Complete characterization of the pareto boundary for the MISO interference channel. *IEEE Trans. Signal Process.*, 56(10):5292–5296, Oct. 2008.

[20] E. A. Jorswieck and R. Mochaourab. Secrecy rate region of MISO interference channel: Pareto boundary and non-cooperative games. In *Proc. of ITG IEEE Workshop on Smart Antennas (WSA)*, 2009.

[21] E. G. Larsson and E. A. Jorswieck. Competition versus collaboration on the MISO interference channel. *IEEE Journal on Selected Areas in Communications*, 26:1059–1069, 2008.

[22] E. G. Larsson and E. A. Jorswieck. *Mathematical Foundations for Signal Processing, Communications and Networking*, chapter Game Theory. CRC Press, 2011.

[23] A. Leshem and E. Zehavi. Bargaining over the interference channel. *Proc. IEEE ISIT*, pages 2225–2229, 2006.

[24] S. Leung-Yan-Cheong and M. Hellman. The Gaussian wire-tap channel. *IEEE Transactions on Information Theory*, Vol. 24(No. 4):451–456, July 1978.

[25] Y. Liang, H. V. Poor, and S. Shamai (Shitz). *Information Theoretic Security*, Vol. 5 of *Foundations and Trends in Communications and Information Theory*, pages 355–580. Now Publishers, 2009.

[26] R. Liu, I. Maric, P. Spasojevic, and R. D. Yates. Discrete memoryless interference and broadcast channels with confidential messages: Secrecy rate regions. *IEEE Trans. on Information Theory*, 54:2439–2507, 2008.

[27] R.-H. Liu and W. Trappe, editors. *Security Wireless Communications at the Physical Layer.* Springer, 2009.

[28] A. MacKenzie and L. DaSilva. *Game Theory for Wireless Engineers.* Morgan & Claypool Publishers, 2006.

[29] A. Mas-Colell, M. D. Whinston, and J. R. Green. *Microeconomic Theory.* Oxford University Press, 1995.

[30] R. Mochaourab and E. A. Jorswieck. Optimal beamforming in interference networks with perfect local channel information. *IEEE Trans. on Signal Processing*, 59(3):1128–1141, Mar. 2011.

[31] R. Mochaourab and E. A. Jorswieck. Exchange economy in two-user multiple-input single-output interference channels. *Selected Topics in Signal Processing, IEEE Journal of*, 6(2):151–164, Apr. 2012.

[32] R. Mochaourab, E. A. Jorswieck, Z. K.-M. Ho, and D. Gesbert. Bargaining and beamforming in interference channels. In *Proc. Asilomar Conference on Signals, Systems, and Computers*, 2010.

[33] M. Nokleby and A. L. Swindlehurst. Bargaining and the MISO interference channel. *EURASIP Journal on Advances in Signal Processing*, ID 368547:13 pages, 2009.

[34] H. J. M. Peters. *Axiomatic Bargaining Game Theory.* Kluwer Academic Publishers, 1992.

[35] H. V. Poor. Information and inference in the wireless physical layer. *IEEE Trans. on Wireless Communications*, 19(1):40–47, Jan. 2012.

[36] S. Salimi, E. Jorswieck, and M. Skoglund. Secret key agreement over an interference channel using noiseless feedback. In *subm. to IEEE ISIT*, 2013.

[37] S. Schaible. Fractional programming. *Zeitschrift für Operations Research*, 27(1):39–54, 1983.

[38] D. Schmidt, C. Shi, R. Berry, M. Honig, and W. Utschick. Distributed resource allocation schemes. *IEEE Signal Processing Magazine*, 26(5):53–63, Sept. 2009.

[39] G. Scutari, D. Palomar, F. Facchinei, and J.-S. Pang. Distributed dynamic pricing for MIMO interfering multiuser systems: A unified approach. In *International conference on NETwork Games, COntrol and OPtimization*, 2011.

[40] G. Scutari, D. P. Palomar, and S. Barbarossa. Competitive design of multiuser MIMO systems based on game theory: A unified view. *IEEE Journal on Selected Areas in Communications*, 26:1089–1103, 2008.

[41] G. Scutari, D. P. Palomar, and S. Barbarossa. Optimal linear precoding strategies for wideband non-cooperative systems based on game-theory—Part i: Nash equilibria. *IEEE Trans. on Signal Processing*, 56(3):1230–1249, Mar. 2008.

[42] C. Shannon. Communication theory of secrecy systems. *Bell Syst. Tech. Journal*, 28:656–715, 1949.

[43] Y.-S. Shiu, S.-Y. Chang, H.-C. Wu, S. C.-H. Huang, and H.-W. Chen. Physical layer security in wireless networks: A tutorial. *IEEE Transactions on Wireless Communications*, 18(2):66–74, Feb. 2011.

[44] H. Uzawa. Walras tâtonnement in the theory of exchange. *The Review of Economic Studies*, 27(3):182–194, 1960.

[45] T. F. Wong, M. Bloch, and J. M. Shea. Secret Sharing over Fast-Fading MIMO Wiretap Channels. *EURASIP Journal on Wireless Communications and Networking*, Article ID 506973, 2009.

[46] A. D. Wyner. The wire-tap channel. *Bell Syst. Tech. Journal*, 54:1355–1387, 1975.

Chapter 11

Ascending Clock Auction for Physical Layer Security

Rongqing Zhang
Peking University

Lingyang Song
Peking University

Zhu Han
University of Houston

Bingli Jiao
Peking University

Physical layer security is an emerging security research area that explores the possibilities of achieving perfect secrecy data transmission between sources and intended destinations, while possible malicious eavesdroppers who intend to eavesdrop the communication links obtain zero information. However, such a security is determined by the wireless channel conditions: if the channel between source and destination is worse than the channel between source and eavesdropper, the secrecy rate is typically zero. To overcome this limitation, cooperative jamming is considered as a promising approach where selected jammers can transmit jamming signals to interfere with the malicious eavesdroppers and thus the secrecy capacity can be effectively improved. In this chapter, we consider the jamming power allocation issue in a jammer-assisted cooperative wireless network. The secrecy rate of the source–destination links can be optimized utilizing a well-chosen amount of jamming power from the friendly jammer, and thus each source that can benefit intends to obtain an optimal jamming power to maximize its secrecy rate for data transmission. Then, the problem comes to how to effectively allocate the limited jamming power owned by the friendly jammer among the sources in demand to achieve an optimized system performance. By considering the friendly jammer as the auctioneer and the sources as the bidders, we formulate this power allocation problem as an auction game model, and introduce three distributed auction-based power allocation schemes, i.e., Power allocation scheme based on Single object pay-as-bid Ascending Clock Auction (P-ACA-S), Power allocation scheme based on Traditional Ascending Clock Auction (P-ACA-T), and Power Allocation Scheme based on Alternative Ascending Clock Auction (P-ACA-A). In addition, we investigate some basic properties of the proposed three auction-based power allocation schemes, i.e., convergence, cheat-proof property, and social welfare maximization.

11.1   Introduction

11.1.1   Cooperative Jamming for Physical Layer Security

Traditionally, security in wireless networks is mainly considered at the upper layers using cryptographic methods. However, recent advances in wireless decentralized and adhoc networking have led to increasing attention on studying physical layer security. The basic idea of physical layer security is to exploit the physical characteristics of wireless channels to provide secure communications. This line of work was pioneered by Wyner [1], who introduced the wiretap channel and showed that when the wiretap channel is a degraded version of the main channel, the two legitimate users can exchange secure messages at a nonzero rate without relying on a private key. In follow-up work [2], Wyner's result was generalized to a nondegraded discrete memoryless broadcast channel with common messages sent to both the receivers and confidential messages sent to only one of the receivers. In [3], the secrecy capacity of the Gaussian wiretap channel was studied, and in [30], the secrecy capacity of the quasistatic fading channel was investigated in terms of the outage probability.

Motivated by the fact that if the wiretap channel is less noisy than the main channel the secrecy capacity will be zero, some recent work has been devoted to overcoming this limitation utilizing relay cooperation, which can mainly be classified into two type methods, i.e., cooperative relaying and cooperative jamming.

Cooperative jamming is considered as a promising approach to improve the secrecy capacity by confusing the eavesdropper with codewords independent of the confidential messages [4]. Several cooperative jamming schemes were then investigated for different scenarios to increase the physical layer security [5,7–10,17,27,28]. In [5], the available secrecy rate of the cooperative schemes, i.e., decode-and-forward (DF), amplify-and-forward (AF), and cooperative jamming (CJ) was analyzed and the corresponding system designs consisting of the determination of relay weights and the allocation of transmit power for each cooperative were proposed. In [7], the problem of secure communication in fading channels with a multiantenna transmitter capable of simultaneous transmission of both the information signal and the artificial noise was studied. In [8], the cooperative jamming problem to increase the physical layer security of a wiretap fading channel via distributed relays was solved using a combination of convex optimization and a one-dimensional search. In [9], cooperative jamming strategies in a two-hop relay network where the eavesdropper can wiretap the relay channels in both hops were investigated. In [10], a cooperative wireless network in the presence of one or more eavesdroppers was considered, and node cooperation for achieving physical layer security based on DF and CJ was studied. In [17], several joint cooperative relay and friendly jammer selection schemes were provided for two-way relay networks. In [27,28], coalition formation game-based cooperation schemes of conventional relays and friendly jammers were investigated in order to improve the physical layer security performance through cooperation in a distributed manner.

Note that although the jamming signals sent from the friendly jammers can help the sources by reducing the leaking data rate from the sources to the malicious eavesdropper, at the same time they also reduce the transmission data rate from the sources to the corresponding destinations. Thus, in such cooperative jamming-assisted networks, the network performance of physical layer security depends very much on the efficient power allocation of the jamming signals and the transmitting signals. In [14], the optimal and suboptimal power allocation solutions were investigated as a centralized optimization problem in a network including one source–destination pair, one malicious eavesdropper, and one trusted relay as friendly jammer. However, when there are multiple source–destination pairs or multiple friendly jammers in the network, the optimal power allocation problem is of great complexity and difficult to handle in a centralized scheme, especially in a decentralized or adhoc network where the number of source–destination pairs or friendly jammers is large.

11.1.2 Game Theory-based Jamming Power Allocation

Game theory [19, 20] offers a formal analytical framework with a set of mathematical tools that can effectively study the complex interactions among interdependent rational players. Throughout the past, game theory has made revolutionary impact on a large number of disciplines ranging from engineering, economics, political science, philosophy, and so on. Recently, there has been significant growth in research activities that use game theory for analyzing communication networks. This is mainly due to the need for developing autonomous, distributed, and flexible mobile networks where the network devices can make independent and rational strategic decisions, and the need for low-complexity distributed algorithms that can efficiently represent competitive or collaborative scenarios between network entities. To address the distributed resource allocation problems is an important application of game theory, and different game theory models can be formulated to analyze the corresponding situations.

As for the jamming power allocation issues for physical layer security, in [11, 12] the authors employed a game theory perspective to investigate the interaction between the source and the friendly jammers in a network including one source–destination pair but multiple friendly jammers, where the game between the source and the friendly jammers was formulated as a Stackelberg leader–follower game, and a distributed evolutionary algorithm was proposed to control the jamming power obtained by the source from each friendly jammer. In [13, 14], two-way relaying scenarios for physical layer security with multiple friendly jammers were investigated, and a Stackelberg game-based jamming power control scheme was proposed to optimize the system secrecy capacity. In [21], several auction-based schemes were provided for a jammer-assisted secure network to utilize the jamming power effectively and efficiently.

11.1.3 Ascending Auctions

The auction is a traditional method for selling commodities that have undetermined or variable values. In an auction, each bidder bids for an item, or items, according to a specific mechanism, and the allocation(s) and price(s) for the item, or items, are determined by specific rules. Auction theory, which can be treated as a type of game theory, was pioneered by William Vickrey, who first gives an analysis from the perspective of the incomplete information game [22]. And this has been well developed and widely used in economic theory [23, 24]. Recently, auction theory has been widely employed into wireless resource allocation issues [25, 26].

In ascending auctions [27], the auctioneer announces a price, the bidders report back the quantities demanded at that price, and the auctioneer raises the price. Items are awarded to the bidders at the current price whenever they are "clinched," and the price is incremented until the market clears. With private values, this (dynamic) auction yields the same outcome as the (sealed-bid) Vickrey auction, but has advantages of simplicity and privacy preservation. With interdependent values, this auction may retain efficiency, whereas the Vickrey auction suffers from a generalized Winner's Curse. As discussed in [28], ascending auctions provide a process of price discovery. Value is socially determined through the escalation of bids. Rarely does a bidder enter an auction with fixed values for the items being sold. Rather the bidders learn from each other's bidding, adjusting valuations throughout the process. This process is especially important when resale is a possibility or more generally when others have information relevant to assessing the item's value. This open competition gives ascending auctions a legitimacy that is not shared by other auctions. Throughout the auction, every bidder is given the opportunity to top the high bid. The auction ends when no bidder is willing to do so. The winners win because they are willing to pay a bit more than the others. Losers are given every opportunity to top the winning bid. They lose stems solely due to their failure to do so.

In this chapter, the distributed jamming power allocation problem is formulated as an ascending clock auction [29], and three auction-based power allocation schemes, i.e., Power allocation scheme based on Single object pay-as-bid Ascending Clock Auction (P-ACA-S), Power allocation scheme based on Traditional Ascending Clock Auction (P-ACA-T), and Power allocation scheme based on Alternative Ascending Clock Auction (P-ACA-A), consider the friendly jammer as the auctioneer and the sources as the bidders. During each auction, the auctioneer first announces an initial price, then the bidders report to the auctioneer their demands at that price, and the auctioneer raises the price until the total demands meet the power supply.

11.1.4   Chapter Outline

The rest of this chapter is organized as follows. In Section 1.2, we describe the investigated cooperative jamming scenario and formulate the jamming power allocation problem by defining the utilities of the sources and the friendly jammer. In Section 1.3, we propose three auction-based jamming power allocation schemes, i.e., P-ACA-S, P-ACA-T, and P-ACA-A. In Section 1.4, we further investigate the properties of the proposed auction-based jamming power allocation schemes including the convergence, the cheat-proof property, the social welfare, as well as the complexity and the communication overhead. In Section 1.5, we conclude the chapter and discuss some open issues.

11.2   System Model and Problem Formulation

11.2.1   System Model

In this chapter, we consider a cooperative network consisting of N sources, N corresponding destinations, one friendly jammer, and one malicious eavesdropper, which are denoted by S_i, D_i, $i = 1, 2, \ldots, N$, J, and E, respectively. The network scenario is illustrated in Figure 12.1. We denote by $\mathcal{N}$ the set of indices $\{1, 2, \ldots, N\}$. All the nodes here are equipped with only a single omnidirectional antenna and operate in a half-duplex manner, i.e., each node cannot receive and transmit simultaneously.

Suppose source S_i transmits with power p_i, $i \in \mathcal{N}$. The channel gains from source S_i to destination D_i and malicious eavesdropper E are g_{S_i,D_i} and $g_{S_i,E}$, respectively. Friendly jammer J transmits with power p_i^J to help improve the secrecy rate of data transmission from source S_i to destination D_i. The channel gains from friendly jammer J to destination D_i and eavesdropper E are g_{J,D_i} and $g_{J,E}$, respectively. Note that the channel gains contain the path loss, as well as the Rayleigh fading coefficient with zero mean and unit variance. For simplicity, a quasistatic fading channel is assumed, i.e., the fading coefficients are constant over one slot, and vary independently from one slot to another. The thermal noise at the destination and eavesdropper nodes satisfies independent Gaussian distribution with zero mean and the same variance denoted by σ^2. The channel bandwidth is W.

The channel capacity for source S_i to destination D_i, denoted by C_1^i, can be written as

$$C_1^i = W \log \left(1 + \frac{p_i g_{S_i,D_i}}{\sigma^2 + p_i^J g_{J,D_i}} \right). \tag{11.1}$$

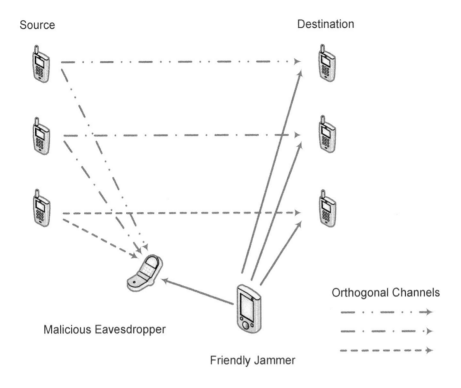

Figure 11.1: System model for the jammer-assisted cooperative network.

Similarly, the channel capacity for source S_i to malicious eavesdropper E, denoted by C_2^i, can be written as

$$C_2^i = W \log \left(1 + \frac{p_i g_{S_i,E}}{\sigma^2 + p_i^J g_{J,E}} \right). \qquad (11.2)$$

Note that here we assume that there is no interference from the other sources, since different sources transmit their messages to the corresponding destinations in orthogonal channels. Then, the secrecy rate for source S_i can be defined as [30]

$$
\begin{aligned}
C_s^i &= \left(C_1^i - C_2^i \right)^+ \\
&= W \left[\log \left(1 + \frac{p_i g_{S_i,D_i}}{\sigma^2 + p_i^J g_{J,D_i}} \right) - \log \left(1 + \frac{p_i g_{S_i,E}}{\sigma^2 + p_i^J g_{J,E}} \right) \right]^+,
\end{aligned} \qquad (11.3)
$$

where $(x)^+$ represents $\max\{x, 0\}$.

From (12.1) and (12.2), we can see that both C_1^i and C_2^i are decreasing and convex functions of the jamming power p_i^J, $i \in \mathcal{N}$. However, if C_2^i decreases faster than C_1^i as p_i^J increases, C_s^i may increase in a certain region of p_i^J. But when p_i^J further increases, both C_1^i and C_2^i approach zero, and thus C_s^i approaches zero. Then, the problems come to whether or not C_s^i can be effectively improved with the help of the friendly jammer, and how to control the jamming power to optimize C_s^i. Comparing the expression of C_1^i with that of C_2^i, we can find that if $\frac{g_{J,D_i}}{g_{S_i,D_i}} < \frac{g_{J,E}}{g_{S_i,E}}$, the gain of C_s^i will be above zero in a certain region of p_i^J. To achieve this improvement effectively in a distributed manner, we propose some auction-based approaches in the following section in this chapter.

11.2.2   Source's Utility Function

To formulate the jamming power allocation problem as an auction, we consider source S_i as one of the bidders, $i \in \mathcal{N}$, while friendly jammer J as the auctioneer. The sources submit bids to compete for the jamming power from the friendly jammer, in order to optimize the secrecy rate of their own data transmission. Source S_i can have a performance gain by successfully getting a proper jamming power. However, in return, it needs to pay for the jamming power offered by the friendly jammer, and the payment is determined by the amount of the jamming power and its unit price. Therefore, the utility function of source S_i can be defined as

$$U_i \left(p_i^J, \lambda \right) = \mathcal{G} \left(p_i^J \right) - \mathcal{P} \left(p_i^J, \lambda \right), \tag{11.4}$$

where $\mathcal{G} \left(p_i^J \right)$ is the performance gain with the jamming power p_i^J, $\mathcal{P} \left(p_i^J, \lambda \right)$ is the cost paid for the friendly jammer, and λ represents the unit price of the jamming power asked by the friendly jammer during the auction.

The performance gain is mainly determined by how much the secrecy rate can be increased with the help of the jamming power. Then $\mathcal{G} \left(p_i^J \right)$ can be written as

$$\mathcal{G} \left(p_i^J \right) = C_s^i - \tilde{C}_s^i, \tag{11.5}$$

where C_s^i and $\tilde{C}_s^i$ represent the secrecy rate with and without the jamming power p_i^J, respectively. C_s^i is given in (12.3), while $\tilde{C}_s^i$ can be obtained by setting $p_i^J = 0$ in (12.3) as

$$\tilde{C}_s^i = W \left[\log \left(1 + \frac{p_i g_{S_i, D_i}}{\sigma^2} \right) - \log \left(1 + \frac{p_i g_{S_i, E}}{\sigma^2} \right) \right]^+. \tag{11.6}$$

Generally speaking, the cost paid for the friendly jammer is higher if the jamming power used is larger. Therefore, the cost function $\mathcal{P} \left(p_i^J, \lambda \right)$ should be monotonically increasing with p_i^J. In the literature, due to its simplicity and efficiency, linear pricing is widely used [29]. Then, the cost function can be written as

$$\mathcal{P} \left(p_i^J, \lambda \right) = \lambda p_i^J, \tag{11.7}$$

where the unit price λ is a constant for all the units of the jamming power, though it may change in different auction rounds.

From (12.4), (12.5), (12.7), and the expressions of C_s^i in (12.3) and $\tilde{C}_s^i$ in (12.6), we can define the utility of source S_i as

$$U_i \left(p_i^J, \lambda \right) = W \left[\log \left(1 + \frac{p_i g_{S_i, D_i}}{\sigma^2 + p_i^J g_{J, D_i}} \right) - \log \left(1 + \frac{p_i g_{S_i, D_i}}{\sigma^2} \right) \right.$$
$$\left. - \log \left(1 + \frac{p_i g_{S_i, E}}{\sigma^2 + p_i^J g_{J, E}} \right) + \log \left(1 + \frac{p_i g_{S_i, E}}{\sigma^2} \right) \right] - \lambda p_i^J, \tag{11.8}$$

which is subject to the transmitting power constraints $0 \leq p_i \leq p_{max}$, $i \in \mathcal{N}$. Note that the optimal bid $p_{i,t}^J$ computed by maximizing the utility of source S_i during each round of the following proposed auction schemes, should also guarantee that $C_s^i \left(p_i, p_{i,t}^J \right)$ in (12.3) is a positive value at some source's transmitting power p_i, which indicates that the jamming power $p_i^J = p_{i,t}^J$ employed in the transmission can indeed lead to a positive gain in the secrecy rate, not just in the utility compared to the no jamming case.

11.2.3 Jammer's Utility Function

The friendly jammer charges the sources for the jamming service at a price λ for every unit of the jamming power. Provided the maximum power is bounded by p_{max}, we can define the utility of friendly jammer J as

$$U_J\left(\{p_i^J\},\lambda\right) = \lambda\sum_i p_i^J, \tag{11.9}$$

which is subject to the total jamming power constraint $0 \le \sum_i p_i^J \le p_{max}$. Besides, there should be a reserve price in the trade, denoted by λ^0, which can be set equal to the average cost of transmitting unit jamming power, i.e., $\lambda^0 = \mathcal{C}/p_{max}$, where $\mathcal{C}$ represents the basic cost of sending the jamming signals at the friendly jammer. Then, we can easily get if the asking price λ is higher than λ^0, the friendly jammer will always benefit from the trade. Otherwise, it will not participate in the trade.

11.3 Auction-based Jamming Power Allocation Schemes

In this section, to solve the power allocation problem in a distributed manner, we propose three auction-based power allocation schemes, i.e., P-ACA-S, P-ACA-T, and P-ACA-A, considering the friendly jammer as the auctioneer and the sources as the bidders based on the ascending clock auction. During each auction, the auctioneer first announces an initial price, then the bidders report to the auctioneer their demands at that price, and the auctioneer raises the price until the total demands meet the power supply.

11.3.1 Power Allocation Scheme based on Single Object Pay-as-Bid Ascending Clock Auction (P-ACA-S)

In this section, P-ACA-S based on the well-known single object pay-as-bid ascending clock auction [27] is proposed, in which the jamming power is sold as a single object and the sources can only bid zero or p_{max} during the auction.

As shown in Table 11.1, before the auction, the friendly jammer sets up the iteration index $t = 0$, the price step $\delta > 0$, as well as the initial asking price λ^0 which is equal to the reserve price given in Section 11.2.3, and then announces λ^0 to all the sources. Each source computes the maximal utility it can obtain if buying the whole jamming power p_{max}

$$U_i^0 = \max_{p_i} U_i\left(p_{max}, \lambda^0\right), \tag{11.10}$$

$$\text{s.t. } C_s^i\left(p_i, p_{max}\right) \ge 0.$$

If the utility U_i^0 is positive, source S_i submits its optimal bid p_{max}. Otherwise, source S_i submits its optimal bid zero. If less than two sources bid p_{max}, the friendly jammer will conclude the auction and choose not to participate in the trade, since its payoff is not more than the basic cost of transmitting the jamming signals. If more than one source bid p_{max}, the friendly jammer continues the auction by raising the asking price $\lambda^{t+1} = \lambda^t + \delta$, increasing the iteration index $t = t + 1$, and announcing λ^t to all the sources. Then, each source resubmits its optimal bid (either zero or p_{max}) by checking the maximal utility

$$U_i^t = \max_{p_i} U_i\left(p_{max}, \lambda^t\right), \tag{11.11}$$

$$\text{s.t. } C_s^i\left(p_i, p_{max}\right) \ge 0.$$

Table 11.1: Algorithm 1: P-ACA-S

Algorithm 1: Power Allocation Scheme based on Single Object Pay-as-Bid Ascending Clock Auction

---

1. Given the available jamming power p_{max}, the price step $\delta > 0$, and the iteration index $t = 0$, friendly jammer J initializes the asking price with the reserve price λ^0.

2. Source S_i computes $U_i^0 = \max\limits_{p_i} U_i \left(p_{max}, \lambda^0 \right)$. If $U_i^0 > 0$, S_i submits its optimal bid p_{max}. Otherwise, S_i submits its optimal bid zero.

3. If less than two sources bid p_{max}, friendly jammer J concludes the auction and chooses not to participate in the trade.

4. Else, set $\lambda^{t+1} = \lambda^t + \delta$, $t = t + 1$, and repeat:
   * Friendly jammer J announces λ^t to all the sources.
   * Source S_i computes $U_i^t = \max\limits_{p_i} U_i \left(p_{max}, \lambda^t \right)$. If $U_i^t > 0$, S_i submits its optimal bid p_{max}. Otherwise, S_i submits its optimal bid zero.
   * If more than one source bid p_{max}, set $\lambda^{t+1} = \lambda^t + \delta$, $t = t + 1$, and continue the auction.
   * Else, conclude the auction. If there still leaves one source that bids p_{max} at the final iteration $T = t$, then allocate the whole jamming power to the source who bids p_{max}. If all the sources bid zero at the final iteration $T = t$, then allocate the whole jamming power to the source who has the maximal utility at the asking price λ^T and make $T = T - 1$.

5. Finally, the utility of source S_i who buys the jamming power is

$$U_i^\star(p_{max}, \lambda^T) = \mathcal{G}\left(p_{max}, p_{i,T}\right) - \lambda^T p_{max},$$

where $p_{i,T} = \text{argmin} \max\limits_{p_i} U_i \left(p_{max}, \lambda^T \right)$ and the expression of $\mathcal{G}$ is given in (12.5).

The auction is repeated until there is not more than one source left bidding p_{max}. In most cases at the final iteration T, there will still leave one source that bids p_{max}, then the whole jamming power is allocated to the source to improve its secrecy rate. However, we should also note that there is a possibility that all the sources bid zero at the final iteration T, then the friendly jammer will allocate the whole jamming power to the source who has the maximal utility at the asking price λ^T but charge it with the unit power price λ^{T-1}.

11.3.2 Power Allocation Scheme based on Traditional Ascending Clock Auction (P-ACA-T)

From Section 1.3.1, we can see that the jamming power offered by the friendly jammer is sold as a single object in P-ACA-S, which may lead to inefficient power allocation since the sources may need only part of rather than the whole jamming power p_{max}. To address this problem, in this section, P-ACA-T based on traditional ascending clock auction [27] is proposed, where each source is allowed to bid any power demand between zero and p_{max} at every iteration.

As shown in Table 11.2, when the friendly jammer gives the initial asking price λ^0, each source submits its optimal bid $p_{i,0}^J$ by computing

$$\left(p_{i,0}^J, p_{i,0}\right) = \text{argmin} \max_{\left(p_i^J, p_i\right)} U_i \left(p_i^J, \lambda^0\right), \tag{11.12}$$

$$\text{s.t. } C_s^i \left(p_{i,0}, p_{i,0}^J\right) \geq 0.$$

Table 11.2: Algorithm 2: P-ACA-T

Algorithm 2: Power Allocation Scheme based on Traditional Ascending Clock Auction

1. Given the available jamming power p_{max}, the price step $\delta > 0$, and the iteration index $t = 0$, friendly jammer J initializes the asking price with the reserve price λ^0.

2. Source S_i computes $\left(p_{i,0}^J, p_{i,0}\right) = \arg\min\limits_{\left(p_i^J, p_i\right)} \max U_i\left(p_i^J, \lambda^0\right)$ and submits its optimal bid $p_{i,0}^J$.

3. Friendly jammer J sums up all the bids from the sources $p_{total,0}^J = \sum\limits_i p_{i,0}^J$ and compares $p_{total,0}^J$ with p_{max}:

   ∗ If $p_{total,0}^J \leq p_{max}$, friendly jammer J concludes the auction and chooses not to participate in the trade.

   ∗ Else, set $\lambda^{t+1} = \lambda^t + \delta$, $t = t + 1$, and repeat:

   ⋆ Friendly jammer J announces λ^t to all the sources.

   ⋆ Source S_i computes $\left(p_{i,t}^J, p_{i,t}\right) = \arg\min\limits_{\left(p_i^J, p_i\right)} \max U_i\left(p_i^J, \lambda^t\right)$ and submits its optimal bid $p_{i,t}^J$.

   ⋆ Friendly jammer J sums up all the bids from the sources $p_{total,t}^J = \sum\limits_i p_{i,t}^J$ and compares $p_{total,t}^J$ with p_{max}:

   • If $p_{total,t}^J > p_{max}$, set $\lambda^{t+1} = \lambda^t + \delta$, $t = t + 1$, and continue the auction.

   • Else, conclude the auction, set $T = t$, and allocate $p_i^{J\star} = p_{i,T}^J + \frac{p_{i,T-1}^J - p_{i,T}^J}{\sum\limits_i p_{i,T-1}^J - \sum\limits_i p_{i,T}^J}\left(p_{max} - \sum\limits_i p_{i,T}^J\right)$ to source S_i.

4. Finally, the utility of source S_i is

$$U_i^\star(p_i^{J\star}, \lambda^T) = \mathcal{G}\left(p_i^{J\star}, p_{i,T}\right) - \lambda^T p_i^{J\star}$$

where $p_{i,T} = \arg\min\limits_{p_i} \max U_i\left(p_{i,T}^J, \lambda^T\right)$ and the expression of $\mathcal{G}$ is given in (12.5).

The friendly jammer sums up all the bids from the sources $p_{total,0}^J = \sum\limits_i p_{i,0}^J$ and compares $p_{total,0}^J$ with p_{max}. If $p_{total,0}^J \leq p_{max}$, the friendly jammer will conclude the auction and choose not to participate in the trade. Otherwise, the friendly jammer sets $\lambda^{t+1} = \lambda^t + \delta$, $t = t + 1$, and announces λ^t to all the sources. Then, each source resubmits its optimal bid $p_{i,t}^J$ by computing

$$\left(p_{i,t}^J, p_{i,t}\right) = \arg\min\limits_{\left(p_i^J, p_i\right)} \max U_i\left(p_i^J, \lambda^t\right), \qquad (11.13)$$

$$\text{s.t. } C_s^i\left(p_{i,t}, p_{i,t}^J\right) \geq 0.$$

Comparing the total bid $p_{total,t}^J = \sum\limits_i p_{i,t}^J$ with p_{max}, if $p_{total,t}^J > p_{max}$, the friendly jammer continues the auction until $p_{total,t}^J \leq p_{max}$. Let the final iteration index be T. As the asking price λ increases discretely every round of the auction, we may have that $p_{total,T}^J < p_{max}$, which does not fully utilize the whole jamming power. To make sure that $p_{total,T}^J = p_{max}$, we modify $p_{i,T}^J$ by introducing proportional rationing [29]. The rationing rule may be specified relatively arbitrarily, but it must satisfy the following monotonicity property: if $p_{i,T}^J < p_{i,T-1}^J$, the expected quantity assigned to source S_i must be strictly greater than its final bid $p_{i,T}^J$; if the final bid $p_{i,T}^J$ of source S_i is increased, with the final bids of all the other sources fixed, the expected quantity assigned to source S_i must increase. Therefore, the final jamming power allocated to source S_i can be given as

$$p_i^{J\star} = p_{i,T}^J + \frac{p_{i,T-1}^J - p_{i,T}^J}{\sum_i p_{i,T-1}^J - \sum_i p_{i,T}^J}\left(p_{max} - \sum_i p_{i,T}^J\right), \tag{11.14}$$

where $\sum_i p_i^{J\star} = p_{max}$.

When the auction concludes, the payment of source S_i is

$$\mathcal{P}_i^\star(p_i^{J\star}, \lambda^T) = \lambda^T p_i^{J\star}. \tag{11.15}$$

11.3.3 Power Allocation Scheme based on Alternative Ascending Clock Auction (P-ACA-A)

As a multiple objects ascending auction, P-ACA-T can achieve efficient jamming power allocation. However, as we will prove in the next section, P-ACA-T is not cheat-proof, which may make the power allocation process out of order. To overcome this problem, P-ACA-A based on alternative ascending clock auction is proposed in this section.

As shown in Table 11.3, the procedures of P-ACA-A are the same as P-ACA-T except that at every iteration t in P-ACA-A, the friendly jammer computes the cumulative clinch [29], which is the amount of the jamming power that each source is guaranteed to win at every iteration. For source S_i at iteration t, it can be expressed as

$$L_i^t = \max\left(0, p_{max} - \sum_{j \text{MRT} i} p_{j,t}^J\right). \tag{11.16}$$

Similar to (11.14), to make sure that $p_{total,T}^J = p_{max}$ at the final iteration T, the final jamming power allocated to source S_i is

$$p_i^{J\star} = p_{i,T}^J + \frac{p_{i,T-1}^J - p_{i,T}^J}{\sum_i p_{i,T-1}^J - \sum_i p_{i,T}^J}\left(p_{max} - \sum_i p_{i,T}^J\right), \tag{11.17}$$

where $\sum_i p_i^{J\star} = p_{max}$.

Note that the key difference between P-ACA-T and P-ACA-A lies in the payment rules of each source for the jamming power it obtains when the auction concludes. In P-ACA-T, the friendly jammer charges each source only by the conditions at the final iteration T, i.e., the final amount of the jamming power allocated to each source and the final unit price λ^T. But in P-ACA-A, the friendly jammer charges each source by the cumulative clinch and the unit price λ^t at every iteration, where the payment of source S_i is

$$\mathcal{P}_i^\star(\{L_i^t\}, \{\lambda^t\}) = \lambda^0 L_i^0 + \sum_{t=1}^T \lambda^t \left(L_i^t - L_i^{t-1}\right). \tag{11.18}$$

The payment design based on the cumulative clinch makes each source to be a truth teller at each iteration during the auction, for the final utility of the source will decrease if it chooses to be a liar when submitting its bid to the friendly jammer. Detailed proof of this cheat-proof property is given in Section 11.4.3. Due to the cheat-proof property, we have that P-ACA-A can guarantee the auction proceeds more orderly compared to P-ACA-T.

Table 11.3: Algorithm 3: P-ACA-A

Algorithm 3: Power Allocation Scheme based on Alternative Ascending Clock Auction

---

1. Given the available jamming power p_{max}, the price step $\delta > 0$, and the iteration index $t = 0$, friendly jammer J initializes the asking price with the reserve price λ^0.

2. Source S_i computes $\left(p_{i,0}^J, p_{i,0}\right) = \arg\min \max_{\left(p_i^J, p_i\right)} U_i \left(p_i^J, \lambda^0\right)$ and submits its optimal bid $p_{i,0}^J$.

3. Friendly jammer J sums up all the bids from the sources $p_{total,0}^J = \sum_i p_{i,0}^J$ and compares $p_{total,0}^J$ with p_{max}:

   * If $p_{total,0}^J \leq p_{max}$, friendly jammer J concludes the auction and chooses not to participate in the trade.

   * Else, set $\lambda^{t+1} = \lambda^t + \delta$, $t = t + 1$, and repeat:

     * Friendly jammer J announces λ^t to all the sources.

     * Source S_i computes $\left(p_{i,t}^J, p_{i,t}\right) = \arg\min \max_{\left(p_i^J, p_i\right)} U_i \left(p_i^J, \lambda^t\right)$ and submits its optimal bid $p_{i,t}^J$.

     * Friendly jammer J sums up all the bids from the sources $p_{total,t}^J = \sum_i p_{i,t}^J$ and compares $p_{total,t}^J$ with p_{max}:

       • If $p_{total,t}^J > p_{max}$, in this algorithm first compute $L_i^t = \max\left(0, p_{max} - \sum_{j \mathrm{MRT} i} p_{j,t}^J\right)$, then set $\lambda^{t+1} = \lambda^t + \delta$, $t = t + 1$, and continue the auction.

       • Else, conclude the auction, set $T = t$, compute $p_i^{J\star} = p_{i,T}^J + \frac{p_{i,T-1}^J - p_{i,T}^J}{\sum_i p_{i,T-1}^J - \sum_i p_{i,T}^J} \left(p_{max} - \sum_i p_{i,T}^J\right)$, and allocate $p_i^{J\star}$ to source S_i.

4. Finally, the utility of source S_i is

$$U_i^\star \left(p_i^{J\star}, \{\lambda^t\}\right) = \mathcal{G}\left(p_i^{J\star}, p_{i,T}\right) - \mathcal{P}_i^\star,$$

where $\mathcal{P}_i^\star$ is the payment of source S_i and can be expressed as

$$\mathcal{P}_i^\star = \lambda^0 L_i^0 + \sum_{t=1}^T \lambda^t \left(L_i^t - L_i^{t-1}\right).$$

11.4 Properties of the Proposed Auction-based Power Allocation Schemes

In this section, we investigate some basic properties of the three proposed auction-based jamming power allocation schemes, i.e., convergence, cheat-proof, and social welfare maximization.

11.4.1 Optimal Jamming Power for Each Source

In this section, we derive the optimal solution of the jamming power for each source during the auction. By differentiating the utility function (12.8) with respect to p_i^J, we have

$$\frac{\partial U_i}{\partial p_i^J} = -\frac{W\gamma_{S_i,D_i}\gamma_{J,D_i}p_i}{\left(1+\gamma_{J,D_i}p_i^J\right)\left(1+\gamma_{S_i,D_i}p_i+\gamma_{J,D_i}p_i^J\right)}$$
$$+ \frac{W\gamma_{S_i,E}\gamma_{J,E}p_i}{\left(1+\gamma_{J,E}p_i^J\right)\left(1+\gamma_{S_i,E}p_i+\gamma_{J,E}p_i^J\right)} - \lambda, \tag{11.19}$$

where $\gamma_{S_i,D_i} \triangleq \frac{g_{S_i,D_i}}{\sigma^2}$, $\gamma_{S_i,E} \triangleq \frac{g_{S_i,E}}{\sigma^2}$, $\gamma_{J,D_i} \triangleq \frac{g_{J,D_i}}{\sigma^2}$, and $\gamma_{J,E} \triangleq \frac{g_{J,E}}{\sigma^2}$, $i \in \mathcal{N}$.

To obtain the optimal solution of the jamming power for source S_i, let $\frac{\partial U_i}{\partial p_i^J} = 0$, and then we have

$$\frac{\lambda}{W} = -\frac{\gamma_{S_i,D_i}\gamma_{J,D_i}p_i}{\left(1+\gamma_{J,D_i}p_i^J\right)\left(1+\gamma_{S_i,D_i}p_i+\gamma_{J,D_i}p_i^J\right)}$$
$$+ \frac{\gamma_{S_i,E}\gamma_{J,E}p_i}{\left(1+\gamma_{J,E}p_i^J\right)\left(1+\gamma_{S_i,E}p_i+\gamma_{J,E}p_i^J\right)}. \tag{11.20}$$

Rearranging the above equation, we can get a fourth order polynomial equation as

$$\left(p_i^J\right)^4 + F_{i,3}\left(p_i^J\right)^3 + F_{i,2}\left(p_i^J\right)^2 + F_{i,1}\left(p_i^J\right) + F_{i,0} = 0, \tag{11.21}$$

where $F_{i,l}$, $l = 0, 1, 2, 3$, are formulae of the constants γ_{S_i,D_i}, $\gamma_{S_i,E}$, γ_{J,D_i}, and $\gamma_{J,E}$, as well as the variables p_i and λ, and can be further calculated as

$$F_{i,3} = \frac{2+\gamma_{S_i,D_i}p_i}{\gamma_{J,D_i}} + \frac{2+\gamma_{S_i,E}p_i}{\gamma_{J,E}}, \tag{11.22}$$

$$F_{i,2} = \frac{(2+\gamma_{S_i,D_i}p_i)(2+\gamma_{S_i,E}p_i)}{\gamma_{J,D_i}\gamma_{J,E}} + \frac{1+\gamma_{S_i,D_i}p_i}{\gamma_{J,D_i}^2} + \frac{1+\gamma_{S_i,E}p_i}{\gamma_{J,E}^2}$$
$$+ \frac{W\gamma_{S_i,D_i}p_i}{\lambda\gamma_{J,D_i}} - \frac{W\gamma_{S_i,E}p_i}{\lambda\gamma_{J,E}}, \tag{11.23}$$

$$F_{i,1} = \frac{(1+\gamma_{S_i,E}p_i)(2+\gamma_{S_i,D_i}p_i)}{\gamma_{J,D_i}\gamma_{J,E}^2} + \frac{(1+\gamma_{S_i,D_i}p_i)(2+\gamma_{S_i,E}p_i)}{\gamma_{J,D_i}^2\gamma_{J,E}}$$
$$+ \frac{2Wp_i(\gamma_{S_i,D_i}-\gamma_{S_i,E})}{\lambda\gamma_{J,D_i}\gamma_{J,E}}, \tag{11.24}$$

$$F_{i,0} = \frac{(1+\gamma_{S_i,D_i}p_i)(1+\gamma_{S_i,E}p_i)}{\gamma_{J,D_i}^2\gamma_{J,E}^2} + \frac{Wp_i\gamma_{S_i,D_i}(1+\gamma_{S_i,E}p_i)}{\lambda\gamma_{J,D_i}\gamma_{J,E}^2}$$
$$- \frac{Wp_i\gamma_{S_i,E}(1+\gamma_{S_i,D_i}p_i)}{\lambda\gamma_{J,D_i}^2\gamma_{J,E}}. \tag{11.25}$$

The solutions of the quartic Equation (11.21) can be expressed in closed form [33], but this is not our primary goal here. The optimal solution to our particular interest can be given as

$$p_i^{J*} = p_i^{J*}\left(\lambda, p_i, \gamma_{S_i,D_i}, \gamma_{S_i,E}, \gamma_{J,D_i}, \gamma_{J,E}\right), \tag{11.26}$$

which is a function of the asking price λ, the source transmitting power p_i, and other channel parameters. Noting that there may be up to four roots of the polynomial Equation (11.21), the selected solution should be a real root and can lead to a higher value of U_i in (12.8) than the other real ones. Subject to the power constraint $0 \leq \sum_i p_i^J \leq p_{max}$ in the auction, we can get the optimal strategy for source S_i, $i \in \mathcal{N}$, as

$$p_{i\_opt}^J (\lambda) = \min \left[p_{max}, \max \left(p_i^{J*}, 0 \right) \right]. \tag{11.27}$$

If there are no real roots of the Equation (11.21), then the optimal strategy will be either $p_{i\_opt}^J = 0$ or $p_{i\_opt}^J = p_{max}$, according to which can achieve a higher utility value U_i when the other parameters are settled.

11.4.2 Convergence

In this section, we prove that all the three auction-based power allocation schemes have the convergence property, i.e., each scheme concludes in a finite number of iterations.

Theorem 11.4.1. *P-ACA-S concludes in a finite number of iterations.*

Proof: From (12.8) and (12.11), we have

$$
\begin{aligned}
U_i^t &= \max_{p_i} U_i \left(p_{max}, \lambda^t \right) \\
&= \max_{p_i} \left\{ W \left[\log \left(1 + \frac{p_i g_{S_i, D_i}}{\sigma^2 + p_{max} g_{J, D_i}} \right) - \log \left(1 + \frac{p_i g_{S_i, D_i}}{\sigma^2} \right) \right. \right. \\
&\quad \left. \left. - \log \left(1 + \frac{p_i g_{S_i, E}}{\sigma^2 + p_{max} g_{J, E}} \right) + \log \left(1 + \frac{p_i g_{S_i, E}}{\sigma^2} \right) \right] - \lambda^t p_{max} \right\}.
\end{aligned} \tag{11.28}
$$

Therefore, we can get

$$U_i^{t+1} - U_i^t = -\delta p_{max} < 0. \tag{11.29}$$

According to Algorithm 11.1, we have that $p_{i,t}^J = p_{max}$ if $U_i^t > 0$, and $p_{i,t}^J = 0$ if $U_i^t \leq 0$. As given in (11.29), $U_i^{t+1} < U_i^t$. Then, with a sufficiently large t, we can get $p_{i,t+1}^J = 0 \leq p_{i,t}^J$, $i \in \mathcal{N}$. Therefore, there exists a finite positive iteration index T satisfying the condition that $\sum_{i=1}^N p_{i,T}^J \leq p_{max}$, which means at most one source bids p_{max} at iteration T, and thus P-ACA-S concludes in a finite number of iterations.

Theorem 11.4.2. *P-ACA-T and P-ACA-A conclude in a finite number of iterations.*

Proof: Subject to the power constraints $0 \leq \sum_i p_i^J \leq p_{max}$ and $0 \leq p_i \leq p_{max}$, we can obtain that the right side of the Equation (11.20) is upper bounded by a nonnegative value, which can be denoted by M_i, $i \in \mathcal{N}$. Then, from (11.19), we have

$$\frac{\partial U_i}{\partial p_i^J} \leq W M_i - \lambda < 0, \quad \text{when } \lambda > W M_i, \tag{11.30}$$

which means the utility U_i is a monotone decreasing function with respect to p_i^J under the condition that $\lambda > W M_i$. Thus, we can obtain that the optimal strategy for source S_i satisfies

$$p_{i\_opt}^J (\lambda) = 0, \quad \text{when } \lambda > W M_i. \tag{11.31}$$

According to Algorithm 11.2 and Algorithm 11.3, we have that the asking price λ increases with a fixed price step $\delta > 0$ until the auction concludes, then λ becomes quite a high value with a sufficiently large t. From (11.31) we have

$$p_{i\_opt}^J (\lambda) = 0, \quad \text{when } t > \frac{W M_i}{\delta}, \tag{11.32}$$

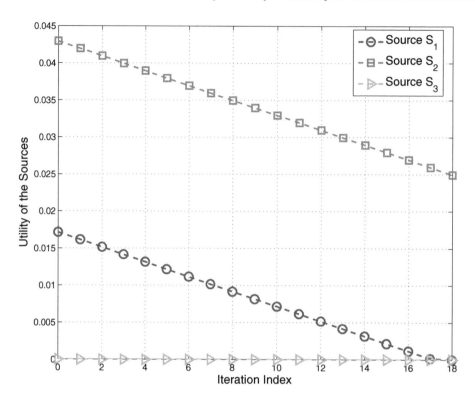

Figure 11.2: Convergence of P-ACA-S.

which means each source will submit its optimal bid $p_{i,t}^J = 0$, $i \in \mathcal{N}$, at the iteration index $t > \frac{WM_{max}}{\delta}$, where $M_{max} = \max\{M_i \mid i \in \mathcal{N}\}$. Therefore, there exists a finite positive iteration index T, $T \leq \frac{WM_{max}}{\delta}$, satisfying the condition that $\sum_{i=1}^{N} p_{i,T}^J \leq p_{max}$, which means P-ACA-T and P-ACA-A conclude in a finite number of iterations.

In addition, from the proofs of Theorem 11.4.1 and Theorem 11.4.2, we can see that the value of the price step δ has a direct impact on the speed of convergence of the proposed auction-based schemes. The proposed schemes conclude fast when δ is large, while they conclude slowly when δ is small. Thus, we can design the value of δ based on the actual demands in a system to achieve a trade-off between the convergence time and the system performance.

Figures 11.2 to 11.4 illustrate the convergence process of P-ACA-S, P-ACA-T, and P-ACA-A, respectively, where there are three independent source–destination pairs in the considered network. From the figures, we can see that no matter in which of the three proposed auction-based schemes, the utility of each source is always a nonincreasing function in terms of the iteration index before the auction concludes, and the utility stays the same value only when it falls to zero. Therefore, each scheme concludes in a finite number of iterations, which means all three proposed schemes can converge. Note that the simulation channel conditions of the three figures are independent, thus the utilities of different schemes here have no comparability. Furthermore, in Figure 11.2, we can see that before one source bids zero and actually quits the auction at one iteration index, the utilities of the source to quit and other sources who also bid p_{max} are parallel with each other. This is because the utilities of the sources that still bid p_{max} and stay in the auction are decreasing linear functions of the variable λ in (12.8) only, where the performance gain of the secrecy rate always stays the same with the jamming power p_{max}. P-ACA-S concludes when there is

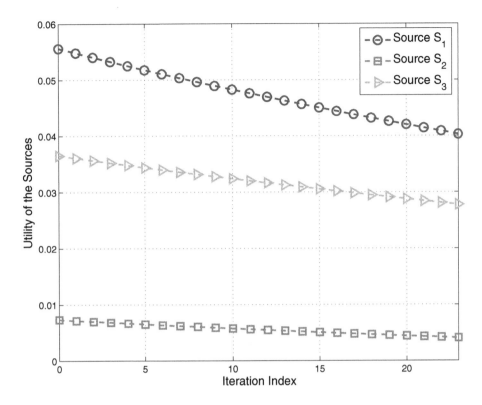

Figure 11.3: Convergence of P-ACA-T.

only one source left to bid in the auction, which means only one source can win through the auction. In Figures 11.3 and 11.4, we can see that all three sources may have a positive performance gain through the auction. In P-ACA-A, the utility of each source is increased at the final iteration index, because when the auction concludes, the final payment of each source is calculated by (11.18) that is slightly less than the usual payment.

11.4.3 Cheat-proof

In this section, we prove that P-ACA-S and P-ACA-A are cheat-proof, while P-ACA-T is not. In our definition, cheat-proof means reporting true optimal demand at every iteration during the auction is a mutually best response for each source. Since the sources are naturally selfish, the property of cheat-proof is crucial for the network performance.

Theorem 11.4.3. *P-ACA-S is cheat-proof.*

Proof: According to Algorithm 11.1, in P-ACA-S, each source has only two choices to bid at every iteration, i.e., either zero or p_{max}. Given that all the other sources report their true optimal demands at every iteration during the auction, we consider source S_i's strategy under the following two cases: if S_i bids zero when its true optimal demand is p_{max}, it will quit the auction and have no performance gain; if S_i bids p_{max} when its true optimal demand is zero and finally wins the whole jamming power, it will get a negative performance gain. From the above analysis, we can see that for source S_i, its optimal strategy is to report its true optimal demand at every iteration since any cheating will lead to a loss in its utility. Note that all the sources are noncollaborative. Therefore, P-ACA-S is cheat-proof.

Theorem 11.4.4. *P-ACA-A is cheat-proof.*

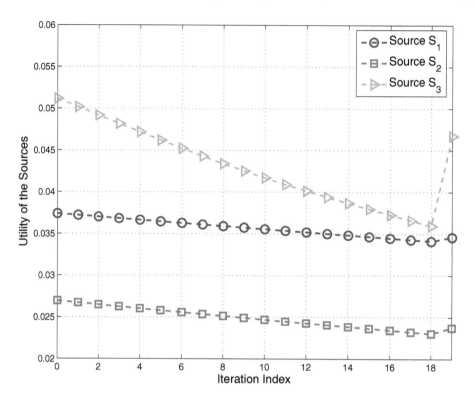

Figure 11.4: Convergence of P-ACA-A.

Proof: Given that all the other sources report their true optimal demands at every iteration during the auction, the auction will conclude at iteration T_1 if source S_i also reports its true optimal demand at every iteration, while the auction will conclude at iteration T_2 if source S_i does not report its true optimal demand at every iteration. The final utility of source S_i is then denoted by $U_i^{T_1}$ and $U_i^{T_2}$, respectively.

According to Algorithm 11.3, we have

$$U_i^{T_j} = \mathcal{G}\left(p_{i,T_j}^J, p_{i,T_j}\right) - \lambda^0 L_i^0 - \sum_{t=1}^{T_j} \lambda^t \left(L_i^t - L_i^{t-1}\right),\ j \in \{1,2\}, \tag{11.33}$$

where the expression of $\mathcal{G}$ is given by (12.3), (12.5), and (12.6).

When δ is sufficiently small, we have

$$L_i^{T_j} = p_{max} - \sum_{k=1, k_{\mathrm{MRT}i}}^{N} p_{k,T_j}^J = p_{i,T_j}^J,\ j \in \{1,2\}. \tag{11.34}$$

- If $T_2 < T_1$, as the asking price increases with the iteration index t, we have $\lambda^{T_1} > \lambda^{T_2}$. Then, we can get

$$\begin{aligned}
U_i^{T_1} - U_i^{T_2} &= \mathcal{G}\left(p_{i,T_1}^J, p_{i,T_1}\right) - \mathcal{G}\left(p_{i,T_2}^J, p_{i,T_2}\right) - \sum_{t=T_2+1}^{T_1} \lambda^t \left(L_i^t - L_i^{t-1}\right) \\
&> \mathcal{G}\left(p_{i,T_1}^J, p_{i,T_1}\right) - \lambda^{T_1} p_{i,T_1}^J - \mathcal{G}\left(p_{i,T_2}^J, p_{i,T_2}\right) + \lambda^{T_1} p_{i,T_2}^J \\
&= U_i\left(p_{i,T_1}^J, p_{i,T_1}, \lambda^{T_1}\right) - U_i\left(p_{i,T_2}^J, p_{i,T_2}, \lambda^{T_1}\right) \\
&\geq 0, \tag{11.35}
\end{aligned}$$

where the last inequality comes from (11.13) that $\left(p_{i,T_1}^J, p_{i,T_1}\right) = \underset{\left(p_i^J, p_i\right)}{\mathrm{argmin}} \max$

$U_i\left(p_i^J, \lambda^{T_1}\right)$.

- If $T_2 \geq T_1$, as the asking price increases with the iteration index t, we have $\lambda^{T_1} \leq \lambda^{T_2}$. Then, we can get

$$
\begin{aligned}
U_i^{T_1} - U_i^{T_2} &= \mathcal{G}\left(p_{i,T_1}^J, p_{i,T_1}\right) - \mathcal{G}\left(p_{i,T_2}^J, p_{i,T_2}\right) + \sum_{t=T_1+1}^{T_2} \lambda^t \left(L_i^t - L_i^{t-1}\right) \\
&> \mathcal{G}\left(p_{i,T_1}^J, p_{i,T_1}\right) - \lambda^{T_1} p_{i,T_1}^J - \mathcal{G}\left(p_{i,T_2}^J, p_{i,T_2}\right) + \lambda^{T_1} p_{i,T_2}^J \\
&= U_i\left(p_{i,T_1}^J, p_{i,T_1}, \lambda^{T_1}\right) - U_i\left(p_{i,T_2}^J, p_{i,T_2}, \lambda^{T_1}\right) \\
&\geq 0, \hspace{6cm} (11.36)
\end{aligned}
$$

where the last inequality comes from (11.13) that $\left(p_{i,T_1}^J, p_{i,T_1}\right) = \underset{\left(p_i^J, p_i\right)}{\mathrm{argmin}} \max$

$U_i\left(p_i^J, \lambda^{T_1}\right)$.

From (11.35) and (11.36), we can obtain that $U_i^{T_1} \geq U_i^{T_2}$. Therefore, given that all the other sources report their true optimal demands at every iteration, the best strategy for source S_i is to report its true optimal demand at every iteration. Since all the sources are noncollaborative, reporting true optimal demand at every iteration is a mutually best response for each source. There is no incentive for the sources to cheat since any cheating may lead to a loss in utility. Therefore, P-ACA-A is cheat-proof. $\blacksquare$

Theorem 11.4.5. *P-ACA-T is not cheat-proof.*

Proof: Given that all the other sources report their true optimal demands at every iteration during the auction, the auction will conclude at iteration T_1 with the asking price λ^{T_1} and the jamming power p_{i,T_1}^J allocated to S_i if source S_i also reports its true optimal demand at every iteration, while the auction will conclude at iteration T_2 with the asking price λ^{T_2} and the jamming power p_{i,T_2}^J allocated to S_i if source S_i does not report its true optimal demand at every iteration. The final utility of source S_i is then denoted by $U_i^{T_1}$ and $U_i^{T_2}$, respectively.

According to Algorithm 11.2, for any fixed p_i, we have

$$
U_i^{T_j} = \mathcal{G}\left(p_{i,T_j}^J, p_i\right) - \lambda^{T_j} p_{i,T_j}^J, \ j \in \{1,2\}. \hspace{2cm} (11.37)
$$

Then, we can get

$$
U_i^{T_1} - U_i^{T_2} = \mathcal{G}\left(p_{i,T_1}^J, p_i\right) - \mathcal{G}\left(p_{i,T_2}^J, p_i\right) - \lambda^{T_1} p_{i,T_1}^J + \lambda^{T_2} p_{i,T_2}^J. \hspace{1cm} (11.38)
$$

From (11.38), we cannot guarantee that $U_i^{T_1} > U_i^{T_2}$, since if $\lambda^{T_1} p_{i,T_1}^J - \lambda^{T_2} p_{i,T_2}^J < \mathcal{G}\left(p_{i,T_1}^J, p_i\right) - \mathcal{G}\left(p_{i,T_2}^J, p_i\right)$, then $U_i^{T_1} < U_i^{T_2}$. Therefore, the sources have the incentive not to report their true optimal demands since it may lead to a greater utility, which means P-ACA-T is not cheat-proof. $\blacksquare$

Figure 11.5 shows the cheat-proof performance of P-ACA-T and P-ACA-A. Here we assume that source S_3 reports a false jamming power demand $\tilde{p}_{3,t}^J$ by scaling the true optimal demand $p_{3,t}^J$ with a cheat factor k, i.e., $\tilde{p}_{3,t}^J = \min\left(p_{max}, \max\left(0, kp_{3,t}^J\right)\right)$. In Figure 11.5, the final utilities of source S_3 as a function of the cheat factor k in P-ACA-T and P-ACA-A are shown, respectively. In P-ACA-T, we can see that source S_3 achieves the maximal utility when k is around 0.7. Since the sources are noncollaborative, all the sources have the

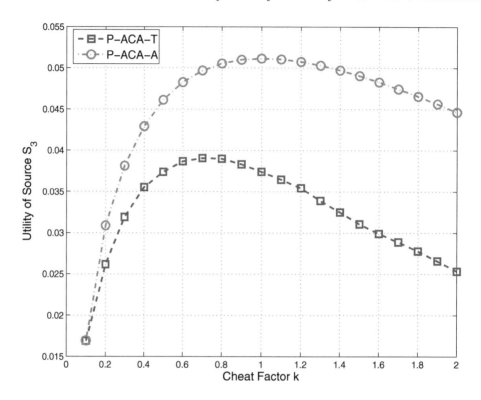

Figure 11.5: Cheat-proof property of P-ACA-T and P-ACA-A.

incentive to report a smaller demand at every iteration. Thus, P-ACA-T is not cheat-proof. In P-ACA-A, we can see that source S_3 achieves the maximal utility when $k = 1$, which means that no source has the incentive to cheat since any cheating will lead to a loss in its utility. Thus, P-ACA-A is cheat-proof. This simulation result verifies Theorem 11.4.4 and Theorem 11.4.5.

11.4.4 Social Welfare Maximization

In this section, we prove that P-ACA-T and P-ACA-A can maximize the social welfare, which is defined as the sum of the sources' and friendly jammer's utilities, in a particularly high interference case, while P-ACA-S achieves a smaller, if not equal, social welfare.

We first define the special high interference case to obtain a simple expression of the optimal jamming power solution for each source. In this special case, we assume that the friendly jammer is close to the malicious eavesdropper but far away from the destinations. Thus, at the eavesdropper the jamming interference from the friendly jammer is much higher than the received signals from the sources as well as the noise, while at the destinations the jamming interference is much lower than the noise. In other words, that means $p_i g_{S_i,E} \ll p_i^J g_{J,E}$, $\sigma^2 \ll p_i^J g_{J,E}$, and $p_i^J g_{J,D_i} \ll \sigma^2$. Therefore, the utility of source S_i can be approximately calculated as

$$
\begin{aligned}
U_i\left(p_i^J, \lambda\right) &\approx W\left[\log\left(1 + \frac{p_i g_{S_i, D_i}}{\sigma^2}\right) - \log\left(1 + \frac{p_i g_{S_i, D_i}}{\sigma^2}\right)\right.\\
&\quad \left. - \log\left(1 + \frac{p_i g_{S_i, E}}{p_i^J g_{J, E}}\right) + \log\left(1 + \frac{p_i g_{S_i, E}}{\sigma^2}\right)\right] - \lambda p_i^J \\
&\approx W\log\left(1 + \frac{p_i g_{S_i, E}}{\sigma^2}\right) - \frac{W p_i g_{S_i, E}}{p_i^J g_{J, E}} - \lambda p_i^J \\
&= \Theta_i - \Phi_i \frac{1}{p_i^J} - \lambda p_i^J,
\end{aligned}
\tag{11.39}
$$

where $\Theta_i = W\log\left(1 + \frac{p_i g_{S_i, E}}{\sigma^2}\right)$, $\Phi_i = \frac{W p_i g_{S_i, E}}{g_{J, E}}$, $i \in \mathcal{N}$, and the second approximation is obtained by the Taylor series expansion $\log(1 + x) \approx x$ when x is sufficiently small. In order to find the optimal jamming power for source S_i, we can calculate

$$
\frac{\partial U_i}{\partial p_i^J} = \frac{\Phi_i}{(p_i^J)^2} - \lambda = 0.
\tag{11.40}
$$

Hence, the optimal closed-form solution can be expressed as

$$
p_{i\_h}^{J}{}^{*} = \sqrt{\frac{\Phi_i}{\lambda}}.
\tag{11.41}
$$

In view of the power constraint, we can obtain the optimal jamming power for source S_i in this special case as

$$
p_{i\_h\_opt}^J = \min\left(p_{i\_h}^{J}{}^{*}, p_{max}\right).
\tag{11.42}
$$

Substituting (11.42) back to the utility function (11.39), we can get the optimal transmitting power $p_{i\_h\_opt}$ of source S_i as

$$
p_{i\_h\_opt} = \arg\min \max_{p_i} \varphi\left(p_i, \lambda\right),
\tag{11.43}
$$

where $\varphi\left(p_i, \lambda\right)$ is defined as

$$
\varphi\left(p_i, \lambda\right) = \begin{cases} \Theta_i - \Phi_i \frac{1}{p_{max}} - \lambda p_{max}, & \text{if } \Phi_i > \lambda p_{max}^2; \\ \Theta_i - 2\sqrt{\Phi_i \lambda}, & \text{if } 0 \le \Phi_i \le \lambda p_{max}^2. \end{cases}
\tag{11.44}
$$

Theorem 11.4.6. *In the high interference case described above, when δ is sufficiently small, P-ACA-T and P-ACA-A converge to $\left(p_1^{J*}, p_1^{\star}, \ldots, p_N^{J*}, p_N^{\star}\right)$, which maximizes the social welfare, i.e., $\left\{p_i^{J*}, p_i^{\star}\right\}$, $i = 1, 2, \ldots, N$, is the solution to the following optimization problem:*

$$
\max_{p_i^J, p_i} \sum_i \left(\Theta_i - \Phi_i \frac{1}{p_i^J}\right),
\tag{11.45}
$$

$$
s.t. \begin{cases} 0 \le p_i \le p_{max}, \forall i \in \mathcal{N}, \\ 0 \le p_i^J \le p_{max}, \forall i \in \mathcal{N}, \\ \sum_i p_i^J = p_{max}. \end{cases}
$$

Proof: From Theorem 11.4.2, we have that P-ACA-T and P-ACA-A conclude in a finite number of iterations, which means each of the two schemes will converge to a solution of jamming power allocation $\left(p_1^{J\star}, p_1^\star, \ldots, p_N^{J\star}, p_N^\star\right)$. From (11.42) and (11.43), we can get

$$p_i^{J\star}(p_i^\star, \lambda) = \min\left(\sqrt{\frac{\Phi_i(p_i^\star)}{\lambda}}, p_{max}\right),\tag{11.46}$$

and

$$p_i^\star(\lambda) = \operatorname*{argmin}_{p_i} \max \varphi(p_i, \lambda),\tag{11.47}$$

where $\varphi(p_i, \lambda)$ is defined in (11.44) and λ is the solution to the following equation:

$$\sum_{i=1}^{N} \min\left(\sqrt{\frac{\Phi_i(p_i^\star)}{\lambda}}, p_{max}\right) = p_{max}.\tag{11.48}$$

On the other hand, noting that the optimization problem in (11.45) is convex in terms of p_i^J for any fixed p_i, we can first find the optimal p_i^J as a function of p_i by solving the Karush-Kuhn-Tucker (KKT) conditions [35]. The Lagrangian of the optimization problem for the social welfare (11.45) can be written as

$$\mathcal{L}\left(p_i^J, p_i, \tilde{\lambda}, \kappa_i, \upsilon_i\right) = -\sum_{i=1}^{N}\left(\Theta_i - \Phi_i \frac{1}{p_i^J}\right) + \tilde{\lambda}\left(\sum_{i=1}^{N} p_i^J - p_{max}\right)$$
$$+ \sum_{i=1}^{N} \kappa_i\left(p_i^J - p_{max}\right) - \sum_{i=1}^{N} \upsilon_i p_i^J.\tag{11.49}$$

The corresponding KKT conditions are as follows:

$$-\frac{\Phi_i}{(p_i^J)^2} + \tilde{\lambda} + \kappa_i - \upsilon_i = 0,\ \forall i \in \mathcal{N};\tag{11.50}$$

$$\tilde{\lambda}\left(\sum_{i=1}^{N} p_i^J - p_{max}\right) = 0;\tag{11.51}$$

$$\kappa_i\left(p_i^J - p_{max}\right) = 0,\ \forall i \in \mathcal{N};\tag{11.52}$$

$$\upsilon_i p_i^J = 0,\ \forall i \in \mathcal{N};\tag{11.53}$$

$$\tilde{\lambda} \geq 0;\tag{11.54}$$

$$\kappa_i \geq 0,\ \forall i \in \mathcal{N};\tag{11.55}$$

$$\upsilon_i \geq 0,\ \forall i \in \mathcal{N}.\tag{11.56}$$

By solving the KKT conditions above subject to the jamming power constraints, we can obtain the optimal solution for p_i^J as

$$\tilde{p}_i^{J\star}\left(p_i, \tilde{\lambda}\right) = \min\left(\sqrt{\frac{\Phi_i(p_i)}{\tilde{\lambda}}}, p_{max}\right),\tag{11.57}$$

where $\sum_{i=1}^{N} \tilde{p}_i^{J\star}\left(p_i, \tilde{\lambda}\right) = p_{max}$.

Substituting $\tilde{p}_i^{J\star}\left(p_i, \tilde{\lambda}\right)$ and the corresponding value of κ_i and υ_i, $i \in \mathcal{N}$, back to (11.49), then the Lagrangian becomes

$$\mathcal{L}\left(\tilde{p}_i^{J\star}, p_i, \tilde{\lambda}, \kappa_i, \upsilon_i\right) = -\sum_{i=1}^{N} \varphi\left(p_i, \tilde{\lambda}\right) + \tilde{\lambda} p_{max},\tag{11.58}$$

where $\varphi\left(p_i, \tilde{\lambda}\right)$ is defined in (11.44). Thus, we can obtain the optimal solution for p_i that minimizes the Lagrangian as

$$\tilde{p}_i^\star = \text{argmin} \max_{p_i} \varphi\left(p_i, \tilde{\lambda}\right). \tag{11.59}$$

In brief, we have the solution to the optimization problem (11.45) as

$$\begin{cases} \tilde{p}_i^{J\star}\left(\tilde{p}_i^\star, \tilde{\lambda}\right) = \min\left(\sqrt{\frac{\Phi_i(\tilde{p}_i^\star)}{\tilde{\lambda}}}, p_{max}\right), \\ \tilde{p}_i^\star\left(\tilde{\lambda}\right) = \text{argmin} \max_{p_i} \varphi\left(p_i, \tilde{\lambda}\right), \end{cases} \tag{11.60}$$

where $\tilde{\lambda}$ is the constant that satisfies

$$\sum_{i=1}^{N} \min\left(\sqrt{\frac{\Phi_i(\tilde{p}_i^\star)}{\tilde{\lambda}}}, p_{max}\right) = p_{max}. \tag{11.61}$$

Comparing the final power allocation $\left\{p_i^{J\star}, p_i^\star\right\}$ based on P-ACA-T and P-ACA-A in (11.46) and (11.47) with the optimal solution in (11.60), we have that $\left(p_1^{J\star}, p_1^\star, \ldots, p_N^{J\star}, p_N^\star\right)$ is just the solution to the optimization problem (11.45). In other words, P-ACA-T and P-ACA-A can maximize the social welfare in the high interference case.

Lemma 11.4.1. *Compared with P-ACA-A, P-ACA-S achieves a smaller, if not equal, social welfare.*

Proof: According to Algorithm 11.1 and Algorithm 11.3, we can see that P-ACA-S is a special case of P-ACA-A with an enhanced power constraint $p_i^J \in \{0, p_{max}\}$. This means with respect to the optimization problem of the social welfare, the feasible jamming power set of P-ACA-S is only a subset of that of P-ACA-A. Therefore, the social welfare of P-ACA-S is smaller than, if not equal to, that of P-ACA-A.

Figure 11.6 indicates the social welfare performance of the three proposed schemes in the special high interference case. In Figure 11.6, the social welfare (i.e., the sum of the sources' and jammer's utilities) as a function of the whole jamming power in P-ACA-S, P-ACA-T, and P-ACA-A are shown, respectively. We can see that P-ACA-T and P-ACA-A have almost the same performance in terms of the social welfare. We can also find that the social welfare of P-ACA-S is the same as that of P-ACA-T and P-ACA-A when the whole jamming power is small, but becomes lower than that of P-ACA-T and P-ACA-A when the whole jamming power goes larger than a certain value. This is because in P-ACA-S each source can only choose either to utilize the whole jamming power from the friendly jammer or not to utilize the jamming power. Therefore, when the whole jamming power is larger than the optimal demand of the source that wins in the end, it will lead to a loss of the optimal system performance. However, in P-ACA-T and P-ACA-A, each source has the chance to utilize just a fraction of the whole jamming power. Hence, the system can always achieve efficient jamming power allocation and obtain optimal system performance in P-ACA-T and P-ACA-A. This simulation result verifies Theorem 11.4.6 and Lemma 11.4.1.

11.4.5 Complexity and Overhead

In this section, we analyze the complexity and communication overhead of the proposed auction-based schemes compared with the centralized scheme, from which we can obtain

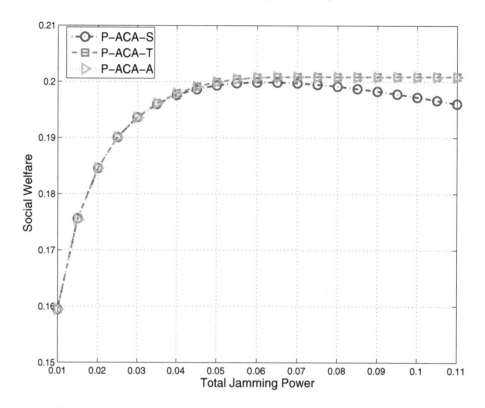

Figure 11.6: Social welfare of P-ACA-S, P-ACA-T, and P-ACA-A.

that the proposed distributed auction-based schemes are more efficient due to their low computational complexity. The centralized scheme is formulated as an optimal jamming power allocation maximizing the system secrecy rate based on the assumption that all the channel information is known at the friendly jammer, which can be expressed as

$$\max_{p_i^J, p_i} \sum_{i=1}^{N} W \left[\log \left(1 + \frac{p_i g_{S_i, D_i}}{\sigma^2 + p_i^J g_{J, D_i}} \right) - \log \left(1 + \frac{p_i g_{S_i, E}}{\sigma^2 + p_i^J g_{J, E}} \right) \right]^+ \quad (11.62)$$

$$\text{s.t.} \begin{cases} 0 \leq p_i \leq p_{max}, \forall i \in \mathcal{N}, \\ 0 \leq p_i^J \leq p_{max}, \forall i \in \mathcal{N}, \\ \sum_i p_i^J = p_{max}. \end{cases}$$

- *Complexity*: By defining the discrete quantification precision of the allocated jamming power as Δp, we can further calculate the computational complexity of the centralized scheme formulated in (11.62) as

$$\mathcal{C}_{centralized} = \mathcal{O} \left(\frac{(K+1)^{N-1}}{[N-1]!} \right), \quad (11.63)$$

where $K = p_{max}/\Delta p$ and $[x]!$ represents the factorial computation of x. From (11.63), we can obtain that the centralized optimal scheme is an NP-hard one, which has high computational complexity to obtain the optimal power allocation solution. In general cases, we have $K \gg N$, thus, when the number of source–destination pairs N grows, the computational complexity of the centralized scheme increases rapidly, which makes it unfeasible when N is large.

In contrast to the centralized scheme, the proposed distributed auction-based schemes transform the centralized global optimization computation into the N-path parallel computation where each distributed path processes only a local optimization problem. From the procedure of the proposed auction-based schemes given in Tables 11.1, 11.2, and 11.3, we can calculate the computational complexity of the proposed auction-based scheme as

$$\mathcal{C}_{proposed} = \mathcal{O}\left((K+N)\frac{(\lambda^T - \lambda^0)}{\delta}\right), \qquad (11.64)$$

where λ^0, λ^T, and δ are the initial price, the final price, and the price step defined in the algorithms. From (11.64), we can see that the computational complexity is well reduced using the proposed auction-based schemes, especially when the number of source–destination pairs N is large. Moreover, the proposed auction-based schemes can greatly reduce the computation load of the friendly jammer, which is crucial to prolonging the life of the battery-powered terminal in this jammer-assisted cooperative network.

- *Overhead*: Compared with the centralized scheme, the main disadvantage of the proposed auction-based schemes is the communication overhead for information exchanging between the friendly jammers and the source–destination pairs during the auction procedure. Actually, the communication overhead cost is the universal problem of most game-theoretical methods since the frequent interactions among the nodes is essential during the convergence procedure. In our proposed auction-based schemes, the communication overhead is mainly determined by the total iteration number when the algorithm converges, i.e., $(\lambda^T - \lambda^0)/\delta$. We have that the value of the price step δ has a direct impact on the speed of convergence of the proposed auction-based schemes and thus the communication overhead. The proposed auction-based schemes conclude fast when δ is large, while they conclude slowly when δ is small. By simulation, we investigate the average number of iterations when the three proposed auction-based schemes converge as the price step δ changes. From the simulation results, we can see that to set a properly larger price step δ can efficiently reduce the convergence time. Therefore, we can design the value of δ based on the actual demands in a practical system to achieve a proper trade-off between the communication overhead and the system performance.

Figure 11.7 shows the average number of iterations of the three proposed auction-based schemes when the price step δ changes. We can see that the average number of iterations of the three proposed auction-based schemes decreases obviously as the price step δ increases. Thus, we can design the value of δ based on the actual demands to control the convergence time of the proposed auction-based schemes and achieve a proper trade-off between the communication overhead and the system performance. Besides, we can also see that P-ACA-S converges faster than P-ACA-T and P-ACA-A. This is because in P-ACA-S, there are only two choices of allocated jamming power (i.e., zero and p_{max}) for each source–destination pair, then the source–destination pairs will quit the auction immediately if the jamming power p_{max} cannot bring a positive utility gain, leading to a faster convergence. P-ACA-T and P-ACA-A have almost the same convergence time since the procedure of P-ACA-A is the same as that of P-ACA-T except that at every iteration an additional cumulative clinch for each source–destination pair is computed.

Figure 12.8 shows the system performance in the three proposed auction-based schemes compared with the no-jamming case and the centralized scheme. The system secrecy rate is defined as the sum secrecy rate of all the source–destination links. We can see that with

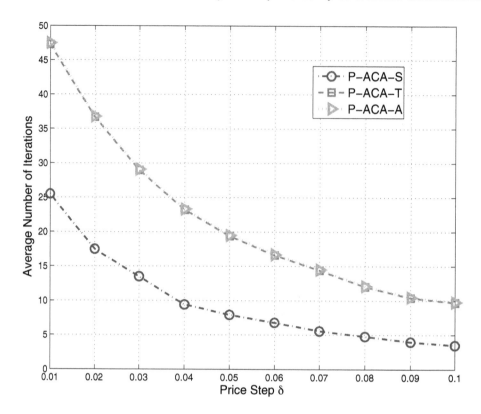

Figure 11.7: Average number of iterations when the price step changes.

the proposed auction-based schemes allocating jamming power from the friendly jammer to the sources, the system can effectively obtain a positive performance gain in the secrecy rate compared with the no-jamming case. We also find that P-ACA-T and P-ACA-A can achieve a comparable performance of the system secrecy rate to the centralized optimal scheme, especially when the whole jamming power is large.

11.5   Conclusions and Open Issues

In this chapter, we have considered the jamming power allocation problem to effectively improve the secrecy rate in a cooperative jamming-based network. By employing the ascending clock auctions in which the friendly jammer serves as the auctioneer and the sources are the bidders, we provided three distributed auction-based power allocation schemes, i.e., P-ACA-S, P-ACA-T, and P-ACA-A. Besides, we investigated the basic properties of the three proposed auction-based power allocation schemes. All three auction-based power allocation schemes can converge in a finite number of iterations. P-ACA-S and P-ACA-A are cheat-proof and can enforce the selfish and noncollaborative sources to report their true optimal demands at every iteration during the auction, while P-ACA-T is not. Moreover, we proved in a special high interference case that P-ACA-T and P-ACA-A can maximize the social welfare, while P-ACA-S may not. Compared with the centralized optimal jamming power allocation scheme, the proposed distributed auction-based schemes have much lower complexity and can be more efficient in practical applications.
Open Issues:

- In [11–13], the jamming power allocation problems are solved based on a Stackel-

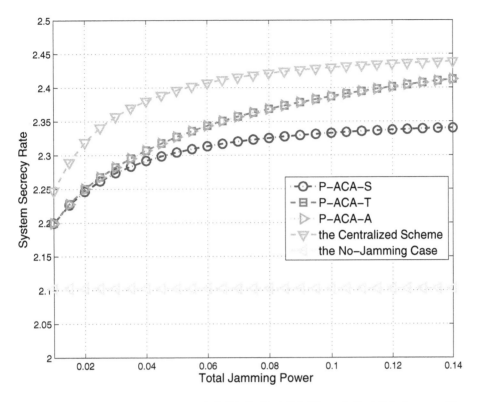

Figure 11.8: System secrecy rate in P-ACA-S, P-ACA-T, and P-ACA-A as well as the no-jamming case and the centralized scheme.

berg game where there is one source–destination pair and multiple friendly jammers (i.e., single buyer and multiple sellers), and in this chapter, the jamming power allocation problem based on ascending clock auctions are investigated where there are multiple source–destination pairs and only one friendly jammer (i.e., multiple bidders and single auctioneer). Then, as for the scenario with multiple source–destination pairs and multiple friendly jammers (i.e., multiple auctioneers and multiple bidders), how to achieve efficient jamming power allocation from a game-theory perspective in such a scenario will be interesting and challenging since the interactions among the source–destination pairs and the friendly jammer are more complicated. Maybe double auction [35] is a good choice to solve such a problem.

- The auction-based resource allocation schemes generally demand obvious communication overhead since the distributed nodes need to continuously interact to exchange information during the auction procedure before converging to a solution. This costs both the feedback channels and the processing time when they are employed into practical networks and lead to a loss of the network efficiency. Then, how to control the iteration to reduce the communication overhead time and meanwhile keep the efficiency to achieve an optimized system performance is important for the application of the auction-based resource allocation schemes in practical networks.

References

[1] A. D. Wyner, "The wire-tap channel," *Bell System Technical Journal*, vol. 54, no. 8, pp. 1355–1387, Oct. 1975.

[2] I. Csiszár and J. Körner, "Broadcast channels with confidential messages," *IEEE Transactions on Information Theory*, vol. 24, no. 3, pp. 339–348, May 1978.

[3] S. K. Leung-Yan-Cheong and M. E. Hellman, "The Gaussian wiretap channel," *IEEE Transactions on Information Theory*, vol. 24, no. 4, pp. 451–456, July 1978.

[4] J. Barros and M. R. D. Rodrigues, "Secrecy capacity of wireless channels," in *Proc. of IEEE International Symposium on Information Theory*, Seattle, USA, July 2006.

[5] L. Lai and H. E. Gamal, "The relay-eavesdropper channel: Cooperation for secrecy," *IEEE Transactions on Information Theory*, vol. 54, no. 9, pp. 4005–4019, Sept. 2008.

[6] L. Dong, Z. Han, A. P. Petropulu, and H. V. Poor, "Improving wireless physical layer security via cooperating relays," *IEEE Transactions on Signal Processing*, vol. 58, no. 3, pp. 1875–1888, Mar. 2010.

[7] X. Zhou and M. R. McKay, "Secure transmission with artificial noise over fading channels: Achievable rate and optimal power allocation," *IEEE Transactions on Vehicular Technology*, vol. 59, no. 8, pp. 3831–3842, Oct. 2010.

[8] G. Zheng, L.-C. Choo, and K.-K. Wong, "Optimal cooperative jamming to enhance physical layer security using relays," *IEEE Transactions on Signal Processing*, vol. 59, no. 3, pp. 1317–1322, Mar. 2011.

[9] J. Huang and A. L. Swindlehurst, "Cooperative jamming for secure communications in MIMO relay networks," *IEEE Transactions on Signal Processing*, vol. 59, no. 10, pp. 4871–4884, Oct. 2011.

[10] J. Li, A. P. Petropulu, and S. Weber, "On cooperative relaying schemes for wireless physical layer security," *IEEE Transactions on Signal Processing*, vol. 59, no. 10, pp. 4985–4997, Oct. 2011.

[11] J. Chen, R. Zhang, L. Song, Z. Han, and B. Jiao, "Joint relay and jammer selection for secure two-way relay networks," *IEEE Transactions on Information Forensics and Security*, vol. 7, no. 1, pp. 310–320, Feb. 2012.

[12] R. Zhang, L. Song, Z. Han, and B. Jiao, "Distributed coalition formation of relay and friendly jammers for secure cooperative networks," in *Proc. IEEE International Conference on Communications (ICC 2011)*, Kyoto, Japan, June 2011.

[13] R. Zhang, L. Song, Z. Han, and B. Jiao, "Relay and jammer cooperation as a coalitional game in secure cooperative wireless networks," in *Proc. the 5th International ICST Conference on Performance Evaluation Methodologies and Tools*, Invited, Cachan, France, May 2011.

[14] L. Dong, H. Yousefi'zadeh, and H. Jafarkhani, "Cooperative jamming and power allocation for wireless relay networks in presence of eavesdropper," in *Proc. IEEE International Conference on Communications (ICC 2011)*, Kyoto, Japan, June 2011.

[15] D. Fudenberg and J. Tirole, *Game Theory*, MIT Press, Cambridge, MA, 1993.

[16] Z. Han, D. Niyato, W. Saad, T. Baçar, and A. Hjørungnes, *Game Theory in Wireless and Communication Networks: Theory, Models and Applications*, Cambridge University Press, New York, 2017.

[17] Z. Han, N. Marina, M. Debbah, and A. Hjørungnes, "Physical layer security game: Interaction between source, eavesdropper and friendly jammer," *EURASIP Journal on Wireless Communications and Networking*, vol. 2009 No. 11, Mar, 2009.

[18] Z. Han, N. Marina, M. Debbah, and A. Hjørungnes, "Physical layer security game: How to date a girl with her boyfriend on the same table," in *Proc. IEEE International Conference on Game Theory for Networks*, Istanbul, Turkey, May 2009.

[19] R. Zhang, L. Song, Z. Han, and B. Jiao, "Physical layer security for two-way untrusted relaying with friendly jammers," *IEEE Transactions on Vehicular Technology*, vol. 61, no. 8, pp. 3693–3704, Oct. 2012.

[20] R. Zhang, L. Song, Z. Han, B. Jiao, and M. Debbah, "Physical layer security for two way relay communications with friendly jammers," in *Proc. IEEE GLOBECOM 2010*, Miami, USA, Dec. 2010.

[21] R. Zhang, L. Song, Z. Han, and B. Jiao, "Improve physical layer security in cooperative wireless network using distributed auction games," in *Proc. IEEE INFOCOM Workshops on Cognitive and Cooperative Networks (INFOCOM WKSHPS 2011)*, Shanghai, China, Apr. 2011.

[22] W. Vickrey, "Counterspeculation, auctions and competitive sealed tenders," *Journal of Finance*, vol. 16, no. 1, pp. 8–37, Mar. 1961.

[23] P. R. Milgrom and R. J. Weber, "A theory of auctions and competitive bidding," *Econometrica*, vol. 50, no. 5, pp. 1089–1122, Sept. 1982.

[24] P. Klemperer, "Auction theory: A guide to the literature," *Journal on Economics Surveys*, vol. 13, no. 3, pp. 227–286, July 1999.

[25] J. Huang, Z. Han, M. Chiang, and H. V. Poor, "Auction-based resource allocation for cooperative communications," *IEEE Journal on Selected Areas in Communications*, vol. 26, no. 7, pp. 1226–1237, July 2008.

[26] Y. Chen, Y. Wu, B. Wang, and K. J. R. Liu, "Spectrum auction games for multimedia streaming over cognitive radio networks," *IEEE Transactions on Communications*, vol. 58, no. 8, pp. 2381–2390, Aug. 2010.

[27] V. Krishna, *Auction Theory*, Academic Press, Burlington, MA, USA, 2002.

[28] P. Cramton, "Ascending auction," *European Economic Review*, vol. 42, pp. 745–756, 1998.

[29] L. M. Ausubel, "An efficient ascending-bid auction for multiple objects," *American Economic Review*, vol. 94, no. 5, pp. 1452–1475, May 2004.

[30] P. Marbach and R. Berry, "Downlink resource allocation and pricing for wireless networks," in *Proc. IEEE INFOCOM 2002*, New York, USA, June 2002.

[31] D. Palomar and M. Chiang, "A tutorial on decomposition methods for network utility maximization," *IEEE Journal on Selected Areas in Communications*, vol. 24, no. 8, pp. 1439–1451, Aug. 2006.

[32] I. N. Stewart, *Galois Theory, 3rd. ed.*, Chapman & Hall/CRC, Boca Raton, FL, 2004.

[33] Z. Han and K. J. R. Liu, *Resource Allocation for Wireless Networks: Basics, Techniques, and Applications*, Cambridge University Press, UK, 2008.

[34] S. Boyd and L. Vandenberghe, *Convex Optimization*, Cambridge University, New York, 2008.

[35] D. Friedman, *The Double Auction Market Institution: A Survey*, Westview Press, 1993.

[36] L. Song and J. Shen, *Evolved Cellular Network Planning and Optimization for UMTS and LTE*, CRC Press, Boca Raton, FL, 2010.

Chapter 12

Relay and Jammer Cooperation as a Coalitional Game

Rongqing Zhang
Peking University

Lingyang Song
Peking University

Zhu Han
University of Houston

Bingli Jiao
Peking University

In this chapter, we investigate the cooperation of conventional relays and friendly jammers for secure cooperative networks. In order to obtain an optimized secrecy rate, the source intends to select several conventional relays and friendly jammers from the intermediate nodes to assist data transmission, and in return, it needs to make a payment. Each intermediate node here has two possible identities to choose, i.e., to be a conventional relay or a friendly jammer, which results in a different impact on the final utility of the intermediate node. After the intermediate nodes determine their identities, they seek to find optimal partners forming coalitions, which improves their chances to be selected by the source and thus to obtain the payoffs in the end. We formulate this cooperation as a coalitional game with transferable utility and study its properties. Furthermore, we define a Max-Pareto order for comparison of the coalition value and construct a distributed merge-and-split coalition formation algorithm for the defined coalition formation game.

12.1   Introduction

12.1.1   Cooperative Relaying and Cooperative Jamming

The literature of physical layer security was pioneered by Wyner [1], who introduced the wiretap channel and showed that when the wiretap channel is a degraded version of the main channel, the two legitimate users can exchange secure messages at a nonzero rate without relying on a private key. In follow-up work [2], Wyner's result was generalized to a nondegraded discrete memoryless broadcast channel with common messages sent to both the receivers and confidential messages sent to only one of the receivers. In [3], the secrecy capacity of a Gaussian wiretap channel was studied.

However, the feasibility of traditional physical layer security approaches is hampered by wireless channels. When the source–destination channel is weaker than the source–eavesdropper channel, a positive secrecy rate cannot be achieved unless multiple transmit antennas are employed. The use of multiple antennas on a node involves a potentially significant hardware cost. Thus, node cooperation is an effective means with low cost that enables single-antenna nodes to enjoy the benefits of multiple-antenna systems [4]. Node cooperation in physical layer security issues can mainly be classified into two types of methods, i.e., cooperative relaying and cooperative jamming. In [5], the available secrecy rates of the cooperative schemes, i.e., decode-and-forward (DF), amplify-and-forward (AF), and cooperative jamming (CJ), were analyzed and the corresponding system designs consisting of the determination of relay weights and the allocation of transmit power for each cooperative were proposed. In [6], the authors focused on obtaining the exact solution or simplified versions of the optimization problem for the DF and CJ schemes. In [7], the problem of secure communication in fading channels with a multiantenna transmitter capable of simultaneous transmission of both the information signal and the artificial noise was studied. In [8], the cooperative jamming problem to increase the physical layer security of a wiretap fading channel via distributed relays was solved using a combination of convex optimization and a one-dimensional search. In [9], cooperative jamming strategies in a two-hop relay network where the eavesdropper can wiretap the relay channels in both hops were investigated. In [10], a cooperative wireless network in the presence of one or more eavesdroppers was considered, and node cooperation for achieving physical layer security based on DF and CJ was studied. In [11–14], several cooperative jamming schemes to achieve an optimized jamming power control were proposed from a game-theoretical perspective.

12.1.2   Relay and Jammer Selection

In a cooperative network, effective opportunistic relay or/and jammer selection with secrecy constraints was recently reported as an interesting research topic. In [15], for a one-way secure relaying network, the authors considered the selection issue of a relay and a jammer with respect to secrecy constraints. The selected relay node performs a DF strategy and assists the source to deliver data to its destination, while the selected jammer node transmits simultaneously with the relaying link in order to create artificial interference to degrade the eavesdropper links. Optimal and suboptimal relay selection schemes were then proposed following the rules that the selected relay increases the perfect secrecy of the relaying link and the selected jammer increases interference at the eavesdropper node while simultaneously protecting the primary destination from interference. In [16, 17], several joint cooperative relay and friendly jammer selection schemes were provided for two-way relay networks. The proposed schemes select two or three intermediate nodes (which act as relays or jammers) to enhance security against the malicious eavesdropper in a two-way secure relaying scenario. The first selected node operates in the conventional relay mode and assists the sources to deliver their data to the corresponding destinations using an AF protocol. The second and third nodes are used in different communication phases (i.e., one for the first phase and the other for the second phase) as jammers in order to create intentional interference upon the malicious eavesdropper. In [18], the authors proposed combined relay selection and cooperative beamforming schemes for physical layer security, with the purpose to achieve an effective trade-off between the number of relays participating in cooperative beamforming to reduce the communication overhead and the performance gain obtained from secure cooperative beamforming.

Different from the above work on relay and jammer selection for physical layer security, in this chapter we attempt to investigate how the relays and jammers can effectively cooperate with each other in a practical decentralized wireless network when optimizing the network performance.

12.1.3   Coalitional Game Theory

Game theory [19, 20] offers a formal analytical framework with a set of mathematical tools that can effectively study the complex interactions among interdependent rational players. Recently, there has been significant growth in research activities that use game theory for analyzing communication networks. Implementing cooperation in large scale communication networks faces several challenges such as adequate modeling, efficiency, complexity, and fairness. Coalitional games prove to be a very powerful tool for designing fair, robust, practical, and efficient cooperation strategies in communication networks [21]. In recent years, coalitional games have been widely employed into wireless networks for analyzing and achieving efficient cooperation, such as spectrum accessing [22] and collaborative spectrum sensing [23] in cognitive radio networks, cooperative routing in vehicular networks [24], and fair user cooperation in decentralized networks [25]. In [26–28], coalitional games are employed into physical layer security issues to improve the network secrecy capacity through efficient cooperation among users or intermediate nodes.

As introduced in [20, 21], coalitional games involve a set of players, denoted by $\mathcal{N} = \{1, 2, \ldots, N\}$, who seek to form cooperative groups, i.e., coalitions, to strengthen their positions in the game. Any coalition represents an agreement between the players in S to act as a single entity. In addition to the player set $\mathcal{N}$, the second fundamental concept of a coalitional game is the coalition value. The coalition value, denoted by v, quantifies the worth of a coalition in the considered coalitional game. The definition of the coalition value determines the form and type of the coalitional game, i.e., coalitional games with transferable utility (TU) and coalitional games with nontransferable utility (NTU). Nonetheless, independent of the definition of the coalition value, a coalitional game is uniquely defined by the pair $(\mathcal{N}, v)$. The most common form of a coalitional game is the characteristic form, whereby the value of a coalition S depends solely on the members of that coalition, with no dependence on how the players in $\mathcal{N} \setminus S$ are structured. Such a category of coalitional games is known as coalition game with transferable utility.

In this chapter, by utilizing a coalitional game, we investigate conventional relay and friendly jammer cooperation for secure data transmission in a distributed cooperative network. The source here can be regarded as a buyer who selects a group of several conventional relays and friendly jammers from the intermediate nodes in order to achieve an optimal secrecy rate, and in return it pays for the "service" offered by the selected nodes. Each intermediate node in the network has two identities to choose, i.e., to be a conventional relay or a friendly jammer, which is mainly determined according to its own channel conditions to the destination and the eavesdropper. After an intermediate node makes a choice of its identity, it starts to find some optimal partners to strengthen its opportunity to be selected by the source that may lead to a satisfactory payoff.

Specifically, we formulate this cooperation as a coalitional game with transferable utility. Some properties of the game are then studied, from which we find that the disjoint coalitions rather than the grand coalition will form in most cases, i.e., the coalitional game defined here can be classified as a coalition formation game. Furthermore, we define a Max-Pareto order for the value comparison between different coalitions, based on which we employ the merge-and-split rules. Then, we devise a distributed merge-and-split coalition formation algorithm of conventional relays and friendly jammers to achieve efficient cooperation in the network.

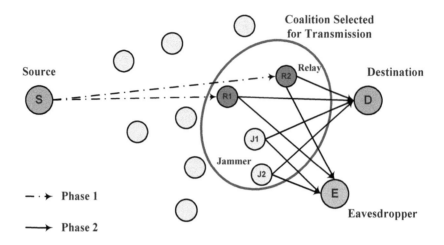

Figure 12.1: System model for the relay and jammer-assisted secure cooperative network.

12.1.4 Chapter Outline

The rest of this chapter is organized as follows. In Section 12.2, we describe the investigated cooperative network consisting of a couple of intermediate nodes and formulate physical layer security issues with cooperative relaying and cooperative jamming in such a scenario. In Section 12.3, we formulate the relay and jammer cooperation as a coalitional game with TU and investigate some basic properties of the proposed game. In Section 12.4, we further propose a merge-and-split-based coalition formation algorithm for the distributed cooperation of the relays and jammers. In Section 12.5, we conclude the chapter and discuss some open issues.

12.2 System Model and Problem Formulation

In this chapter, we consider a conventional cooperative network, as shown in Figure 12.1. The source node, the corresponding destination node, the malicious eavesdropper node, and N intermediate nodes are denoted by S, D, E, and T_i, $i = 1, 2, \ldots, N$, respectively. We denote by $\mathcal{T}$ the whole intermediate nodes set. All the nodes here are equipped with only a single omnidirectional antenna and operate in a half-duplex way, i.e., each node cannot receive and transmit simultaneously. Then, the complete data transmission can be divided into two phases: (1) in the broadcasting phase, the source transmits its messages to the intermediate nodes, and the intermediate nodes who can successfully decode the signals from the source form a decoding node subset $\mathcal{T_D} \subseteq \mathcal{T}$, and (2) in the cooperative relaying phase, several selected intermediate nodes transmit their signals toward both the destination and the eavesdropper with respect to secrecy constraints. Some of the selected nodes denoted by R_m, $m = 1, 2, \ldots, M$, operate as conventional relays, and forward the source messages to the destination. The other selected intermediate nodes denoted by J_k, $k = 1, 2, \ldots, K$, operate as friendly jammers who can transmit jamming signals to interfere with the malicious eavesdropper in order to improve the secrecy rate of data transmission. We denote by $\mathcal{M}$ and $\mathcal{K}$ the set of indices $\{1, 2, \ldots, M\}$ and $\{1, 2, \ldots, K\}$, respectively.

Note that relay R_m, $m \in \mathcal{M}$, must belong to the decoding subset $\mathcal{T_D}$, while friendly jammer J_k, $k \in \mathcal{K}$, can be any node of the node set $\mathcal{T}$ as it need not decode the source messages. Furthermore, we assume that there are no direct links between source and destination as well as between source and eavesdropper. As a result, the data transmission can only be performed through the intermediate nodes, and the eavesdropper cannot overhear the broadcast channels between source and intermediate nodes as the source–eavesdropper channel is blocked.

Suppose the selected relay R_m, $m \in \mathcal{M}$, transmits with power p_m^R. The channel gains from relay R_m to destination D and malicious eavesdropper E are $g_{R_m,D}$ and $g_{R_m,E}$, respectively. Friendly jammer J_k, $k \in \mathcal{K}$, transmits with power p_k^J to help improve the secrecy rate of data transmission from relay R_m to destination D. The channel gains from friendly jammer J_k to destination D and eavesdropper E are $g_{J_k,D}$ and $g_{J_k,E}$, respectively. Note that here the channel gain contains the path loss, as well as the Rayleigh fading coefficient with zero mean and unit variance. For simplicity, we assume that the fading coefficients are constant over one slot, and vary independently from one slot to another. The thermal noise at the destination and eavesdropper nodes satisfies independent Gaussian distribution with zero mean and the same variance denoted by σ^2. The channel bandwidth is W.

First, with friendly jammers out of consideration, using maximal-ratio combining (MRC) at the receiver, the channel capacity for data transmission from the selected relays to destination D, denoted by $C_{R \to D}(\{R_m\})$, $m = 1, 2, \ldots, M$, can be written as

$$C_{R \to D}(\{R_m\}) = \frac{W}{2} \log \left(1 + \frac{\sum\limits_m p_m^R g_{R_m,D}}{\sigma^2} \right). \tag{12.1}$$

Similarly, the channel capacity for data leakage from the selected relays to malicious eavesdropper E, denoted by $C_{R \to E}(\{R_m\})$, can be written as

$$C_{R \to E}(\{R_m\}) = \frac{W}{2} \log \left(1 + \frac{\sum\limits_m p_m^R g_{R_m,E}}{\sigma^2} \right). \tag{12.2}$$

The direct links $S \to D$ and $S \to E$ are not available in our assumptions, and thus, security of data transmission concerns only the cooperative relaying channels. Then, the secrecy rate for data transmission only via the selected relays can be defined as

$$\begin{aligned} C_s(\{R_m\}) &= (C_{R \to D}(\{R_m\}) - C_{R \to E}(\{R_m\}))^+ \\ &= \frac{W}{2} \left[\log \left(1 + \frac{\sum\limits_m p_m^R g_{R_m,D}}{\sigma^2} \right) \right. \\ &\quad \left. - \log \left(1 + \frac{\sum\limits_m p_m^R g_{R_m,E}}{\sigma^2} \right) \right]^+, \end{aligned} \tag{12.3}$$

where $(x)^+$ represents $\max\{x, 0\}$.

Taking friendly jammers into consideration, i.e., with the help of the friendly jammers transmitting jamming signals to interfere with the malicious eavesdropper, the channel capacity for data transmission from the selected relays to destination D, denoted by $C_{R \to D}(\{R_m\}, \{J_k\})$, $m = 1, 2, \ldots, M$, $k = 1, 2, \ldots, K$, can be written as

$$C_{R \to D}(\{R_m\}, \{J_k\}) = \frac{W}{2} \log \left(1 + \frac{\sum\limits_m p_m^R g_{R_m,D}}{\sigma^2 + \sum\limits_k p_k^J g_{J_k,D}} \right). \tag{12.4}$$

Note that in our assumptions destination D is not able to mitigate the artificial interference from the friendly jammers, as the jamming signals are unknown to the destination.

Similarly, the channel capacity for data leakage from the selected relays to malicious eavesdropper E, denoted by $C_{R \to E}(\{R_m\}, \{J_k\})$, can be written as

$$C_{R \to E}(\{R_m\}, \{J_k\}) = \frac{W}{2} \log \left(1 + \frac{\sum\limits_{m} p_m^R g_{R_m, E}}{\sigma^2 + \sum\limits_{k} p_k^J g_{J_k, E}} \right). \tag{12.5}$$

Then, the secrecy rate for data transmission with the help of friendly jammers can be defined as

$$
\begin{aligned}
C_s(&\{R_m\}, \{J_k\}) \\
&= (C_{R \to D}(\{R_m\}, \{J_k\}) - C_{R \to E}(\{R_m\}, \{J_k\}))^+ \\
&= \frac{W}{2} \left[\log \left(1 + \frac{\sum\limits_{m} p_m^R g_{R_m, D}}{\sigma^2 + \sum\limits_{k} p_k^J g_{J_k, D}} \right) \right.\\
&\quad \left. - \log \left(1 + \frac{\sum\limits_{m} p_m^R g_{R_m, E}}{\sigma^2 + \sum\limits_{k} p_k^J g_{J_k, E}} \right) \right]^+.
\end{aligned}
\tag{12.6}
$$

12.3 Relay and Jammer Cooperation as a Coalitional Game

In this cooperative system, we consider the source as a buyer who wants to send its messages to the corresponding destination and to optimize the secrecy rate of data transmission with the help of conventional relays and friendly jammers, while in return it needs to pay for the "service." However, only one coalition consisting of several conventional relays and friendly jammers from the node set $\mathcal{T}$ can be selected by the source for secure data transmission. Thus, there exists competition among the intermediate nodes as each node wants to be selected and get the payment from the source. The payment is determined by the amount of secrecy rate offered by the selected conventional relays and friendly jammers. Moreover, we assume that the source is sufficiently rich, i.e., the source cares much more about the secrecy rate of data transmission than the payment.

Then, in this section, we formulate the conventional relay and friendly jammer cooperation for secure data transmission described above as a coalitional game. In the proposed coalitional game, there involves a set of players, i.e., the intermediate nodes in the cooperative network, who seek to find optimal partners forming cooperative groups, in order to increase their counters in the selective transmission and to be selected by the source in the end, maximizing their utilities.

12.3.1 Coalitional Game Definition

We define the intermediate nodes cooperation as a coalitional TU-game $(\mathcal{T}, v)$, where $\mathcal{T}$ denotes the set of players, i.e., the intermediate nodes $\{T_1, T_2, \ldots, T_N\}$, and $v(\mathcal{S})$ is the utility of a coalition $\mathcal{S}$, $\mathcal{S} \subseteq \mathcal{T}$. The value v of a coalition $\mathcal{S}$ can be given as

$$v(\mathcal{S}) = \frac{W}{2} \left[\log \left(1 + \frac{\varepsilon_r \sum_{m=1}^{M_\mathcal{S}} p_m^R g_{R_m,D}}{\sigma^2 + \sum_{k=1}^{K_\mathcal{S}} p_k^J g_{J_k,D}} \right) \right.$$
$$\left. - \log \left(1 + \frac{\varepsilon_r \sum_{m=1}^{M_\mathcal{S}} p_m^R g_{R_m,E}}{\sigma^2 + \sum_{k=1}^{K_\mathcal{S}} p_k^J g_{J_k,E}} \right) \right]^+, \qquad (12.7)$$

where ε_r is a switching factor that $\varepsilon_r = 1$ when there exists at least one conventional relay in $\mathcal{S}$ and $\varepsilon_r = 0$ when there is no conventional relay in $\mathcal{S}$. $M_\mathcal{S}$ and $K_\mathcal{S}$ represent the number of conventional relays and friendly jammers in the coalition, respectively, satisfying $M_\mathcal{S} + K_\mathcal{S} = |\mathcal{S}|$, where $|\mathcal{S}|$ denotes the total number of members in the coalition.

Each node in the node set $\mathcal{T}$ has two identities to choose, to be a conventional relay or to be a friendly jammer. To choose a proper identity is quite important for an intermediate node, for it has great effects on whether the intermediate node can find optimal partners to form a coalition and to be selected by the source in the end. The source will choose the coalition that can provide the highest secrecy rate for data transmission. Suppose that each intermediate node is selfish and rational, whose objective is to maximize its own utility. We denote by $\mathcal{N}$ the set of indices $\{1, 2, \ldots, N\}$. The channel gains from the intermediate node T_i to destination D and malicious eavesdropper E are $g_{T_i,D}$ and $g_{T_i,E}$, respectively, $i \in \mathcal{N}$.

During the cooperation course, each intermediate node will first choose its proper identity, i.e., to be a conventional relay or a friendly jammer, mainly according to its own channel conditions. Then, the intermediate nodes are divided into two groups, i.e., the relay group and the jammer group. We denote the relay group and the jammer group by $\mathcal{R} = \{R_1, R_2, \ldots, R_t\}$ and $\mathcal{J} = \{J_1, J_2, \ldots, J_l\}$, respectively, where $\mathcal{R} \cup \mathcal{J} = \mathcal{T}$.

Consider a coalition $\mathcal{S}$ with only one member T_i, $T_i \in \mathcal{T}$, either a conventional relay or a friendly jammer. Then, from the coalition value defined in (12.7), we can obtain the self-utility of T_i, i.e., the coalition value of $\mathcal{S} = \{T_i\}$, as

$$U(T_i)$$
$$= v(\mathcal{S} = \{T_i\})$$
$$= \begin{cases} \left[\log \left(1 + \frac{p_{T_i}^R g_{T_i,D}}{\sigma^2} \right) - \log \left(1 + \frac{p_{T_i}^R g_{T_i,E}}{\sigma^2} \right) \right]^+, & T_i \in \mathcal{R}, \\ 0, & T_i \in \mathcal{J}. \end{cases} \qquad (12.8)$$

In addition, we define a payoff division rule of a coalition, which is to divide the extra utility equally among the members. Thus, the payoff utility of member T_i in a coalition $\mathcal{S}$ can be given as

$$\phi_i^v = \frac{1}{|\mathcal{S}|} \left(v(\mathcal{S}) - \sum_{T_j \in \mathcal{S}} U(T_j) \right) + U(T_i), \qquad (12.9)$$

where $U(T_i)$ and $U(T_j)$ are the self-utility of intermediate nodes T_i and T_j defined in (12.8).

Theorem 1: For the intermediate node T_i, $i \in \mathcal{N}$, we have that if $g_{T_i,D} > g_{T_i,E}$ and $T_i \in \mathcal{T}_\mathcal{D}$, T_i will choose to be a conventional relay, otherwise, it will choose to be a friendly jammer.

Proof: First, we consider the intermediate nodes which cannot successfully decode the source messages during the broadcasting phase. If the intermediate node $T_i \notin \mathcal{T}_\mathcal{D}$, then the only choice for it is to be a friendly jammer as it does not satisfy the necessary condition of being a conventional relay.

Then, we consider the intermediate nodes which belong to the decoding subset $\mathcal{T}_\mathcal{D}$. For the case that $g_{T_i,D} > g_{T_i,E}$, if T_i chooses to be a conventional relay, from (12.8), we have that it can get a positive self-utility. If T_i chooses to be a friendly jammer, from (12.8), we have that the self-utility of it is zero. In addition, no one will want to cooperate with it since with a positive jamming power it can only decrease the secrecy rate achieved by a coalition due to (12.6). Therefore, the optimal choice of T_i under the channel condition that $g_{T_i,D} > g_{T_i,E}$ is to be a conventional relay. For the case that $g_{T_i,D} \leq g_{T_i,E}$, from (12.8), the self-utility will be zero either T_i chooses to be a conventional relay or a friendly jammer. However, if T_i chooses to be a friendly jammer, it may find some potential partners to form a stronger coalition and get a positive payoff utility from the formed coalition due to (12.9) for that it can increase the secrecy rate achieved by a coalition with a proper positive jamming power. Therefore, the optimal choice of T_i under the channel condition that $g_{T_i,D} \leq g_{T_i,E}$ is to be a friendly jammer.

12.3.2   Properties of the Proposed Coalitional Game

Definition 1: A coalitional game $(\mathcal{T}, v)$ with transferable utility is said to be *superadditive* if for any two disjoint coalitions $\mathcal{S}_1, \mathcal{S}_2 \subset \mathcal{T}$, $v(\mathcal{S}_1 \cup \mathcal{S}_2) \geq v(\mathcal{S}_1) + v(\mathcal{S}_2)$.

   Property 1: The proposed coalitional game $(\mathcal{T}, v)$ is nonsuperadditive.

   Proof: Consider two disjoint coalitions $\mathcal{S}_1 \subset \mathcal{T}$ and $\mathcal{S}_2 \subset \mathcal{T}$ in the network with their corresponding utilities $v(\mathcal{S}_1)$ and $v(\mathcal{S}_2)$ when they do not cooperate with each other. We assume that there is one node belonging to $\mathcal{R}$ in $\mathcal{S}_1$, which is denoted by R, while all the nodes belong to $\mathcal{J}$ in $\mathcal{S}_2$. Then, we have that $\varepsilon_r = 1$ in the coalition value $v(\mathcal{S}_1 \cup \mathcal{S}_2)$ and $v(\mathcal{S}_1)$, while $v(\mathcal{S}_2) = 0$ as $\varepsilon_r = 0$ for the value of coalition $\mathcal{S}_2$. From the expressions of (12.6) and (12.7), when $\varepsilon_r = 1$, in our assumptions we have that $v(\mathcal{S}) = (C_{R \to D}(R, \{J_k\}) - C_{R \to E}(R, \{J_k\}))^+$, where both $C_{R \to D}(R, \{J_k\})$ and $C_{R \to E}(R, \{J_k\})$ are decreasing and convex functions of $|\mathcal{S}| - 1$ which is the number of friendly jammers in the coalition $\mathcal{S}$ consisting of relay R and several friendly jammers. However, $C_{R \to E}(R, \{J_k\})$ decreases faster than $C_{R \to D}(R, \{J_k\})$ as the number of friendly jammers in $\mathcal{S}$ increases, due to the channel conditions that $g_{J_k,D} < g_{J_k,E}, \forall J_k \in \mathcal{J}$. Thus, $v(\mathcal{S})$ will increase in some region of $|\mathcal{S}| - 1$. But when $|\mathcal{S}| - 1$ further increases, both $C_{R \to D}(R, \{J_k\})$ and $C_{R \to E}(R, \{J_k\})$ approach zero; as a result, $v(\mathcal{S})$ will approach zero. From the above analysis, we can get that there exists an optimal number of friendly jammers i_{opt}, which can maximize the coalition value $v(\mathcal{S})$. Therefore, in the case that $||\mathcal{S}_1 \cup \mathcal{S}_2| - 1 - i_{opt}| > ||\mathcal{S}_1| - 1 - i_{opt}|$, $v(\mathcal{S}_1 \cup \mathcal{S}_2) < v(\mathcal{S}_1) + v(\mathcal{S}_2)$, which means that the proposed coalitional game is not superadditive.

   Definition 2: Given the grand coalition $\mathcal{T}$ consisting of all the intermediate nodes, a payoff vector $\phi^v = (\phi_1^v, \phi_2^v, \ldots, \phi_N^v)$ for dividing the coalition value $v(\mathcal{T})$ is said to be *group rational* if $\sum_{i=1}^N \phi_i^v = v(\mathcal{T})$, and to be *individually rational* if each player can obtain a benefit no less than acting alone, i.e., $\phi_i^v \geq v(\{T_i\}), \forall i \in \mathcal{N}$. An *imputation* is a payoff vector satisfying the above two conditions.

   Having defined an imputation, the core can be defined as

$$\mathcal{C}_{TU} = \left\{ \phi^v : \sum_{T_i \in \mathcal{T}} \phi_i^v = v(\mathcal{T}) \text{ and } \sum_{T_i \in \mathcal{S}} \phi_i^v \geq v(\mathcal{S}), \forall \mathcal{S} \subseteq \mathcal{T} \right\}. \qquad (12.10)$$

In other words, the core is the set of imputations where the players have no incentive to reject the proposed payoff allocation, deviate from the grand coalition $\mathcal{T}$, and form a coalition $\mathcal{S}$ instead. A nonempty core means that the players have an incentive to form the grand coalition.

   Property 2: The core of the proposed coalitional game $(\mathcal{T}, v)$ is not always nonempty.

Proof: For this property, we consider a special case that there are only a few conventional relays but a large number of friendly jammers in $\mathcal{T}$, then from the proof to Property 1 we can get that the grand coalition value $v(\mathcal{T})$ approaches zero due to the high interference of friendly jammers, which will lead to no group rational imputation. Therefore, at least under this case the core of the proposed coalitional game is not nonempty.

Briefly, as a result of the nonsuperadditivity as well as the possible core's emptiness of the proposed game, the grand coalition of all the intermediate nodes will seldom form. Instead, several intermediate nodes will deviate from the grand coalition and form independent disjoint coalitions. Hence, in the next section, we will devise an algorithm for coalition formation that can characterize these disjoint coalitions.

12.4 Coalition Formation Algorithm

Based on the above analysis, the proposed game can be classified as a *coalition formation game* [21]. In this section, using the game-theoretical techniques from coalition formation games, we devise a distributed coalition formation algorithm.

12.4.1 Coalition Formation Concepts

For constructing a coalition formation process suitable to the proposed game, we require several definitions as follows.

Definition 3: A *collection* of coalitions, denoted by $\mathcal{P}$, is defined as a set $\mathcal{P} = \{\mathcal{S}_1, \mathcal{S}_2, \ldots, \mathcal{S}_p\}$ of mutually disjoint coalitions $\mathcal{S}_i \subset \mathcal{T}$. In other words, a collection is any arbitrary group of disjoint coalitions $\mathcal{S}_i$ of $\mathcal{T}$ not necessarily spanning all the players of $\mathcal{T}$. If a collection $\mathcal{P}$ spans all the players of $\mathcal{T}$, i.e., $\bigcup_{i=1}^{p} \mathcal{S}_i = \mathcal{T}$, then the collection is recognized as a *partition* of $\mathcal{T}$.

Definition 4: Consider two collections $\mathcal{P} = \{\mathcal{S}_1, \ldots, \mathcal{S}_p\}$ and $\mathcal{L} = \{\mathcal{S}_1^*, \ldots, \mathcal{S}_l^*\}$ which are partitions of the same subset $\mathcal{A} \subseteq \mathcal{T}$ (i.e., the same players in $\mathcal{P}$ and $\mathcal{L}$). Then, a *comparison relation* $\triangleright$ is defined as that $\mathcal{P} \triangleright \mathcal{L}$ implies the way $\mathcal{P}$ partitions $\mathcal{A}$ is preferred to the way $\mathcal{L}$ partitions $\mathcal{A}$.

Various well-known orders can be used as comparison relations [31]. There are two main categories of these orders, which are coalition value orders and individual value orders. Coalition value orders compare two collections using the value of the coalitions in these collections such as the utilitarian order in which $\mathcal{P} \triangleright \mathcal{L}$ implies $\sum_{i=1}^{p} v(\mathcal{S}_i) > \sum_{i=1}^{l} v(\mathcal{S}_i^*)$. Individual value orders perform the comparison using the individual payoffs of each player such as the Pareto order. Given that two collections $\mathcal{P}$ and $\mathcal{L}$ have the same players and the payoffs of a player T_j in $\mathcal{P}$ and $\mathcal{L}$ are denoted by $\phi_j^v(\mathcal{P})$ and $\phi_j^v(\mathcal{L})$, respectively. The Pareto order can be written as

$$\mathcal{P} \triangleright \mathcal{L} \Leftrightarrow \left\{ \phi_j^v(\mathcal{P}) \geq \phi_j^v(\mathcal{L}), \ \forall T_j \in \mathcal{P}, \mathcal{L} \right\}, \tag{12.11}$$

with at least one strict inequality ($>$) for a player T_k. The Pareto order implies a collection $\mathcal{P}$ is preferred to $\mathcal{L}$, if at least one player is able to improve its payoff when the coalition structure changes from $\mathcal{P}$ to $\mathcal{L}$ without decreasing other players' payoffs.

In this cooperative network, different coalitions of conventional relays and friendly jammers not only have chances for cooperation, but also suffer competition from each other as only one coalition can be selected by the source and gain the payoff in the end. Two disjoint coalitions are preferred to cooperate with each other if they can form a stronger one instead, as the players of them have more chance to win the payoffs from the source. Here, we are concerned more about the coalition who has the highest value and whether the players in it can always benefit from a positive payoff. Then, based on the Pareto order described in (12.11), we define a new comparison relation with respect to our system as follows.

Definition 5: Consider two collections $\mathcal{P} = \{\mathcal{S}_1, \ldots, \mathcal{S}_p\}$ and $\mathcal{L} = \{\mathcal{S}_1^*, \ldots, \mathcal{S}_l^*\}$ with the same players in them. Then, the *Max-Pareto order* is defined as

$$\mathcal{P} \succ \mathcal{L} \Leftrightarrow \{\max\{v(\mathcal{S}_1), \ldots, v(\mathcal{S}_p)\} \geq \max\{v(\mathcal{S}_1^*), \ldots, v(\mathcal{S}_l^*)\},$$
$$\text{and } \phi_j^v(\mathcal{P}) \geq \phi_j^v(\mathcal{L}), \ \forall T_j \in \mathcal{P}, \mathcal{L}\}, \tag{12.12}$$

with at least one strict inequality ($>$) for a player T_j in the individual payoff comparison.

12.4.2   Merge-and-Split Coalition Formation Algorithm

Using the coalition formation concepts in the previous section, we construct a distributed coalition formation algorithm for self-organization in the cooperative network based on two simple rules denoted as "merge" and "split" which permit us to modify a partition of $\mathcal{T}$ [31].

- **Merge Rule:** Merge any set of coalitions $\{\mathcal{S}_1, \ldots, \mathcal{S}_l\}$ whenever the merged form is preferred by the players, i.e., where $\left\{\bigcup_{j=1}^l \mathcal{S}_j\right\} \succ \{\mathcal{S}_1, \ldots, \mathcal{S}_l\}$, then, $\{\mathcal{S}_1, \ldots, \mathcal{S}_l\} \longrightarrow \left\{\bigcup_{j=1}^l \mathcal{S}_j\right\}$.

- **Split Rule:** Split any coalition $\left\{\bigcup_{j=1}^l \mathcal{S}_j\right\}$ whenever a split form is preferred by the players, i.e., where $\{\mathcal{S}_1, \ldots, \mathcal{S}_l\} \succ \left\{\bigcup_{j=1}^l \mathcal{S}_j\right\}$, then, $\left\{\bigcup_{j=1}^l \mathcal{S}_j\right\} \longrightarrow \{\mathcal{S}_1, \ldots, \mathcal{S}_l\}$.

According to the above rules, multiple coalitions can merge into a larger coalition if merging yields a preferred partition based on the Max-Pareto order. This implies that a group of players can agree to form a larger coalition, if the merged coalition has no less value than any of the disjoint coalitions and at least one of the players can improve its payoff without decreasing the utilities of any other player. Similarly, an existing coalition can decide to split into smaller coalitions if splitting yields a preferred partition based on the Max-Pareto order. The merge-and-split rules is a coalition formation algorithm with partially reversible agreements [32], where once the players sign a binding agreement to form a coalition through the merge operation, the coalition can be split only when all the players approve (i.e., the splitting does not decrease the payoff of any coalition member). Using the rules of merge and split is suitable for the proposed coalition game, where each merge or split decision can be taken in a distributed manner by each individual relay node or by each form coalition.

As shown in Table 12.1, we construct a distributed coalition formation algorithm based on the merge-and-split rules. The three phases of the coalition formation algorithm are repeated periodically during the network operation, allowing a topology that is adaptive to environmental changes such as mobility. During the adaptive coalition formation phase, $\mathcal{T}_t$ denotes the partition of the intermediate nodes set at clock t. The merging operation **Merge**($\mathcal{T}_t$) spans all the current coalitions in $\mathcal{T}_t$ orderly, resulting in a final partition $\mathcal{P}_t$ at clock t. Following the merge process, the coalitions in the resulting partition $\mathcal{P}_t$ are next subject to the splitting operation **Split**($\mathcal{P}_t$), if any is necessary. The process will be repeated until there is no need for any merging or splitting in the current partition. Note that the merge or split decisions can be taken in a distributed manner by each individual coalition without relying on any centralized entity.

In Figure 12.2, we show the network scenarios before and after coalition formation based on our proposed merge-and-split algorithm. We can see that before coalition formation each intermediate node chooses its optimal identity (to be a conventional relay or a friendly

Table 12.1: The Proposed Distributed Merge-and-Split Coalition Formation Algorithm

---

∗ Initial State:

The set of intermediate nodes is partitioned by $\mathcal{T}_0 = \{T_1, T_2, \ldots, T_N\}$ at the beginning with noncooperation.

∗ Coalition Formation Algorithm:

⋆ *Phase 1—Identity Choice:*

The source transmits its data to the intermediate nodes. Each intermediate node chooses its optimal identity (to be a conventional relay or a friendly jammer) based on its channel conditions to the destination and the eavesdropper.

For an intermediate node T_i, if $g_{T_i,D} > g_{T_i,E}$, then it will choose to be a conventional relay. Else, it will choose to be a friendly jammer. The intermediate node who cannot decode the source messages correctly has no choice but to be a friendly jammer.

⋆ *Phase 2—Adaptive Coalition Formation:*

Each intermediate node with an identity can be regarded as a disjoint independent coalition, then coalition formation using merge-and-split rules occurs:

− Clock index $t = 0$.

− **Repeat**

(a) $\mathcal{P}_t = \mathbf{Merge}(\mathcal{T}_t)$: coalitions in $\mathcal{T}_t$ decide to merge based on the merge rules.

(b) $\mathcal{L}_t = \mathbf{Split}(\mathcal{P}_t)$: coalitions in $\mathcal{P}_t$ decide to split based on the split rules.

(c) Update clock index $t = t + 1$.

(d) Value assignment $\mathcal{T}_t = \mathcal{L}_t$.

Until merge-and-split terminates.

− Each coalition reports the secrecy rate of data transmission that it can achieve, i.e, the coalition value as we defined in (12.7), to the source.

⋆ *Phase 3—Secure Transmission:*

The source chooses the coalition that can provide the highest secrecy rate for secure data transmission and pays for it.

---

jammer; in the figure △ represents conventional relay and ○ represents friendly jammer) according to its own channel conditions to the destination and the eavesdropper. Then, the conventional relays and friendly jammers start to find optimal partners in order to improve the chance to be selected by the source, using the merge-and-split rules to help them make decisions. After the coalition formation process, several disjoint independent coalitions are formed. The coalition with the highest value will be selected by the source for secure data transmission and obtain the payoff it deserves. Overall, this figure shows how the distributed conventional relays and friendly jammers can self-organize into disjoint independent coalitions based on the proposed merge-and-split algorithm. In addition, from the figure, we can find that all the conventional relays will always join in one coalition in the end.

In Figure 12.3, we show the performance in terms of the final secrecy rate for data transmission as a function of the transmitting power p^R in two different cases. Here, the centralized relay and jammer selection algorithm is the one proposed as the optimal selection with jamming (OSJ) scheme in [15]. The comparison between the two performance results implies the efficiency of our proposed distributed coalition formation algorithm with respect to secure data transmission in a cooperative network.

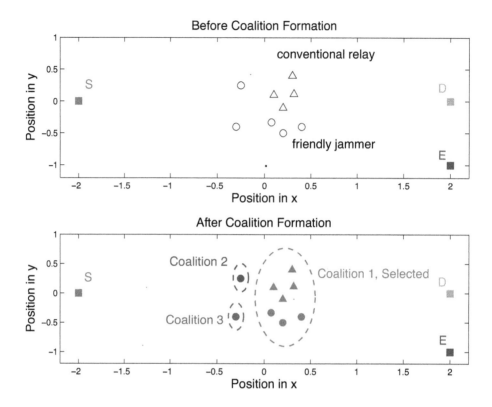

Figure 12.2: Network scenarios before and after coalition formation.

12.5   Conclusions and Open Issues

In this chapter, we have investigated conventional relay and friendly jammer cooperation in secure cooperative networks. In order to obtain an optimized secrecy rate, the source intends to select several conventional relays and friendly jammers from the intermediate nodes to assist data transmission, and in return, it needs to make a payment. Each intermediate node here has two possible identities to choose, i.e., to be a conventional relay or a friendly jammer, which results in a different impact on the final utility of the intermediate node. After the intermediate nodes determine their identities, they seek to find optimal partners forming coalitions, which improves their chances to be selected by the source and thus to obtain the payoffs in the end. We formulated this cooperation as a coalitional game with transferable utility and studied its properties. Besides, we defined a Max-Pareto order and provided a distributed merge-and-split coalition formation algorithm to achieve efficient cooperation between conventional relays and friendly jammers.

As for the future work, two-way relaying and multihop relaying scenarios will be interesting but more complex when investigating the cooperation among intermediate nodes to improve the network secure performance. In such scenarios, the intermediate nodes that can behave as conventional relays or friendly jammers cannot decide their identities based on their own channel conditions in advance. Thus, the node cooperation can no longer be formulated as a coalitional game with transferable utility but should be the one with nontransferable utility [21]. Then, the analysis of node cooperation and the corresponding coalition formation process will be more challenging.

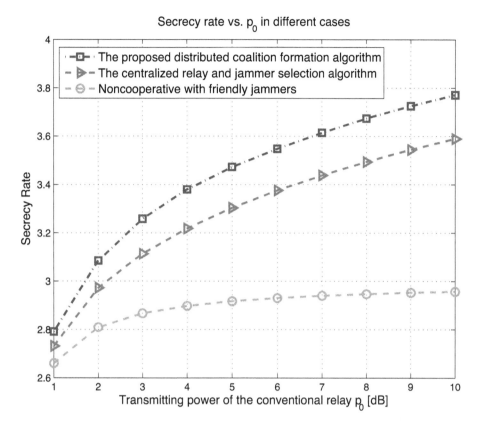

Figure 12.3: Secrecy rate for data transmission in different cases.

References

[1] A. D. Wyner, "The wire-tap channel," *Bell System Technical Journal*, vol. 54, no. 8, pp. 1355–1387, Oct. 1975.

[2] I. Csiszár and J. Körner, "Broadcast channels with confidential messages," *IEEE Transactions on Information Theory*, vol. 24, no. 3, pp. 339–348, May 1978.

[3] S. K. Leung-Yan-Cheong and M. E. Hellman, "The Gaussian wiretap channel," *IEEE Transactions on Information Theory*, vol. 24, no. 4, pp. 451–456, July 1978.

[4] L. Lai and H. E. Gamal, "The relay-eavesdropper channel: Cooperation for secrecy," *IEEE Transactions on Information Theory*, vol. 54, no. 9, pp. 4005–4019, Sept. 2008.

[5] L. Dong, Z. Han, A. P. Petropulu, and H. V. Poor, "Improving wireless physical layer security via cooperating relays," *IEEE Transactions on Signal Processing*, vol. 58, no. 3, pp. 1875–1888, Mar. 2010.

[6] J. Li, A. P. Petropulu, and S. Weber, "On cooperative relaying schemes for wireless physical layer security," *IEEE Transactions on Signal Processing*, vol. 59, no. 10, pp. 4985–4997, Oct. 2011.

[7] X. Zhou and M. R. McKay, "Secure transmission with artificial noise over fading channels: Achievable rate and optimal power allocation," *IEEE Transactions on Vehicular Technology*, vol. 59, no. 8, pp. 3831–3842, Oct. 2010.

[8] G. Zheng, L.-C. Choo, and K.-K. Wong, "Optimal cooperative jamming to enhance physical layer security using relays," *IEEE Transactions on Signal Processing*, vol. 59, no. 3, pp. 1317–1322, Mar. 2011.

[9] J. Huang and A. L. Swindlehurst, "Cooperative jamming for secure communications in MIMO relay networks," *IEEE Transactions on Signal Processing*, vol. 59, no. 10, pp. 4871–4884, Oct. 2011.

[10] J. Li, A. P. Petropulu, and S. Weber, "On cooperative relaying schemes for wireless physical layer security," *IEEE Transactions on Signal Processing*, vol. 59, no. 10, pp. 4985–4997, Oct. 2011.

[11] Z. Han, N. Marina, M. Debbah, and A. Hjørungnes, "Physical layer security game: Interaction between source, eavesdropper and friendly jammer," *EURASIP Journal on Wireless Communications and Networking*, vol. 2009, no. 11, Mar. 2009.

[12] Z. Han, N. Marina, M. Debbah, and A. Hjørungnes, "Physical layer security game: How to date a girl with her boyfriend on the same table," in *Proc. IEEE International Conference on Game Theory for Networks*, Istanbul, Turkey, May 2009.

[13] R. Zhang, L. Song, Z. Han, and B. Jiao, "Physical layer security for two-way untrusted relaying with friendly jammers," *IEEE Transactions on Vehicular Technology*, vol. 61, no. 8, pp. 3693–3704, Oct. 2012.

[14] R. Zhang, L. Song, Z. Han, B. Jiao, and M. Debbah, "Physical layer security for two way relay communications with friendly jammers," in *Proc. IEEE GLOBECOM 2010*, Miami, USA, Dec. 2010.

[15] I. Krikidis, J. Thompson, and S. McLaughlin, "Relay selection for secure cooperative networks with jamming," *IEEE Trans. Wireless Commun.*, vol. 8, no. 10, pp. 5003–5011, Oct. 2009.

[16] J. Chen, R. Zhang, L. Song, Z. Han, and B. Jiao, "Joint relay and jammer selection for secure two-way relay networks," in *Proc. IEEE ICC 2011*, June 2011.

[17] J. Chen, R. Zhang, L. Song, Z. Han, and B. Jiao, "Joint relay and jammer selection for secure two-way relay networks," *IEEE Transactions on Information Forensics and Security*, vol. 7, no. 1, pp. 310–320, Feb. 2012.

[18] J. Kim, A. Ikhlef, and R. Schober, "Combined relay selection and cooperative beamforming for physical layer security," *Journal of Communications and Networks*, vol. 14, no. 4, pp. 364–373, Aug. 2012.

[19] D. Fudenberg and J. Tirole, *Game Theory*, MIT Press, Cambridge, MA, 1993.

[20] Z. Han, D. Niyato, W. Saad, T. Başar, and A. Hjørungnes, *Game Theory in Wireless and Communication Networks: Theory, Models and Applications*, Cambridge University Press, New York, 2017.

[21] W. Saad, Z. Han, M. Debbah, A. Hjørungnes, and T. Başar, "Coaltional game theory for communication networks: A tutorial," *IEEE Signal Processing Magazine*, vol. 26, no. 5, pp. 77–97, Sept. 2009.

[22] D. Li, Y. Xu, X. Wang, and M. Guizani, "Coalitional game theoretic approach for secondary spectrum access in cooperative cognitive radio networks," *IEEE Transactions on Wireless Communications*, vol. 10, no. 3, pp. 844–856, Mar. 2011.

[23] W. Saad, Z. Han, M. Debbah, A. Hjørungnes, and T. Baçsr, "Coalitional games for distributed collaborative spectrum sensing in cognitive radio networks," in *Proc. IEEE INFOCOM 2009*, Apr. 2009.

[24] T. Chen, L. Zhu, F. Wu, and S. Zhong, "Stimulating cooperation in vehicular ad hoc networks: A coalitional game theoretic approach," *IEEE Transactions on Vehicular Technology*, vol. 60, no. 2, pp. 566–579, Feb. 2011.

[25] W. Saad, Z. Han, M. Debbah, and A. Hjørungnes, "A distributed coalition formation framework for fair user cooperation in wireless networks," *IEEE Transactions on Wireless Communications*, vol. 8, no. 9, pp. 4580–4593, Sept. 2009.

[26] W. Saad, Z. Han, T. Baçsr, M. Debbah, and A. Hjørungnes, "Physical layer security: Coalitional games for distributed cooperation," in *IEEE WiOPT 2009*, June 2009.

[27] R. Zhang, L. Song, Z. Han, and B. Jiao, "Distributed coalition formation of relay and friendly jammers for secure cooperative networks," in *Proc. IEEE ICC 2011*, Japan, June 2011.

[28] R. Zhang, L. Song, Z. Han, and B. Jiao, "Relay and jammer cooperation as a coalitional game in secure cooperative wireless networks," in *Proc. the 5th International ICST Conference on Performance Evaluation Methodologies and Tools*, Invited, Cachan, France, May 2011.

[29] P. Marbach and R. Berry, "Downlink resource allocation and pricing for wireless networks," in *Proc. IEEE INFOCOM 2002*, New York, USA, June 2002.

[30] J. Barros and M. R. D. Rodrigues, "Secrecy capacity of wireless channels," in *Proc. of IEEE International Symposium on Information Theory*, Seattle, USA, July 2006.

[31] K. Apt and A. Witzel, "A generic approach to coaltion formation," in *Proc. International Workshop on Computational Social Choice (COMSOC)*, Amsterdam, Netherlands, Dec. 2006.

[32] D. Ray, *A Game-theoretic Perspective on Coalition Formation*, New York: Oxford University Press, 2007.

[33] I. N. Stewart, *Galois Theory, 3rd. ed.*, Chapman & Hall/CRC Mathematics, Boca Raton, FL, 2004.

[34] Z. Han and K. J. R. Liu, *Resource Allocation for Wireless Networks: Basics, Techniques, and Applications*, Cambridge University Press, UK, 2008.

[35] S. Boyd and L. Vandenberghe, *Convex Optimization*, Cambridge University Press, New York, 2004.

[36] L. Song and J. Shen, *Evolved Cellular Network Planning and Optimization for UMTS and LTE*, CRC Press, Boca Raton, FL, 2010.

Chapter 13

Stochastic Geometry Approaches to Secrecy in Large Wireless Networks

Xiangyun Zhou
The Australian National University

Martin Haenggi
University of Notre Dame

In this chapter, we turn our attention to large-scale wireless networks and borrow tools from stochastic geometry to study the physical layer security performance. We consider the legitimate users and the eavesdroppers to be randomly located over a large geographical area according to some probability distributions. The network connectivity and throughput performance are discussed based on two analytical approaches. Specifically, the secrecy graph, as a graph-theoretic approach, is introduced to study the connectivity properties among the legitimate users of the network. It characterizes the existence of connection with perfect secrecy between any two legitimate users. The second approach is the development of a performance metric named secrecy transmission capacity. It considers concurrent transmissions between all the legitimate links and gives a mathematically tractable measure on the achievable network throughput with a given secrecy requirement.

13.1 Introduction

13.1.1 Motivation

For networks with potentially a large number of legitimate users as well as malicious eavesdroppers, the spatial configuration of the nodes is an important parameter affecting the network performance. Commonly used spatial models range from highly structured node placement to completely random node deployment. In cellular networks, for example, the base stations are traditionally placed with careful planning to ensure some minimum separation between neighboring base stations. In future cellular networks, there tends to be more than one tier of base stations, creating a heterogenous architecture. While the macrocell base stations are usually placed with careful planning, picocell and femtocell base stations are likely to have much more randomness in their locations. Besides that, many decentralized networks tend to have random node locations due to the random deployment or substantial mobility of nodes. Apart from the spatial distributions of nodes, the knowledge of the node locations is another important factor, especially when secrecy is taken into account. Typically in decentralized networks, it is difficult to enable any mechanism to track the locations of the eavesdroppers, even if they are ordinary users belonging to the same network as the legitimate nodes. Hence, the assumption of knowing the eavesdropper locations, which is often used in the literature of physical layer security, needs to be relaxed.

In this chapter, we focus on large-scale decentralized networks. Indeed, it is often difficult or costly to enable secret key exchange and management for encrypted transmissions in decentralized networks and hence, physical layer security becomes especially important. As we have seen in the previous chapters, the metrics used to study physical layer security performance of a single link include the maximum achievable secrecy rate (or secrecy capacity) and the outage probability for transmission at a prescribed secrecy rate. In this chapter, we build on these metrics to come up with more suitable measures on the secrecy performance in large-scale wireless networks, explicitly taking into account the spatial locations of both the legitimate nodes and the eavesdroppers. To this end, tools from stochastic geometry are used for the modeling and analysis.

13.1.2   Stochastic Geometry Approaches

Traditionally, the studies on wireless networks often assume a fixed geometry. As a result, the performance analyses merely produced results applicable for that particular network topology. In order to obtain results that are more generally valid for networks with node mobility or otherwise random topology, two ingredients are needed: first, stochastic models that describe the possible realizations of the network geometry; second, a mathematical theory that permits the averaging of the performance metrics over these realizations, weighted by their likelihood of occurrence. Stochastic geometry, in particular point process theory, solves both these problems—it provides the models and the mathematical tools for the analysis [1, 2].

The simplest yet most important model is the homogenous Poisson point process (PPP). A homogenous PPP in an n-dimensional (usually two-dimensional) space roughly means that all nodes are randomly located inside the network according to a uniform distribution. It is completely characterized by the constant intensity parameter λ. Specifically, the value of λ gives the average number of nodes located inside a unit volume in the n-dimensional space. The location of every node is independent from the locations of all other nodes. Hence, knowing the locations of some nodes gives no information about the locations of others. Of course, more complex point processes, which are often derived from the homogenous PPP, have also been used to better model wireless networks with specific features, such as clustering or repulsion between nodes. The analytical tractability is usually reduced as the model of the network geometry becomes more complex.

Using the stochastic geometry approach, initial studies on the physical layer security performance of large wireless network have been carried out from various viewpoints, such as connectivity [3–8], coverage [9, 10], and throughput [11–15]. The majority of these works modeled the locations of both the legitimate nodes and the eavesdroppers as homogeneous PPPs. In this chapter, we present two important frameworks used in the literature, namely, the secrecy graph [3, 7] and the secrecy transmission capacity [13]. The former is concerned with the connectivity properties of the legitimate nodes in the network, while the latter studies the average per-link throughput.

13.2   Secrecy Graph

In decentralized networks, communication is usually initiated in an ad hoc manner with loose or completely random medium access control (MAC). The message transmitted from a source to a destination requires multiple intermediate relays when the source and destination are separated by a large distance. Therefore, a high level of connectivity becomes a very important prerequisite for reliable communications over the entire network.

Connectivity is a basic concept in graph theory. Within a graph-theoretic framework, the location of each node in the network[1] is called a "vertex" and a successful communication between two nodes is represented by the existence of an "edge" between the two corresponding vertices. A graph representing the network is denoted as $G = (\Phi, E)$, where Φ denotes the set of all the vertices, i.e., all node locations, and E denotes the set of all the edges. For two nodes located at x and y, if the edge $xy \in E$, successful communication can be made between the two nodes; or simply, these two nodes are connected. The number of edges represents the level of connectivity of the network. If x can talk to y but not vice versa, we can model this with a directed edge $\overrightarrow{xy}$, which results in a directed graph.

The network we are interested in consists of both the legitimate nodes as well as eavesdroppers. Whether there exists a secure communication link between any two legitimate nodes depends not only on the locations and channel quality of these two nodes, but also on the locations and channel qualities of all the eavesdroppers. In this section, we introduce the concept of the secrecy graph, which extends the conventional graph-theoretic approach for network connectivity analysis to include secrecy considerations.

13.2.1   Network and Graph Model

Let the locations of both the legitimate nodes and the eavesdroppers follow independent homogeneous PPPs over a two-dimensional space[2], denoted as Φ_l and Φ_e, respectively. The intensities of Φ_l and Φ_e are λ_l and λ_e. The transmit power is fixed to the same value P, for all legitimate nodes. The thermal noise power is the same for both legitimate nodes and the eavesdroppers, which is normalized to one. The wireless channels between all nodes are modeled by a large-scale path loss component with exponent α and a small-scale fading component.

Define the *directed secrecy graph* as $\overrightarrow{G} = (\Phi_l, \overrightarrow{E})$ where Φ_l is the set of vertices representing the locations of legitimate nodes and $\overrightarrow{E}$ is the set of edges representing the existence of a secure connection between any legitimate node pair [3]. We say that there exists a secure connection from node at x to node at y when a nonzero secrecy rate is achievable for message transmission from x to y in the presence of all eavesdroppers, i.e., if

$$R_s = \left[\log_2 \left(1 + P \|x - y\|^{-\alpha} S_{xy} \right) - \log_2 \left(1 + P \|x - z\|^{-\alpha} S_{xz} \right) \right] > 0, \ \forall z \in \Phi_e, \quad (13.1)$$

where S_{xy} (S_{xz}) is the fading gain of the channel between the transmitting node at x and the receiving node at y (the eavesdropper at z) and $\|x - y\|$ ($\|x - z\|$) is the distance between the transmitting node at x and the receiving node at y (the eavesdropper at z). In other words, the edge $\overrightarrow{xy}$ exists in the set $\overrightarrow{E}$ when the condition in (13.1) holds. Note that $\overrightarrow{xy}$ means the link from the transmitter at x to the receiver at y. Clearly, the secrecy rate of

---

[1] In this chapter, we denote the location of a node as a vertex. Another convention denotes the node itself as a vertex. Both conventions are widely used in the literature of graph-theoretic study on wireless networks.

[2] For simplicity, we consider a two-dimensional space, while the secrecy graph formulation is applicable to the general case of n-dimensional space.

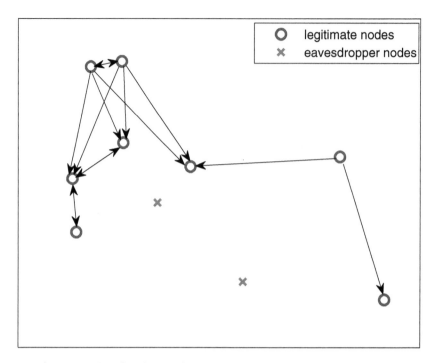

Figure 13.1: An example of a directed secrecy graph. The arrows between the legitimate nodes show the existence of direct edges in the secrecy graph.

the $\overrightarrow{xy}$ link is generally different from the secrecy rate of the $\overrightarrow{yx}$ link. This is why the edges in $\overrightarrow{E}$ have directions. An illustration of the directed secrecy graph is shown in Figure 13.1.

From the directed secrecy graph, we can also define undirected graphs. For instance, the *basic secrecy graph* is defined as $\overline{G} = (\Phi_l, \overline{E})$, where the set of edges $\overline{E}$ is

$$\overline{E} = \{xy : \overrightarrow{xy} \in \overrightarrow{E} \text{ and } \overrightarrow{yx} \in \overrightarrow{E}\}. \tag{13.2}$$

Hence, an edge exists in the basic secrecy graph if and only if the bidirectional links between the two nodes can both support some nonzero secrecy rates. Clearly, this is a stronger condition than that of the directed secrecy graph.

Another example of undirected graphs is called the *enhanced secrecy graph* defined as $G' = (\Phi_l, E')$, where the set of edges E' is

$$E' = \{xy : \overrightarrow{xy} \in \overrightarrow{E} \text{ or } \overrightarrow{yx} \in \overrightarrow{E}\}. \tag{13.3}$$

This means that the existence of a secure connection between two nodes only requires a nonzero secrecy rate in one direction but not necessarily in both directions. The enhanced secrecy graph is relevant in the scenarios where a one-time pad can be transmitted in the intrinsically secured direction to enable encrypted transmission in the other direction.

In what follows, we focus on the directed secrecy graph only and discuss some of its connectivity properties. Readers interested in undirected secrecy graphs can find relevant discussions in [3, 8].

13.2.2   Local Connectivity Properties

Connectivity measures can be broadly categorized into local connectivity and global connectivity. The former focuses on the single-hop connections from/to a typical node of interest, while the latter looks at the multihop connections over the entire network. Here, we study the single-hop connectivity of the directed secrecy graph and use two metrics named the *in-degree* and *out-degree*. Assume a typical node located at the origin of the two-dimensional space. From Slivnyak's Theorem [2], the spatial distribution of all the other legitimate nodes still follows the same homogeneous PPP with intensity λ_l. The in-degree is defined as the total number of secure connections made from all other legitimate nodes (as the transmitters) to the typical node. The out-degree is defined as the total number of secure connections made from the typical node to all other legitimate nodes (as the receivers).

Let us first consider a special channel model of path loss only; the impact of fading will be discussed next. With path loss only, the condition of existence of a secure connection from x to y in (13.1) reduces to

$$\|x - y\| < \|x - z\|, \ \forall z \in \Phi_e, \tag{13.4}$$

where we have set $S_{xy} = 1$ and $S_{xz} = 1$ in (13.1). This simplified condition has the following geometrical meaning: if we draw a circle centered at the transmitting node with radius equal to the distance to the receiving node, then there should be no eavesdropper located inside (or on) the circle.

Out-Degree: We start by characterizing the out-degree of the typical node at the origin. The maximum distance up to which the typical node is able to establish a secure connection is limited by the nearest eavesdropper. If we draw a circle centered at the typical node with radius equal to the distance to the nearest eavesdropper, then any legitimate node(s) located inside the circle is able to receive a secure connection from the typical node. Denote the out-degree of the typical node by N_{out}. If we consider a sequence of its nearest neighbors in the combined PPP $\Phi_l \cup \Phi_e$, then $N_{\text{out}} = n$ means that the closest n neighbors are all in Φ_l while the $(n+1)$-th neighbor is in Φ_e. Due to the independence between Φ_l and Φ_e, it is not hard to show that the probability distribution of the out-degree is given by

$$\mathbb{P}\Big(N_{\text{out}} = n\Big) = \frac{\lambda_e}{\lambda_l + \lambda_e} \Big(\frac{\lambda_l}{\lambda_l + \lambda_e}\Big)^n, \tag{13.5}$$

which is a geometric distribution.

From (13.5), one can easily derive the probability of (out-)isolation as

$$\mathbb{P}\Big(N_{\text{out}} = 0\Big) = \frac{\lambda_e}{\lambda_l + \lambda_e}, \tag{13.6}$$

and the average out-degree as

$$\mathbb{E}\Big\{N_{\text{out}}\Big\} = \frac{\lambda_l}{\lambda_e}. \tag{13.7}$$

In-Degree: Now let us treat the typical node at the origin as the receiver and derive its in-degree, denoted as N_{in}, i.e., the number of legitimate nodes that can make a secure connection to the typical node. For a legitimate node at x, a secure connection to the origin o exists if there is no eavesdropper located within $\|x - o\| = \|x\|$ from x. The expression of probability distribution of N_{in} is generally unavailable. Nevertheless, the probability of isolation and the average in-degree can be obtained as

$$\mathbb{P}\Big(N_{\text{in}} = 0\Big) = \mathbb{E}\Big\{ \exp\Big(-\frac{\lambda_l}{\lambda_e}\tilde{A}\Big)\Big\}, \tag{13.8}$$

and

$$\mathbb{E}\Big\{N_{\text{in}}\Big\} = \frac{\lambda_l}{\lambda_e}, \tag{13.9}$$

where $\widetilde{A}$ in (13.8) stands for the area of a typical Voronoi cell induced by a unit-intensity PPP [2]. Although there is no closed-form expression of the probability distribution of $\widetilde{A}$, existing studies have shown that the gamma distribution can be used to provide an accurate approximation [16], hence (13.8) can be reduced to a closed-form approximation.

One important observation from comparing the results between out-degree and in-degree is that the average node degree is the same in both directions. The equality in the average node degree of both directions is a general result which holds for any directed random geometric graph, since each directed edge contributes one to both the sum of all out-degrees and all in-degrees over the entire network, so the two sums are equal. On the other hand, the probability distribution is not the same, which can be seen by comparing the probability of isolation results. As shown in [7], the probability of isolation is higher in the out-direction than the in-direction. An intuitive explanation is as follows: in order to make the typical node isolated in the out-direction, we only need one eavesdropper located close to it so that it is not able to transmit securely. While if we want to make the typical node isolated in the in-direction, there needs to be multiple close-by eavesdroppers surrounding the typical node from different angles, so that no legitimate nodes can transmit securely to the typical node from any angle. Therefore, from the eavesdropper's perspective, it is easier to make the typical node isolated in the out-direction than the in-direction.

Effect of Small-Scale Fading: So far, our discussion on the local connectivity properties assumed a simple path loss channel model. When the small-scale fading is also taken into account, an extra source of randomness is added into the condition for the existence of a secure link from x to y, given as

$$\|x - y\| S_{xy} < \|x - z\| S_{xz}, \ \forall \ z \in \Phi_e, \tag{13.10}$$

where S_{xy} and all S_{xz} are independent random variables following some probability distribution $f_S(s)$ (where $s > 0$). The impact of fading on the connectivity is not trivial and a detailed analysis can be found in [7]. Here we give a summary of some important results:

- The probability distribution of the out-degree is the same regardless of the presence of any type of small-scale fading. In other words, the equation in (13.5) holds for any valid $f_S(s)$.

- The probability distribution of the in-degree depends on the specific form of $f_S(s)$, hence is generally different from the path loss-only case.

In practical wireless networks, fading always exists and hence needs to be taken into account in order to obtain more accurate network design and analysis. The above list of results suggest that sometimes one can carry out the analysis in a simple path loss channel model and the obtained results will also be applicable in more practical scenarios with fading. However, one should do so with care and pay attention to the underlying assumptions and formulations from which the above results arrive. For example, if the legitimate nodes utilize the knowledge of the fading channel gains in some smart way, the probability distribution of the out-degree is likely to be different from that in the path loss-only case.

13.2.3   Global Connectivity Properties

While the distribution of the node degrees provides some insight into the connectivity of a network, it does not capture its global properties. The most pertinent question is whether a significant part of the network is connected, i.e., whether there exists a *giant component* in the network whose nodes can all reach each other via multihop routes. A giant component in an infinite network is a component that itself contains infinitely many nodes; in other words, it contains a positive fraction of the nodes. The results on the node degrees imply that for any nonzero intensity of eavesdroppers λ_e, the network will be disconnected almost surely, since each legitimate node is isolated with positive probability, both in the in- and the out-directions. However, we can still hope for a giant (connected) component to exist for $\lambda_e > 0$. The mathematical technique to analyze for which λ_e such a component exists is *percolation theory*. Originally applied to the question of the connectivity of lattices with random edges [17], percolation theory has become an important tool also for random geometric graphs where the vertices are points in the Euclidean space, typically $\mathbb{R}^2$ [18]. Applications of percolation theory to wireless networks in general are discussed in Part 2 of [2].

To apply percolation theory to secrecy graphs, we first need to specify what type of components we are interested in. Adding a node at the origin, and focusing on the directed case, the in-component is the set of nodes the origin can receive messages from; conversely, the out-component is set of nodes that can be reached by the origin, all via directed paths. Alternatively, we may consider the components in the undirected secrecy graphs (basic and enhanced). Defining $\xi = \lambda_e/\lambda_l$, we denote by $\theta(\xi)$ the probability that the origin's component is infinite as a function of the eavesdropper's and legitimate nodes' relative density. The critical (relative) density then follows as

$$\xi_c = \inf\{\xi \colon \theta(\xi) = 0\} = \sup\{\xi \colon \theta(\xi) > 0\}.$$

In words, it is the smallest density that ensures that the origin's component is finite, or, equivalently, the largest density for which the component still has a positive probability of being infinite. The graph is said to *percolate* if $\xi < \xi_c$, since $\theta(\xi) > 0$ implies that there exists an infinite component somewhere in the network. This follows from Kolmogorov's 0-1 law.

In the following, we present several bounding techniques for ξ_c for the case of out-percolation, i.e., for the critical density pertaining to an infinite out-component.

Branching techniques. From the theory of (independent) branching processes, it is known that the out-component is finite as soon as the mean out-degree is one or smaller. So, from (13.7), if $\xi^{-1} \leq 1$, there is no percolation, hence $\xi_c < 1$. A more detailed analysis yields the sharper result $\theta(\xi) \leq \max\{0, 1 - \xi\}$.

Lattice percolation. In percolation models on lattices, vertices form a lattice, for example the square lattice $\mathbb{Z}^2$, and edges exist between all pairs of nearest neighbors. Edges are declared *open* (passable) independently with probability p. For the square lattice, it is known that the critical probability for the emergence of a giant component is $p_c = 1/2$ [17]. A connection between a continuum model, such as the secrecy graph, and a lattice model can be established by identifying a suitable geometric condition in the continuum model that determines whether a certain edge is open or closed in the lattice model. However, in the case of the secrecy graph, while such methods can be used to show that $\xi_c > 0$, they have led to extremely loose bounds only. The reason is the strong correlation between the existence of edges of nodes that are geometrically close.

Rolling ball method. This method results in better bounds, since it is explicitly designed for dependent percolation models. It exploits a connection to a so-called one-dependent lattice percolation model, where an edge state may depend on the state of adjacent edges. Using this method, it can be shown that $\xi_c > 0.0008$, which is not tight (the exact value is assumed to be around 0.15, as simulations in [3] suggest), but orders of magnitude better than bounds obtained with other methods.

All approaches are described in detail in [8].

13.2.4   Connectivity Enhancements

There are various existing wireless technologies that can be used to enhance the connectivity of the secrecy graph. Here, we briefly discuss the use of multiple antennas at the legitimate nodes for transmission. Two commonly used techniques are considered, namely, directional antenna transmission and eigen-beamforming. The former technique requires the location information of the intended receiving nodes, while the latter requires the fading channel gains of the receiving nodes to be known at the transmitting node. Naturally the channel model considered here includes both path loss and small-scale fading. In particular, Rayleigh fading is assumed. Our analysis still focuses on the typical node located at the origin and for simplicity, we study the average out-degree to demonstrate the benefits of the multi-antenna transmission techniques. That is to say, we treat the typical node as the transmitter and compute the average number of legitimate nodes to which the typical node can establish secure connections. The number of antennas at the typical node is assumed to be M.

13.2.4.1   Directional Antenna Transmission

Directional antenna transmission is usually accomplished by using an array of antenna elements with different phase shifts such that the intensity of energy radiation reaches the maximum in the desired direction. We consider a simplified model for directional antenna transmission [19] in which there is a main-lobe of gain M and spread angle $2\pi\omega$, as well as a side-lobe of gain κM and spread angle $2\pi(1-\omega)$. Note that $\kappa < 1$ represents the relative power attenuation in the side-lobe as compared to the main-lobe. The values of κ and ω are not chosen freely but should be representative of the antenna gain pattern of some practical array, such as uniform linear array (ULA) and uniform circular array (UCA).

With this antenna model, the received power at a legitimate node at x is given by $PMS_x\|x\|^{-\alpha}$, where P is the transmit power of the typical node, S_x is the fading gain of the channel from the typical node to the receiving node at x, and $\|x\|$ is the distance between them. Here we have implicitly assumed that the typical node steers its main-lobe towards the intended receiving node. When multiple receiving nodes are considered, a time-division multiplexing transmission scheme is assumed so that one receiving node is served at a time using the main-lobe. For the eavesdroppers, the received power depends on whether the eavesdropper is located inside the main-lobe or the side-lobe. For the main-lobe eavesdropper at y, the received power is given by $PMS_y\|y\|^{-\alpha}$, while for a side-lobe eavesdropper at z, the received power is given by $P\kappa MS_z\|z\|^{-\alpha}$. Clearly, the main-lobe eavesdroppers received the same benefit from directional antenna transmission as the legitimate receivers, whereas the side-lobe eavesdroppers are disadvantaged by a factor of κ in their received power. Note that directional antenna transmission does not utilize the knowledge of fading channel gain, hence the result on the out-degree is generally valid for any fading distributions as well as for the path-loss only case. For notational convenience, we define $\delta = 2/\alpha$.

Omitting the detailed derivations (which can be found in [6]), the average out-degree of the typical node with directional antenna transmission is given by

$$\mathbb{E}\left\{N_{\text{out}}^{\text{DAT}}\right\} = \frac{1}{\omega + \kappa^\delta - \omega\kappa^\delta} \frac{\lambda_l}{\lambda_e}. \tag{13.11}$$

The values of ω and κ depend on the particular antenna array pattern as well as the number of antenna elements in the array, i.e., M (see [19] for the description on the exact relations). Therefore, the leading fraction in (13.11) quantifies the improvement on the average out-degree from directional antenna transmission.

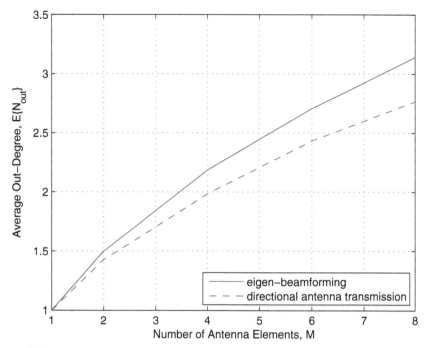

Figure 13.2: The average out-degree versus the number of antenna elements for transmission. The ratio of intensities is $\lambda_l/\lambda_e = 1$ and the path loss exponent is $\alpha = 4$.

13.2.4.2 Eigen-Beamforming

Eigen-beamforming is a popular technique for achieving spatial diversity. It requires the knowledge of the fading gain of the intended receiver available at the transmitter. By transmitting linearly weighted copies of the same signal on different antenna elements, the signal arrived at the intended receiver has the maximum power. When multiple receiving nodes are considered, a time-division multiplexing transmission scheme is assumed so that one receiving node is served at a time with eigen-beamforming according to its fading channel gain. Assuming the Rayleigh fading channels between all nodes are independent, the signal power arrived at the intended receiver follows a gamma distribution due to eigen-beamforming, while the signal power arrived at any other nodes, including the eavesdroppers, is still exponentially distributed.

Omitting the detailed derivations (which can be found in [6]), the average out-degree of the typical node with M-antenna eigen-beamforming is given by

$$\mathbb{E}\left\{N_{\text{out}}^{\text{EBF}}\right\} = \frac{\Gamma(M+\delta)}{\Gamma(1+\delta)\Gamma(M)} \frac{\lambda_l}{\lambda_e}, \qquad (13.12)$$

where $\Gamma(.)$ denotes the gamma function. From well-known properties of the gamma function, it is easily shown that

$$M^\delta \frac{\lambda_l}{\lambda_e} \leq \mathbb{E}\left\{N_{\text{out}}^{\text{EBF}}\right\} \leq \frac{M^\delta}{\Gamma(1+\delta)} \frac{\lambda_l}{\lambda_e},$$

which shows the scaling behavior and gives simple bounds on the out-degree. In particular, we see that M^δ quantifies how the average out-degree scales with the number of antennas with eigen-beamforming.

Figure 13.2 shows the improvement in the average out-degree from multiantenna transmission. In the case of directional antenna transmission, we have used a UCA and computed the values of ω and κ accordingly (see details in [6]). We have set the intensities of legitimate nodes and eavesdroppers to be equal, i.e., $\lambda_l = \lambda_e$, hence the average out-degree is one in the case of single-omnidirectional antenna transmission. Clearly, both directional antenna transmission and eigen-beamforming achieve a significant improvement in the average out-degree even with a relatively small number of antennas. This improvement, however, is not linear and the increment reduces as the number of antennas gets larger.

13.3 Secrecy Transmission Capacity

In this section, we turn our attention from connectivity to network throughput of secure transmission. Even without secrecy considerations, the characterization of the maximum achievable throughput over a large network is a long-standing challenge in the research of wireless communication. While the information theory for point-to-point communications was established in the 1940s, no such "network information theory" is found for wireless networks [20].

Various approaches have been developed toward understanding the network throughput behavior, including stochastic geometry. The basic idea behind the stochastic geometry approach is as follows: although it is extremely difficult to directly characterize the throughput performance limit of a network, it is often tractable to study the performance of a typical communication pair, taking into account the interaction from other nodes in the network. Furthermore, if all nodes have roughly the same properties, e.g., transmit power, code rate, mobility, etc., then the average performance of the typical communication pair represents the average performance of all the links in the network.

Following the success of the existing works on applying stochastic geometry to study network throughput without secrecy consideration [1,21], we present a framework for analyzing the throughput of large-scale decentralized networks with physical layer security considerations. The central part of this framework is the metric termed *secrecy transmission capacity*, which measures the achievable rate of successful transmission of conditional messages per unit area for given constraints on two important quality of service (QoS) parameters to be defined later. In this section, after first presenting the network model, we give the formal definition of secrecy transmission capacity, followed by an example to illustrate the use of such a metric.

13.3.1 Network Model

Let us consider a decentralized wireless network in a two-dimensional space. Homogeneous PPPs are used to model the locations of both the legitimate nodes and the eavesdroppers. Specifically, the locations of the legitimate transmitters follow a homogeneous PPP, denoted by Φ_l, with intensity λ_l. Each transmitter has an intended receiver at a fixed distance r in a random direction. This is sometimes referred to as a *Poisson bipolar model*. The locations of eavesdroppers follow an independent homogeneous PPP, denoted by Φ_e, with intensity λ_e. An example of a network snapshot is shown in Figure 13.3.

13.3.2 Capacity Formulation

In order to send confidential messages, each legitimate transmitter–receiver pair employs a wiretap code, such as the classical Wyner code [22]. Unlike most studies in physical layer security which assume the availability of the channel state information (CSI) at both the receiver and the transmitter, we consider a more practical scenario where only the receiver has CSI but not the transmitter. Furthermore, no feedback link exists between the receiver

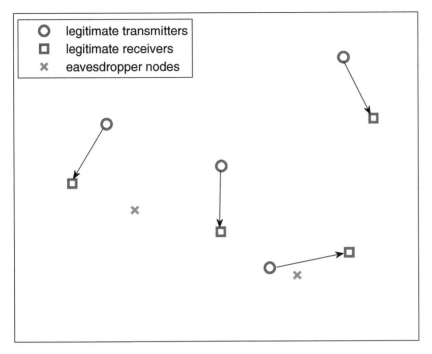

Figure 13.3: An example of a network snapshot consisting of a number of transmit–receiver pairs and eavesdroppers.

and the transmitter. As a result, the transmission of confidential messages can only be done with a fixed code rate. Different from any conventional code where there is a one-to-one correspondence between the message and the codeword, a wiretap code effectively associates one message with multiple codewords. Hence, there are two rate parameters of any wiretap code, namely, the codeword rate, R_t, and the confidential data rate, R_s. The codeword rate is simply the rate at which the codewords are transmitted, which is larger than the actual rate of confidential message transmission. The rate difference, denoted by $R_e = R_t - R_s$, represents the redundancy added into the code. This redundancy introduces randomness into the code structure for providing physical layer security in the message transmission. Roughly, a larger R_e gives a higher secrecy level. A detailed description of wiretap code construction can be found in [23, 24]

With such a wiretap code in place, we can define the following two outage events for the confidential message transmission between a legitimate transmitter–receiver pair:

- Connection Outage: The event that the capacity of the channel from the transmitter to the intended receiver is below the codeword rate R_t. The probability of this event happening is referred to as the connection outage probability, denoted by P_{co}.

- Secrecy Outage: The event that the capacity of the channel from the transmitter to at least one eavesdropper is above the rate redundancy R_e. The probability of this event happening is referred to as the secrecy outage probability, denoted by P_{so}.

These outage definitions are similar to the ones used in [24]. The connection outage is easy to understand, as the receiver cannot decode the message correctly when the codeword rate exceeds the channel capacity. On the other hand, the secrecy outage can be caused by any eavesdropper inside the network. When the rate redundancy is larger than the channel capacity for a particular eavesdropper, this eavesdropper is not capable of resolving the randomness added into the wiretap code, and hence, perfect secrecy is achieved against this eavesdropper. However, there is more than one eavesdropper in the network. To have perfect secrecy, we need the rate redundancy to exceed the channel capacity for all eavesdroppers. Otherwise, the message transmission fails to provide perfect secrecy, in which case a secrecy outage is deemed to occur. Since the rate parameters of the wiretap code are constant, it is impossible to avoid either outage events all the time. Hence, we have defined the corresponding outage probabilities, which can be treated as QoS measures on reliability and secrecy.

Now, we introduce the secrecy transmission capacity, which measures the achievable rate of successful transmission of confidential messages per unit area of the network, with constraints on the connection outage probability and the secrecy outage probability. In other words, the secrecy transmission capacity measures the network throughput in terms of the area spectral efficiency with some target QoS requirements on both the reliability of the communication over the legitimate links and the secrecy level of the communication against all the eavesdroppers. Mathematically, denoting the constraints on the reliability and secrecy as $P_{co} = \sigma$ and $P_{so} = \epsilon$, respectively, the secrecy transmission capacity is defined as

$$\tau = R_s(1 - \sigma)\lambda_l. \tag{13.13}$$

In the above equation, R_s is the data rate of the confidential message transmission, $(1 - \sigma)$ is the probability of reliable transmission, and λ_l is the number of message transmissions per unit area at any point in time. It is important to note that $R_s = R_t - R_e$ cannot be chosen freely, but is a function of σ and ϵ. Intuitively, if a higher reliability (smaller σ) is required, we need to reduce the codeword rate R_t to reduce the connection outage probability. Similarly, if a higher secrecy level (smaller ϵ) is required, we need to increase the rate redundancy R_e to make the message transmission more difficult to be intercepted. Here, we also have an implicit assumption of $R_s \geq 0$, which simply means that the data rate cannot be negative. Therefore, for operational purpose, we redefine $R_s = [R_t - R_e]^+$, where $[a]^+ = \max\{0, a\}$. Whenever the required reliability and secrecy constraints result in $R_s = 0$, we claim that transmission of confidential messages is not possible.

It is important to note that the secrecy transmission capacity formulation given in (13.13) is conceptually different from the usual "transmission capacity" framework. Specifically, the key step in deriving the usual transmission capacity of a network is to find the achievable transmission intensity (i.e., the number of concurrent transmissions per unit area) for a given data rate and outage probability. In contrast, the key step in deriving the secrecy transmission capacity is to find the achievable data rate of the confidential message transmission, R_s, for prescribed outage probabilities, σ and ϵ, and given node intensities, λ_l and λ_e. When the intensity of legitimate nodes λ_l is also a controllable parameter, the secrecy transmission capacity τ can be further maximized by optimally choosing the value of λ_l.

Also note that the secrecy transmission capacity defined in (13.13) depends on the distance r between the legitimate transmitter–receiver pair. For simplicity, we do not consider the more practical case where r is different for different pairs. A discussion on how to incorporate variable distance into the capacity formulation can be found in [13].

13.3.3   Illustrative Example

Here, we provide an example to illustrate the use of the secrecy transmission capacity framework for the analysis and design of secure transmissions in large-scale decentralized wireless networks. The following assumptions are made:

- All legitimate nodes and eavesdroppers in the network are equipped with a single omnidirectional antenna.

- Gaussian signaling is assumed at the transmitters in order to obtain tractable capacity expressions.

- The wireless channels between all nodes are modeled by large-scale path loss with exponent α and small-scale Rayleigh fading. The channels between different nodes experience independent fading.

- Both the legitimate receivers and the eavesdroppers do not apply any multiuser decoding techniques and treat the interference from concurrent transmissions as noise.

- The network is interference limited, i.e., the thermal noise is negligible compared to the aggregate interference from concurrent transmissions.

Note that the second last assumption is practical for the legitimate receivers in a decentralized wireless network as the receivers are only interested in the message from their associated transmitters. On the other hand, the eavesdroppers may have a common objective and try to eavesdrop any legitimate link. Hence, the assumption of no multiuser decoding at the eavesdroppers may underestimate the eavesdropping capability. An alternative treatment, which allows the eavesdroppers to have any arbitrary multiuser decoding technique, can be found in [15].

In an interference-limited network, the capacity of any legitimate or eavesdropping link is determined by the signal-to-interference ratio (SIR). Let us add an arbitrary legitimate transmitter–receiver pair, with the receiver placed at the origin and the transmitter located at distance r away in an arbitrary direction. We call them the *typical* transmitter–receiver pair, on which our analysis will be performed. For the signal reception at the typical receiver, all the other legitimate transmitters in Φ_l act as interferers. Hence, Φ_l also denotes the locations of all interferers.

In what follows, we first derive expressions of the connection outage probability and secrecy outage probability, which will then be used to obtain the secrecy transmission capacity. For notational convenience, we have defined $\delta = 2/\alpha$.

13.3.3.1 Connection Outage Probability

The SIR at the typical receiver can be expressed as

$$\mathrm{SIR}_0 = \frac{S_0 r^{-\alpha}}{\sum_{x \in \Phi_l} S_x \|x\|^{-\alpha}}, \tag{13.14}$$

where S_0 is the fading gain of the channel between the typical transmitter and receiver, S_x is the fading gain of the channel from the interferer located at x and the typical receiver, and $\|x\|$ denotes the distance from the interferer located at x and the typical receiver. Note that the summation on the denominator is over all interferer locations and forms a two-dimensional shot noise process [25]. According to the outage definitions given in Section 13.3.2, a connection outage occurs when $\log_2(1 + \mathrm{SIR}_0) < R_t$. Defining $\beta_t = 2^{R_t} - 1$, the connection outage probability can be written as

$$\mathrm{P_{co}} = \mathbb{P}\left(\mathrm{SIR}_0 < \beta_t\right) = \mathbb{P}\left(\frac{S_0 r^{-\alpha}}{\sum_{x \in \Phi_l} S_x \|x\|^{-\alpha}} < \beta_t\right). \tag{13.15}$$

The random variables in the above probability calculation include the fading gains, i.e., S_0 and S_x, as well as the PPP Φ_l (i.e., the locations of all interferers). Computation of the statistical properties of a shot noise process is a classical problem in stochastic geometry, and there are many existing studies looking at the computation of probability in the above form, e.g., [21, 26]. With the fading gains having exponential distributions (due to Rayleigh fading), the connection outage probability can be computed in a closed form as

$$P_{co} = 1 - \exp\left[-\lambda_l \pi r^2 \beta_t^\delta \Gamma(1-\delta)\Gamma(1+\delta)\right],\tag{13.16}$$

where $\Gamma(.)$ denotes the gamma function.

13.3.3.2 Secrecy Outage Probability

After obtaining the connection outage probability, we carry on with the derivation of the secrecy outage probability. For ease of analysis, we now shift the coordinates such that the legitimate transmitter is located at the origin. Note that shifting the coordinates does not change the distribution of either the interferer locations Φ_l or the eavesdropper locations Φ_e. Considering an eavesdropper located at z in Φ_e, the received SIR is given by

$$\text{SIR}_z = \frac{S_z \|z\|^{-\alpha}}{\sum_{y \in \Phi_l} S_{yz} \|y-z\|^{-\alpha}},\tag{13.17}$$

where S_z is the fading gain of the channel between the typical transmitter and the eavesdropper at z, S_{yz} is the fading gain of the channel from the interferer located at y and the eavesdropper at z, $\|z\|$ denotes the distance from the typical transmitter to the eavesdropper at z, and $\|y-z\|$ denotes the distance from the interferer at y and the eavesdropper at z. Unlike the connection outage event which is only concerned with one link, the secrecy outage event is affected by all the eavesdropping "links." More precisely, a secrecy outage occurs when $\log_2(1 + \text{SIR}_z) < R_e$ holds for at least one z in Φ_e. In other words, perfect secrecy is achieved when $\log_2(1 + \text{SIR}_z) > R_e$ for all z in Φ_e. Defining $\beta_e = 2^{R_e} - 1$, the secrecy outage probability can be written as

$$P_{so} = 1 - \mathbb{E}_{\Phi_l}\left\{\mathbb{E}_{\Phi_e}\left\{\prod_{z \in \Phi_e}\left(\mathbb{P}\left(\frac{S_z \|z\|^{-\alpha}}{\sum_{y \in \Phi_l} S_{yz} \|y-z\|^{-\alpha}} < \beta_e \Big| z, \Phi_l, \Phi_e\right)\right)\right\}\right\},\tag{13.18}$$

where the probability term $\mathbb{P}(.)$ in the above expression stands for the probability of perfect secrecy against the eavesdropper located at z in a given network realization. Since the fading gains are independent for all eavesdroppers, the probability of perfect secrecy against all eavesdroppers in a given network realization equals the product of individual probabilities. This is why we have a product over Φ_e in the above expression. The double expectations are needed to average over all possible network realizations. Unfortunately, the secrecy outage probability in (13.18) does not admit a closed-form expression. Nevertheless, using techniques in [27] a very accurate upper bound can be found as

$$P_{so}^{UB} = 1 - \exp\left[-\frac{\lambda_e}{\lambda_l \beta_e^\delta \Gamma(1-\delta)\Gamma(1+\delta)}\right].\tag{13.19}$$

In the following, we will use the expression in (13.19) as the exact secrecy outage probability for network analysis.

13.3.3.3 Secrecy Transmission Capacity

Now, we are ready to derive the secrecy transmission capacity as defined in (13.13). Specifically, we need to obtain an expression of $R_s = [R_t - R_e]^+$ as a function of the outage

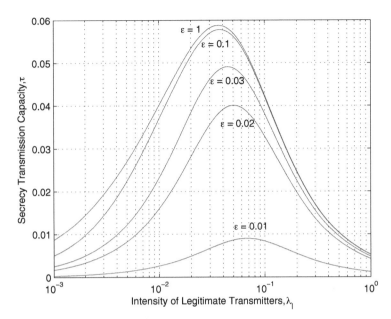

Figure 13.4: The secrecy transmission capacity τ versus the intensity of legitimate transmitters λ_l. Results are shown for networks with different secrecy outage constraints, i.e., $\epsilon = 0.01, 0.02, 0.03, 0.1$, as well as no secrecy constraint, i.e., $\epsilon = 1$. The other system parameters are $r = 1$, $\alpha = 4$, $\sigma = 0.3$, and $\lambda_e = 0.001$.

constraints $\mathrm{P_{co}} = \sigma$ and $\mathrm{P_{so}} = \epsilon$. This is done by inverting the outage probability expressions in (13.16) and (13.19), such that R_t is expressed in terms of $\mathrm{P_{co}} = \sigma$ and R_e is expressed in terms of $\mathrm{P_{so}} = \epsilon$. After obtaining the expressions of R_t and R_e in terms of σ and ϵ, respectively, the expression of the confidential data rate R_s is readily obtained. Using (13.13) and with some algebraic manipulation, the secrecy transmission capacity with both outage constraints is given by

$$\tau = (1-\sigma)\lambda_l \left[\log_2 \left(\frac{1 + \left[\frac{\ln \frac{1}{1-\sigma}}{\lambda_l \pi r^2 \Gamma(1-\delta)\Gamma(1+\delta)} \right]^{\frac{1}{\delta}}}{1 + \left[\frac{\lambda_l}{\lambda_e} \Gamma(1-\delta)\Gamma(1+\delta) \ln \frac{1}{1-\epsilon} \right]^{-\frac{1}{\delta}}} \right) \right]^+ . \tag{13.20}$$

This expression helps network designers to understand the impacts of QoS requirements and various network parameters on the throughput of secure communication over the entire network. The use of the secrecy transmission capacity expression is further elaborated in the following numerical results.

Figure 13.4 shows the secrecy transmission capacity τ as the intensity of legitimate transmitter λ_l varies. Clearly, there is an optimal value of λ_l. Below this optimal value, there are too few concurrent transmissions resulting in an inefficient spatial reuse. On the other hand, above this optimal value, there is too much interference limiting the achievable data rate. Comparing the five curves with different levels of secrecy requirement, we see that the optimal intensity of legitimate transmitters increases as the secrecy requirement increases, i.e., as ϵ reduces. This shows the benefit of having more interference in the network as a secrecy enhancement. Perhaps, a more important observation made from comparing the five curves is the throughput cost of achieving a certain level of secrecy. Specifically, the

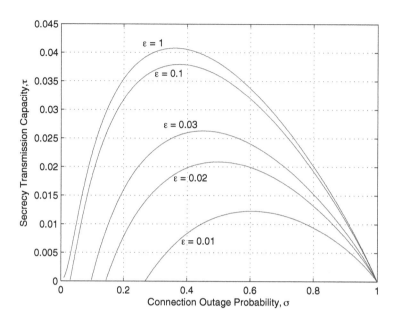

Figure 13.5: The secrecy transmission capacity τ versus the connection outage probability σ. Results are shown for networks with different secrecy outage constraints, i.e., $\epsilon = 0.01, 0.02, 0.03, 0.1$, as well as no secrecy constraint, i.e., $\epsilon = 1$. The other system parameters are $r = 1$, $\alpha = 4$, $\lambda_l = 0.01$, and $\lambda_e = 0.001$.

reduction in the achievable throughput from the network with no secrecy requirement (i.e., $\epsilon = 1$) to the network with a relatively moderate secrecy requirement (e.g., $\epsilon = 0.1$) is often insignificant. On the other hand, when the network already has a relatively high secrecy level, any further increment in the secrecy level (e.g., from $\epsilon = 0.02$ to $\epsilon = 0.01$) results in a huge reduction in the achievable throughput.

Figure 13.5 illustrates the impact of the connection outage probability on the secrecy transmission capacity. We see that the optimal connection outage probability is usually not very small and it increases as the secrecy requirement increases. This shows the trade-off between the reliability and secrecy, i.e., when the secrecy requirement increases, it is wise to sacrifice more on the reliability performance from the network throughput point of view. Furthermore, secure transmissions become impossible when the reliability requirement is too high, i.e., when σ is too small. The minimum acceptable connection outage probability needs to be increased, i.e., sacrificing the reliability, if one wants to increase the secrecy requirement whilst still transmitting at a nonzero rate.

13.4   Current Limitations and Future Directions

In this chapter, we have introduced two stochastic geometry approaches for understanding the physical layer security implications in large-scale wireless networks. In particular, the secrecy graph and its connectivity properties have been discussed in detail. A major limitation in the current secrecy graph formulation is the absence of interference. For networks, typically centralized ones, with tight MAC, interference can be negligible. In decentralized networks, however, only simple MAC layer protocols can be used. In this case, interference caused by the transmissions of neighboring nodes has a nonnegligible impact on the connectivity properties of the secrecy graph, which needs to be understood.

On the other hand, the secrecy transmission capacity formulation explicitly takes the interference into account and still gives mathematically tractable results on the network throughput. Hence, it is very convenient to be used by network designers to understand the roles of different system parameters as well as the feasibility and impact of QoS requirements. The main limitation of this formulation is that it only considers one-hop transmissions. In order to truly understand the end-to-end throughput performance in large wireless networks, multi-hop communication needs to be considered, which is largely an open research problem even without secrecy considerations.

Another limitation in the current studies is the simplistic spatial model of network geometry, which applies to both secrecy graph and secrecy transmission capacity. In particular, the homogeneous PPP is commonly used to model the locations of both the legitimate nodes and the eavesdroppers in most existing studies with few exceptions. On the one hand, homogeneous PPPs can well approximate ad hoc networks with unplanned node deployment or with substantial mobility. Hence, it is a suitable model for many practical scenarios of the legitimate network. On the other hand, the assumption of the eavesdroppers being located accordingly a homogeneous PPP often underestimates the potential security threat when the eavesdroppers are truly malicious. The impact of eavesdroppers' distribution on both connectivity and throughput performance is an interesting problem to investigate.

References

[1] M. Haenggi, J. G. Andrews, F. Baccelli, O. Dousse, and M. Franceschetti, "Stochastic geometry and random graphs for the analysis and design of wireless networks," *IEEE J. Sel. Areas Commun.*, vol. 27, no. 7, pp. 1029–1046, Sep. 2009.

[2] M. Haenggi, *Stochastic Geometry for Wireless Networks*, Cambridge University Press, 2012.

[3] M. Haenggi, "The secrecy graph and some of its properties," in *Proc. IEEE Int. Symp. Inf. Theory (ISIT)*, Toronto, Canada, July 2008, pp. 539–543.

[4] P. C. Pinto and M. Z. Win, "Continuum percolation in the intrinsically secure communications graph," in *Proc. IEEE Int. Symp. Inf. Theory and Its Applications (ISITA)*, Taichung, Taiwan, Oct. 2010.

[5] S. Goel, V. Aggarwal, A. Yener, and A. R. Calderbank, "The effect of eavesdroppers on network connectivity: A secrecy graph approach," in *IEEE Trans. Info. Forensics and Security*, vol. 6, no. 3, pp. 712–724, Sep. 2011.

[6] X. Zhou, R. K. Ganti, and J. G. Andrews, "Secure wireless network connectivity with multi-antenna transmission," *IEEE Trans. Wireless Commun.*, vol. 10, no. 2, pp. 425–430, Feb. 2011.

[7] P. C. Pinto, J. Barros, and M. Z. Win, "Secure communication in stochastic wireless networks—Part I: Connectivity," *IEEE Trans. Info. Forensics and Security*, vol. 7, no. 1, pp. 125–138, Feb. 2012.

[8] A. Sarkar and M. Haenggi, "Percolation in the secrecy graph," *Discrete Applied Mathematics*. Accepted for publication.

[9] A. Sarkar and M. Haenggi, "Secrecy coverage," in *Proc. Asilomar Conf. Signals, Systems, and Computers (ACSSC)*, Pacific Grove, CA, Nov. 2010.

[10] A. Sarkar and M. Haenggi, "Secrecy coverage," *Internet Mathematics*. Accepted for publication.

[11] S. Vasudevan, D. Goeckel, and D. Towsley, "Security-capacity trade-off in large wireless networks using keyless secrecy," in *Proc. ACM Int. Symp. Mobile Ad Hoc Networking and Computing (MobiHoc)*, Chicago, IL, Sep. 2010, pp. 21–30.

[12] Y. Liang, H. V. Poor, and L. Ying, "Secrecy throughput of MANETs under passive and active attacks," in *IEEE Trans. Inf. Theory*, vol. 57, no. 10, pp. 6692–6702, Oct. 2011.

[13] X. Zhou, R. K. Ganti, J. G. Andrews, and A. Hjørungnes, "On the throughput cost of physical layer security in decentralized wireless networks," *IEEE Trans. Wireless Commun.*, vol. 10, no. 8, pp. 2764–2775, Aug. 2011.

[14] O. O. Koyluoglu, C. E. Koksal, and H. El Gamal, "On secrecy capacity scaling in wireless networks," *IEEE Trans. Inf. Theory*, vol. 58, no. 5, pp. 3000–3015, May 2012.

[15] X. Zhou, M. Tao, and R. A. Kennedy, "Cooperative jamming for secrecy in decentralized wireless networks," in *Proc. IEEE Int. Conf. Commun. (ICC)*, Ottawa, Canada, June 2012.

[16] D. Weaire, J. P. Kermode, and J. Wejchert, "On the distribution of cell areas in a Voronoi network," *Philosophical Mag. Part B*, vol. 53, no. 5, pp. L101–L105, 1986.

[17] G. Grimmett, *Percolation*, Springer, New York 1999.

[18] R. Meester and R. Roy, *Continuum Percolation*, Cambridge University Press, New York 1996.

[19] J. Yu, Y. D. Yao, A. Molisch, and J. Zhang, "Performance evaluation of CDMA reverse links with imperfect beamforming in a multicell environment using a simplified beamforming model," *IEEE Trans. Veh. Technol.*, vol. 55, no. 3, pp. 1019–1031, 2006.

[20] J. G. Andrews, N. Jindal, M. Haenggi, R. Berry, S. Jafar, D. Guo, S. Shakkottai, R. Heath, M. Neely, S. Weber, and A. Yener, "Rethinking information theory for mobile ad hoc networks," *IEEE Commun. Mag.*, vol. 46, pp. 94–101, Dec. 2008.

[21] S. Weber, J. G. Andrews, and N. Jindal, "An overview of the transmission capacity of wireless networks," *IEEE Trans. Commun.*, vol. 58, no. 12, pp. 3593–3604, Dec. 2010.

[22] A. Wyner, "The wire-tap channel," *Bell Syst. Tech. J.*, vol. 54, no. 8, pp. 1355–1387, Oct. 1975.

[23] A. Thangaraj, S. Dihidar, A. R. Calderbank, S. W. McLaughlin, and J.-M. Merolla, "Applications of LDPC codes to the wiretap channel," *IEEE. Trans. Inf. Theory*, vol. 53, no. 8, pp. 2933–2945, Aug. 2007.

[24] X. Tang, R. Liu, P. Spasojević, and H. V. Poor, "On the throughput of secure hybrid-ARQ protocols for Gaussian block-fading channels," *IEEE. Trans. Inf. Theory*, vol. 55, no. 4, pp. 1575–1590, Apr. 2009.

[25] J. Venkataraman, M. Haenggi, and O. Collins, "Shot noise models for outage and throughput analyses in wireless ad hoc networks," in *Proc. IEEE Military Commun. Conf. (MILCOM)*, Washington, DC, Oct. 2006, pp. 1–7.

[26] F. Baccelli, B. Błaszczyszyn, and P. Mühlethaler, "An Aloha protocol for multihop mobile wireless networks," *IEEE Trans. Inform. Theory*, vol. 52, no. 2, pp. 421–436, Feb. 2006.

[27] R. K. Ganti and M. Haenggi, "Single-hop connectivity in interference-limited hybrid wireless networks," in *Proc. IEEE Int. Symp. Inf. Theory (ISIT)*, Nice, France, June 2007, pp. 366–370.

Chapter 14

Physical Layer Secrecy in Large Multihop Wireless Networks

Dennis Goeckel
University of Massachusetts - Amherst

Cagatay Capar
University of Massachusetts - Amherst

Don Towsley
University of Massachusetts - Amherst

Wireless networks provide flexible ubiquitous user communications but, at the same time, provide the same access to an attacker or eavesdropper trying to intercept private messages. Hence, security has become a primary concern in wireless networks, and there has been a resurgence of interest in physical layer forms of secrecy based on Shannon's original information-theoretic formulation and Wyner's consideration of the wiretap formulation of such. In this chapter, we venture beyond the prototypical three-node Alice-Bob-Eve security scenario to consider physical layer forms of secrecy in wireless *networks,* where the need to protect the message over multiple hops opens up system vulnerabilities, but the addition of system nodes beyond the communicating parties also offers opportunities. In particular, the scaling of secrecy in (asymptotically) large multihop networks will be considered, and we will focus, in particular, on how the flexibility offered by the large number of system nodes in the network can facilitate secure communication in the face of a large number of eavesdroppers, even if the location of those eavesdroppers is unknown. The transmission techniques that are employed in the conclusive results for secrecy scaling in large networks with noncollaborating eavesdroppers also have clear analogs that should prove effective in finite networks, but there are challenges to be addressed in such an adaptation.

14.1 Introduction

Wireless networks have revolutionized the way that people communicate and carry out their daily lives. Wireless networks offer convenience and flexibility to users in the developed world, and they often offer the only reliable access in the developing world or in the face of natural disasters when much of the wired communications infrastructure is inadequate. Rapid advances in commercial wireless networks are resulting in a generation growing up in a world where fast, reliable network connectivity is taken as a given, regardless of location.

In addition, military users rely on the rapid set-up and employment of wireless networks in missions extending beyond their current sphere of control, and wireless sensor networks are becoming more and more widespread to track and monitor everything from animal populations to structural integrity.

The enabling feature of all of the preceding applications is the ability to have access to the signal regardless of the user's location. But this necessary availability of the signal requires naturally that it is not tightly physically constrained, as it was in a traditional wired network, and this presents security vulnerabilities. In particular, as with the user, a potential eavesdropper can have potentially easy access to the signal, and, in a large ad hoc network, it is likely that the system will need to deal with eavesdroppers intermingled with the system nodes. This creates quite a challenge. Thus, the need to secure wireless communications networks has taken on great prominence in the twenty-first century.

There are two classes of security generally considered for wireless (and other) networks. Traditionally, security has been obtained by cryptography, where the trusted nodes share a key that allows for efficient message decryption, while the eavesdropper, without a key, must solve a "hard" problem, which is assumed to be beyond his/her current and future computational capabilities [1]. However, advances in computation (e.g., quantum computers [2,3]), the susceptibility of many implemented cryptographic schemes to attacks, and the ability of the adversary to record and continue to work on the signal [4], motivates a type of security that is provably everlasting. This is the realm of information-theoretic security [5,6] and motivates this text.

Information-theoretic security guarantees that the eavesdropper cannot decode the message—regardless of his/her current or future computational capabilities. Much of the current research is based on the contribution from Wyner in the form of the wiretap channel [6], where he showed, perhaps surprisingly, that even under the strict requirements of information-theoretic security that, if the channel from the transmitter (Alice) to the eavesdropper (Eve) is a degraded version of the channel from Alice to the receiver (Bob), a positive rate of secret information could flow from Alice to Bob. This scheme is described briefly in Section 14.2 and more thoroughly in the first chapter of this book and in recent texts [7,8].

Although there were a number of advances over the next twenty-five years, notably the generalization of the wiretap channel and the introduction of public discussion methods [9,10], it was not until the turn of the century that security in wireless networks became an extremely hot topic. As discussed in detail below in Section 14.2, whereas the wiretap channel of [6] fit the wireless Alice-Bob-Eve scenario well, it also immediately revealed the vulnerabilities of the approach: speaking roughly, if the distance from Alice to Eve is less than the distance from Alice to Bob, secret communication is at best challenging and at worst impossible. And, more troubling is the fact that we generally cannot know the location of a passive eavesdropper and thus must guard against very close eavesdroppers, thus making it very difficult to send any message and have some guarantee that its security is not compromised. It is this vulnerability that has led to many in the security community to completely disregard information-theoretic secrecy, arguing quite convincingly that they would rather have the long-term computational risk of cryptographic approaches rather than the very near-term possibility of the security of a message being compromised in information-theoretic approaches. In this chapter, we keep this stinging criticism of information-theoretic security in mind as we consider how to overcome these limitations by exploiting the possibilities of communication in an (asymptotically) large multihop network.

In moving from the Alice-Bob-Eve scenario to large wireless networks, there are challenges and opportunities. Consider a large ad hoc network, where communication takes place wirelessly in many hops across the network between various source and destination pairs. Now, suppose that there are a large number of eavesdroppers intertwined with the

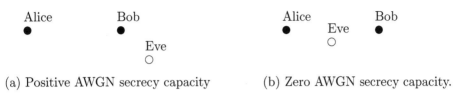

(a) Positive AWGN secrecy capacity   (b) Zero AWGN secrecy capacity.

Figure 14.1: Scenarios for information secrecy on the additive white Gaussian noise (AWGN) wiretap channel. In (a), the received signal at Bob is stronger than that at Eve, leading to the degraded wiretap channel formulation of [6]. In (b), the signal strength ordering is reversed, and the secrecy capacity under a wiretap formulation is zero.

system nodes. The challenge to the system designer is immediately apparent: not only must the secrecy of the message be maintained over many hops, thus yielding many locations where an eavesdropper might be located and still capture the message, but it is also almost certain that there will be many Alice-Bob-Eve scenarios in such a network where Eve is between a transmitter and a receiver, thus making the provisioning of physical layer security difficult if not impossible. However, what is likely less clear to the reader is that the network also opens up a number of opportunities not present in the Alice-Bob-Eve scenario. In particular, the main theme of this chapter is the description of techniques, ranging from multiuser diversity to cooperative jamming to clever routing to network coding, that exploit features of large wireless networks to robustly obtain physical layer security in the presence of many eavesdroppers—even if the location of those eavesdroppers is completely unknown to the system nodes.

With the goal of considering how to provision secrecy in very large multihop networks, this chapter is organized as follows. Section 14.2 briefly reviews the basic wireless Alice-Bob-Eve scenario and its relation with Wyner's wiretap scheme, with a focus on the challenges that arise. We then turn to considering the two main metrics for network performance—connectivity (Section 14.3) and capacity (Section 14.4)—in large wireless networks with many eavesdroppers. It is Section 14.4 that forms the core of this chapter, starting with a consideration of schemes that assume knowledge of eavesdropper locations (Section 14.4.2), but then turning to the authors' main contribution to the field: that of considering secrecy scaling in large wireless networks when eavesdropper locations are unknown. Section 14.4.3 revisits the arc of our thinking as we considered various techniques, ranging from cooperative jamming to routing to network coding (and combinations), before arriving at a conclusive result in the case of noncollaborating eavesdroppers. Section 14.5 presents the conclusions and future work.

14.2   Background: Physical Layer Security in One-hop Networks

Consider the Alice (transmitter), Bob (receiver), and Eve (eavesdropper) scenarios shown in Figure 14.1. If we suppress the time index, the standard mathematical model for a narrowband wireless communication channel between Alice and Bob yields a received signal at Bob of:

$$y_B = \frac{h_{A,B}}{d_{A,B}^{\frac{\alpha}{2}}} \sqrt{E_s} x_A + n_B$$

where $h_{A,B}$ is the multipath fading (if any) between Alice and Bob, $d_{A,B}$ is the distance between nodes A and B, α is the path-loss exponent, E_s is the transmitted energy per symbol, x_A is the transmitted signal, and n_B is a zero mean complex Gaussian random variable (i.e., noise) with $E[|n_B|^2] = N_0$. Defining the analogous quantities for the link from Alice to Eve in the obvious manner yields the received signal at Eve as:

$$y_E = \frac{h_{A,E}}{d_{A,E}^{\frac{\alpha}{2}}} \sqrt{E_s} x_A + n_E.$$

Conditioned on the fading and path-loss variables, this is a Gaussian wiretap channel [6,11]. Defining the *secrecy* capacity as the number of bits per symbol that can be sent without any leakage of information to the eavesdropper, the secrecy capacity of this channel is given by [11]:

$$C_s = \min\left(\frac{1}{2}\log\left(1 + \frac{|h_{A,B}|^2}{d_{A,B}^\alpha}\frac{E_s}{N_0}\right) - \frac{1}{2}\log\left(1 + \frac{|h_{A,E}|^2}{d_{A,E}^\alpha}\frac{E_s}{N_0}\right),\ 0\right). \qquad (14.1)$$

The challenges of information-theoretic security, which drive this work, become immediately apparent from (14.1), resulting in the skepticism expressed by the cryptographic community for such an approach. In particular, if the channel gain encompassing both fading and path loss to the eavesdropper Eve is better than that to the intended recipient Bob, the secrecy capacity is zero. Furthermore, more concerning is that $h_{A,E}$ and $d_{A,E}$ are generally unknown except under the most optimistic set of assumptions; hence, the secrecy capacity is unknown to the transmitter and a message transmitted at any positive secrecy rate has the potential to be compromised. There have been approaches to try to exploit whatever knowledge of the fading is available (e.g., [12]), but the problem remains that, without knowledge of the location of the eavesdropper, there is no way to choose a secrecy rate below which the message is certainly protected with high probability from compromise. Two-way schemes, such as public discussion [9,10], provide some alleviation of this problem, although it is still challenging to pick a rate without knowing the eavesdropper characteristics. However, two-way schemes are useful in combination with other network tools, as will be considered in Section 14.4.3.

Unless otherwise stated, we will assume throughout this work that Rayleigh fading is present. That is, $|h_{A,B}|^2$ is exponentially distributed with $E[|h_{A,B}|^2] = 1$. We hasten to note, however, that wireless systems generally have to operate in a variety of fading environments, ranging from Rayleigh faded through Rician to additive white Gaussian noise (AWGN), and it is important to remember that a system should be robust to the type of fading encountered.

The simple wiretap channel and its limitations will have far-reaching consequences on the two main metrics for networks: connectivity and capacity. These are addressed in the next two sections, respectively.

14.3   Secure Connectivity: The Secrecy Graph

Before turning to the main topic of our interest (throughput capacity), we first briefly review results in the secure connectivity of large networks, referring the reader to the previous chapter (or [13] and [14]) for an extensive treatment of the topic.

A classical topic in networking is the study of the connectivity of large networks through percolation theory [15]. In particular, consider an infinite two-dimensional plane with nodes distributed randomly according to a Poisson point process with intensity $\lambda > 0$ nodes per unit square. Suppose that each node can communicate directly with any other node within a radius r, and, hence, draw an edge (line) between any pair of nodes that are less than r distant apart. After doing this for all pairs, an undirected graph is obtained. Now, through multihop communications, any two nodes connected in the graph by an unbroken sequence of edges can have a communication flow between them, and, if every node can reach every other node through such a sequence of edges, the network is fully connected. However, since there will be some isolated nodes in the infinite plane, the question generally asked is whether a "giant" component containing a large (infinite) number of nodes arises in the network. The interesting result is that there is a threshold, termed the "percolation threshold" such that, when λr^2 exceeds that threshold, the giant component exists with probability one, and, when λr^2 is less than that threshold, the giant component exists with probability zero. In other words, there is a clear threshold effect.

The secrecy graph formulation of [13] studies *secure* connectivity from a similar perspective. The system set-up is changed by assuming a number of eavesdroppers are intertwined with legitimate users in the network. Then, the "secrecy graph" is constructed in two steps. First, the standard communication graph is constructed as described in the previous paragraph: an undirected edge is added between any pair of nodes that are less than a distance of r apart. Next, each undirected edge is split into two directed edges. Finally, the directed edges are thinned by removing those that are "insecure"; that is, assuming a Gaussian wiretap channel, remove edges where the distance from the transmitter to any eavesdropper is less than the distance from the transmitter to the receiver. For the scenarios given in Figure 14.1, assuming that Alice and Bob are less than a distance of r apart, the edge from Bob to Alice would need to be removed as insecure in the scenario of Figure 14.1(a), and both edges (Alice to Bob, Bob to Alice) would need to be removed as insecure for the scenario in Figure 14.1(b). Thus, it is clear that the thinning of the edges is not necessarily symmetric, and the result is the directed security graph. We can then ask questions about the secure connectivity of the network. This was considered in [13] and has been an area of significant research interest since that time, as reviewed in [14].

14.4   Secure Capacity

The major line of research presented in this chapter is the design and analysis of algorithms that maximize the secure per-session throughput between pairs of nodes in large wireless networks. Stated loosely, we are interested in the rate at which randomly distributed sources and destinations in a large wireless network can exchange information securely in the presence of other transmissions (which can cause interference) and eavesdroppers (which can compromise security). After reviewing the prominent work in the networking field over the last decade addressing the per-session (insecure) throughput of large wireless networks, we then consider how that throughput can be secured in the presence of a large number of eavesdroppers. This latter endeavor will start with work that assumes eavesdropper locations are known, in which case the network can simply route around the eavesdroppers, but then addresses in much more detail the case where eavesdropper locations are unknown.

14.4.1   Background: Throughput Scaling in Large Wireless Networks

Consider a network of n nodes that are uniformly distributed on the region $[0, \sqrt{n}] \times [0, \sqrt{n}]$ as shown in Figure 14.2. Each node is the source for one flow and and also the destination for one flow, and the (source, destination) pairs are assigned randomly. The transmission from the source to the destination will generally take place in a multihop fashion, consisting of a number of wireless transmissions. For each of these wireless transmissions, assume for now

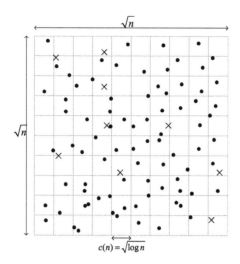

Figure 14.2: The two-dimensional network consists of legitimate nodes (represented by dots) and eavesdroppers (represented by crosses) placed in the square $[0, \sqrt{n}] \times [0, \sqrt{n}]$. The whole region is divided into square cells of size $c(n) \times c(n)$, with $c(n) = \sqrt{\log n}$, as part of the signaling construction of [16].

only path loss between the transmitter and receiver; that is, whenever a node A transmits with some transmit power P, the received power at node B is modeled as

$$P_{\text{rcv},B} = P/d_{AB}^\alpha,$$

where, as introduced in Section 14.2, d_{AB} is the distance between nodes A, B, and α is the path loss exponent. The transmission will be said to be successful if the received signal-to-interference-plus-noise ratio (SINR) exceeds a threshold; that is,

$$\text{SINR}_B = \frac{P_{\text{rcv},B}}{N_0 + I_B} > \gamma, \tag{14.2}$$

where I_B is the variance of the interference received at node B due to other transmissions in the network.

The main throughput scaling question of interest over the last decade has been: what per-session throughput can be maintained in the random network for each source–destination pair as $n \to \infty$? This question was first addressed in the seminal paper of Gupta and Kumar [16], where they showed that a per-pair throughput of $O(1/\sqrt{n \log n})$ was achievable and that the per-pair throughput was upper bounded by $O(1/\sqrt{n})$. An important interpretation of this result was that it was the interference that limited the number of simultaneous transmissions in the network, and, hence, ultimately the capacity. Since a certain transmission radius was needed to guarantee network connectivity, this lower bounded the system interference and hence limited the overall network throughput. Later, the work of Franceschetti et al. [17] demonstrated that the upper bound of $O(1\sqrt{n})$ was indeed achievable if a more sophisticated routing strategy ("percolation highways") was employed; in particular, the percolation highways allow a lower transmission power for connectivity than that employed in [16], and, hence, a higher throughput.

Since the results of [16] and [17], there has been significant further work. In some cases, $O(1)$ per-session throughput was obtained by changing the system model (e.g., adding mobility [18]) or introducing a complicated physical layer (e.g., the hierarchical scheme of [19]), but, for reasons of practical implementation, we will restrict ourselves to the original network and physical layer assumptions of [16]. Hence, our interest is in asking whether the per-pair throughputs of $O(1/\sqrt{n \log n})$ and $O(1/\sqrt{n})$ can be achieved securely, and, if so, in the presence of how many eavesdroppers.

14.4.2   Secrecy Scaling with Known Eavesdropper Locations

An early major contribution on secrecy scaling in wireless networks is given in [20]. Starting with the work of [17], the density of randomly located eavesdroppers that can be added to the network while still maintaining the same per-pair throughput of $O(1/\sqrt{n})$ is considered. A critical assumption is that the eavesdropper locations are known; hence, in essence, the network is able to route around the locations where the eavesdroppers are located. More precisely, the network nodes cannot transmit if they are too close to an eavesdropper; hence, the presence of an eavesdropper leads to dead spots for transmission in the network, and, with enough eavesdropper density, can cause the percolation highways required for the results of [17] to break down. Stated differently, one can view this problem as starting with a standard (insecure) network as described in [17] and punching a hole in the network wherever an eavesdropper (of known location) prohibits communication[1]. The authors in [20] show that secure communication breaks down at a number of eavesdroppers that is roughly $O(n/(\log n)^2)$, and, importantly, some nodes with eavesdroppers nearby are unable to source their transmissions. As noted previously in Section 14.2, this problem of the "near eavesdropper" is a major problem of information-theoretic secrecy and one that drives much of the design in the next section.

Finally, an important result of independent interest in secure multihop networks is used as part of the main proof in [20]; in particular, [20] shows that it is sufficient to secure each link individually to achieve end-to-end security of the entire multihop transmission.

14.4.3   Secrecy Scaling with Unknown Eavesdropper Locations

The bulk of our contribution to the literature addresses secrecy scaling in large wireless networks when the eavesdropper locations are completely unknown to the system nodes. In particular, we will generally consider securing a per-session throughput of $O(1/\sqrt{n \, \log n})$ bits/second and ask how many randomly located and noncollaborating eavesdroppers can be present in the network while this throughput is kept secure with high probability. As noted in Section 14.2, an eavesdropper near the source can make the secrecy capacity of the wiretap channel zero, and, if the eavesdropper location is unknown, there is essentially no rate at which transmissions can be made such that they can be guaranteed to be secure from an eavesdropper (since the eavesdropper might be very near the source). However, in this section, we consider how to exploit the many nodes present in a large-scale network to achieve security in the face of many eavesdroppers of unknown location. We present a number of techniques in succession: cooperative jamming, network "coloring," and then network coding, with each improving the security performance of the network. The final result of this section (and our sequence of papers) is that any number of arbitrarily placed (as long as they are not right on top of a system node) and noncollaborating eavesdroppers can be tolerated while maintaining a secure per-session throughput of $O(1/\sqrt{n \, \log n})$.

---

[1]It is interesting to note that this is mathematically analogous to similar work in cognitive radio networks, where the primary networks create exclusion zones for secondary network transmission and hence lead to similar scaling laws.

Figure 14.3: Wiretap scenario with a cooperative jammer.

14.4.3.1   Cooperative Jamming

As noted in Section 14.2, the secrecy capacity of the Gaussian wiretap channel is zero when the eavesdropper is closer than the destination to the source. Hence, although we can (and will) exploit fading in our work, it is also important to be able to alter the Gaussian wiretap channel using other system nodes. One way to alter the characteristics of the physical layer channel is to employ nodes not directly in the transmission as cooperative jammers. Because of its promise, this has been an active area of research with many of the schemes, including ours, tracing back to [21,22]. In particular, suppose that there exists a (noncommunicating) node located in the environment as shown in Figure 14.3. Then, using the same notation as in Section 14.2, but now with the node quantities denoted with a superscript "J" for jammer, the secrecy capacity is given by:

$$C_s = \min\left(\frac{1}{2}\log\left(1 + \frac{|h_{A,B}|^2}{d_{A,B}^\alpha}\frac{E_s}{N_0 + \frac{E_J|h_{J,B}|^2}{d_{J,B}^\alpha}}\right) - \frac{1}{2}\log\left(1 + \frac{|h_{A,E}|^2}{d_{A,E}^\alpha}\frac{E_s}{N_0 + \frac{E_J|h_{J,E}|^2}{d_{J,E}^\alpha}}\right), 0\right)$$

$$(14.3)$$

Hence, roughly, such cooperative jamming will improve the secrecy capacity of the link if the jammer is located close to the eavesdropper and far from the receiver. If the eavesdropper location is unknown, then only nodes that are far from the receiver (or whose signals are in some other way blocked from the receiver) can jam. Whereas we are more interested in theoretical results, we note that this results in a practical protocol that can be employed in a wireless network to help thwart unwanted listeners.

Algorithm: Cooperative Jamming in a Finite Network

1. Transmission Slot 1: The transmitter sends a request-to-send (RTS) message.

2. Transmission Slot 2: The receiver responds with a clear-to-send (CTS) message. All other nodes, besides the transmitter and receiver, use the CTS message as a pilot to measure the strength of the channel between them and the receiver.

3. Transmission Slot 3: The transmitter transmits the message. With the exception of the transmitter and receiver, any system node *who heard the CTS weakly or not at all*, indicating a weak channel between that node and the receiver, transmits random noise into the air.

Such a protocol is practical, does not cause undue interference at the receiver, and can inhibit the ability of unwanted listeners from overhearing messages.

However, our main interest here is on algorithms whose performance can be established analytically for large wireless networks. Turning to this more theoretical approach, we note

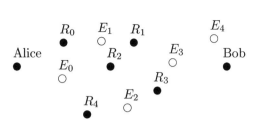

Figure 14.4: Local Communication: The source node Alice wishes to communicate securely with destination node Bob with the assistance of intervening relays $R_0, R_1, \cdots, R_{N-1}$ ($N = 5$ in the figure) in the presence of passive eavesdroppers $E_0, E_1, \cdots, E_{M-1}$ of unknown locations ($M = 5$ in the figure).

two problems, in general, with (14.3). First, because the signal from the jammer to the eavesdropper is faded, it is once again difficult to pick a secrecy rate, and this would be true even if we knew the eavesdropper location(s). Second, the eavesdroppers and desired recipient are roughly on the same playing field. However, both of these problems can be solved asymptotically by exploiting characteristics of the wireless channel fading and network structure [23,24].

Before considering large multihop networks, let's first focus in on a local transmission as shown in Figure 14.4. In particular, consider Alice and Bob attempting to communicate secretly in the presence of N other system nodes and M eavesdroppers of unknown location. To gain an advantage against the eavesdroppers, we exploit multipath fading through multiuser diversity. In particular, we find which of the N system nodes other than the transmitter and receiver receives a strong signal from each of them, hence providing a significant signal-to-noise ratio (SNR) gain against the average eavesdropper. The following protocol can then be employed.

Algorithm: Standard Multiuser Diversity for Two-hop Security

1. The source Alice transmits a pilot. All relay nodes measure the strength of the signal received.

2. The destination Bob transmits a pilot. All relay nodes measure the strength of the signal received.

3. Any relay node with gain $\log N$ greater than the average for its path loss (i.e., where the multipath fading adds constructively) attempts to grab the channel. With positive probability, one and only one relay is successful (see [24]).

4. The source transmits with the very small energy $E_s/\log N$, where E_s would be the energy normally required to reach the relay successfully.

5. The relay transmits with the very small power $E_r/\log N$, where E_r would be the power normally required to reach the destination successfully.

This straightforward form of multiuser diversity allows a reduction in system node transmission powers, hence making it much more difficult for a randomly located eavesdropper to intercept the transmission. In fact, it can be shown [24] that roughly $M = (\log n)^\beta$ noncollaborating eavesdroppers, for some $\beta > 0$ can be tolerated. But, with cooperative jamming, we can do even better. Consider the following enhanced multiuser diversity protocol that performs intelligent cooperative jamming using similar ideas to the protocol for finite networks given above.

Algorithm: Enhanced multi-user diversity with cooperative jamming for two-hop security

1. The source Alice transmits a pilot. All relay nodes measure the strength of the signal received.

2. The destination Bob transmits a pilot. All relay nodes measure the strength of the signal received.

3. Any relay node with gain $\log N$ greater than the average for its path loss (i.e., where the multipath fading adds constructively) attempts to grab the channel. With positive probability, one and only one relay is successful (see [24]). *All relay nodes measure the strength of the signal received from the relay grabbing the channel.*

4. The source Alice transmits with her standard power E_s. *Relay nodes that measured a poor channel to the relay broadcast random noise.*

5. The relay transmits with its standard energy E_r. *Relay nodes that measured a poor channel to the destination broadcast random noise.*

Essentially, nodes whose transmissions do not interfere significantly with the message relaying establish a background of white noise, adding to the straightforward multiuser diversity that allows for the main message to be heard above that noise; both effects significantly inhibit the ability of the eavesdropper to pick up the message. It can be shown [24] that this results in the ability to handle any number of noncollaborating eavesdroppers M up to $o(N)$, thus demonstrating that intelligent cooperative jamming provides a significant gain over employing standard multiuser diversity approaches.

This idea for two-hop networks can then be embedded in large multihop networks, the details of which can be found in [25]. In particular, the network is set up exactly the same as in the standard scaling literature [16] with the same routes along fixed vertical and horizontal cells. But now, a transmission from one cell in the network to the next is done by selecting a node to whom the multipath fading provides a significant gain, and other system nodes set up a background of jamming as described above. Assuming n nodes in this large network, [25] shows that the $O(1/\sqrt{n \ \log n})$ per-session throughput of [16] can be maintained *securely* in the presence of $O((\log n)^\beta)$ eavesdroppers, where β is a constant between zero and one.

14.4.3.2 Network Coloring

At the secure Gupta–Kumar throughput of $O(1/\sqrt{n \ \log n})$, one sees that the result from the previous section is quite pessimistic relative to its counterpart for when eavesdropper locations are known [20]. In particular, only $O((\log n)^\beta)$ eavesdropper can be tolerated in the construction of [25], which is far fewer than the $O(n/(\log n)^2)$ of [20] when eavesdropper locations are known. One wonders whether this is simply the price to be paid for a lack of knowledge of the eavesdropper locations. In particular, one notes that, since the eavesdropper locations are unknown, [25] does not change the routing of the original [16] work and pays a stiff penalty from potentially having to charge right over the eavesdroppers, rather than routing around them as done in [20]. The question is then apparent: how does one "route around" eavesdroppers of unknown location?

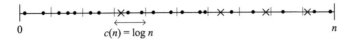

0 $c(n) = \log n$ n

Figure 14.5: The one-dimensional network consists of legitimate nodes (represented by dots) and eavesdroppers (represented by crosses) placed in the interval $[0, n]$, divided into cells of length $c(n) = \log n$, as part of the signaling construction.

The problem was solved and the number of eavesdroppers of unknown location that can be tolerated was greatly increased by a technique that we term network "coloring" [26, 27]. Consider the desire to communicate from a source to a destination in the presence of eavesdroppers of unknown location. In particular, to make the approach clear, consider the most difficult case: that of a one-dimensional network as shown in Figure 14.5.

Without the help of other nodes providing jamming support to nodes transmitting the messages, even a single eavesdropper in the middle of the one-dimensional (1-D) network easily disconnects all of the nodes to its left from all of the nodes to its right in a standard additive white Gaussian noise channel, as follows. For any transmission from a source on the left of the eavesdropper to a destination on the right of the eavesdropper, the distance between the source and the eavesdropper is less than that between the source and destination, yielding a secrecy capacity of zero. But cooperative jamming yields some hope. In the same way that somebody sitting next to you in a large auditorium can talk to you and keep you from hearing the speaker, while users in the back of the room can still hear that same speaker, a cooperative jammer placed near the eavesdropper can keep him/her from obtaining the message while allowing nodes further down the 1-D network to hear the message. In essence, the presence of a cooperative jammer near the eavesdropper can allow the message to securely "jump" the eavesdropper. But, without knowing the location of the eavesdropper, how do we know which cooperative jammer to employ? That is where the idea of network "coloring" is employed, as described next.

Suppose we do not have to protect from eavesdroppers in all locations with every transmission. Rather, suppose we are able for each transmission to *hypothesize* a set of eavesdropper locations and only protect against those locations. Then, we could "turn on" jammers near those hypothesized locations for message transmission and communicate securely from source to destination—given that our hypothesis is correct. For example, suppose that we knew that the eavesdropper is in one of the boxes in the set $\Gamma_1(n)$ depicted on the first line of Figure 14.6. It turns out [27] that it is then relatively straightforward to communicate along the 1-D network by turning on jammers in the cells that are shaded.

But what of eavesdroppers in other regions? For example, what about eavesdroppers in the sets $\Gamma_2(n), \Gamma_3(n), \ldots \Gamma_{10}(n)$ in Figure 14.6? It is not possible to jam hypothesized eavesdroppers in all of the sets simultaneously. Fortunately, it turns out that it is also not necessary. Here is where the use of secret sharing [28] completes the construction. Consider a secret information packet $\underline{w}$ of length L. Generate $t - 1$ random binary strings $\underline{w}_1, \underline{w}_2, \ldots \underline{w}_{t-1}$, each of length L, which will be the "keys"[2]. Next, calculate the bitwise exclusive-or of the information packet with all of the keys to form an encoded message $\underline{w}_t$ of length L. Now, the critical observation is that any node in possession of all $\underline{w}_1, \underline{w}_2, \ldots, \underline{w}_t$ can decode the entire information packet; *however*, any node missing even one of these t messages gets no information about the information packet. Hence, the system need only guarantee that each eavesdropper misses at least one message to guarantee secrecy. The idea is then to protect $\underline{w}_i$ from eavesdroppers in $\Gamma_i(n)$ by multiplexing transmission of the messages along the line, where, during any transmission involving w_i, the appropriate

---

[2]Note that true random numbers can be generated efficiently through a variety of methods (e.g., [29]).

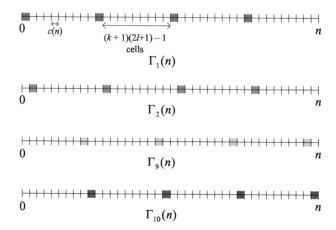

Figure 14.6: The network is partitioned into regions (colors), where each region is a collection of cells regularly sampled in the linear grid. All of the eavesdroppers in a given region can be simultaneously thwarted. The parameters k and l are set based on detailed technical arguments that guarantee the physical layer secrecy (see [27]). The network is shown here with ten regions, with four of these ten regions highlighted.

cooperative jammers are activated to prevent eavesdroppers in $\Gamma_i(n)$ from eavesdropping on the message. This technique is effective in the 1-D network, allowing us to tolerate $m = o(n/\log n)$ noncollaborating and randomly distributed eavesdroppers [27].

Next, consider the extension to 2-D networks, where the use of secret sharing allows a form of routing diversity to be effective against any eavesdroppers of unknown location—as long as they are not near the source. In particular, proceed as follows: generate three random binary keys $\underline{w}_1, \underline{w}_2, \underline{w}_3$ for the first three messages, as above, and form the fourth message to be transmitted by taking the exclusive-or of all of the keys and the information packet to be protected. Next, form four paths from the source to the destination as shown in Figure 14.7. Then, send one message along each path, while noting that the messages are not protected from eavesdroppers near the source node. This construction provides a secure per-session throughput of $O(1/\sqrt{n \log n})$ in the presence of $o(n/\log n)$ noncollaborating and randomly distributed eavesdroppers, thus roughly matching the results from [20] with known eavesdropper locations (although we hasten to note that the per-session data rates are slightly different). This consistency makes sense: the secret sharing approach allows us to hypothesize known eavesdropper locations and employ a routing/jamming scheme that is effective against eavesdroppers in those locations. This allows a number of eavesdroppers to be tolerated that is similar to the case of known eavesdropper locations, and, since the number of keys is finite, there is no loss in per-session throughput order. Also, as is the case in [20], it is still difficult to handle eavesdroppers right near a source node. This last limitation is finally mitigated in the next section.

14.4.3.3   Network Coding: Solving PHY Layer Problems a Layer Up

As discussed in the previous section, the employment of secret sharing allows the system without eavesdropper knowledge to roughly perform the same as that where eavesdroppers locations are known. Thus, having achieved without eavesdropper knowledge roughly the same result as [20] (although at a slightly lower per-session throughput rate) one might think that we have hit the fundamental limit. In particular, as noted in the previous section,

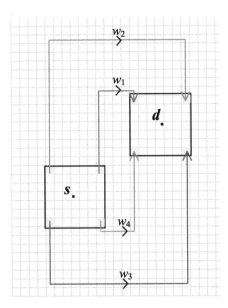

Figure 14.7: The source and the destination regions are connected with four paths, each carrying one of the packets. The paths have the same minimum spacing throughout the route; hence, no eavesdropper can be close enough to all four paths at once.

both schemes suffer from what seems to be an intractable problem - eavesdroppers near the source. One wonders whether this is simply the (steep) price of information-theoretic secrecy. However, it turns out the problem of eavesdroppers near the source can indeed be solved through a form of network coding. The result is striking and conclusive: the Gupta–Kumar multiple unicast per-session throughput of $O(1/\sqrt{n \log n})$ can be achieved *securely* in a large wireless network in the presence of an arbitrary number of noncollaborating eavesdroppers of arbitrary location, as long as no eavesdropper lies exactly on top of a system node.

Consider the case of a two-dimensional network. Per the previous section, employing network coloring and routing diversity, we know how to maintain security once we are reasonably far from the source. Next, consider how to maintain security near the source. First, consider again the scenario of Figure 14.2(a). In Figure 14.2(a), Bob is not able to transmit securely to Alice under the standard wiretap channel formulation because the distance from Bob to Eve is shorter than the distance from Bob to Alice. But suppose that we employ two-way communications in the following way [13]. Employing wiretap coding, Alice securely sends to Bob an L-bit "key"—a random binary string with the same length as the information packet. Next, Bob then takes the exclusive-or of that key with the information packet, and then sends the result via wiretap coding back to Alice. The key observation is that an eavesdropper needs to get both messages to decode the signal, and any eavesdropper who is not between the source and the destination will not be able to do such. In essence, any eavesdropper on the far side of the source from the eventual destination is blocked from obtaining this message.

Next, consider the extension of this simple two-way scheme for securing the Bob to Alice communication in Figure 14.2(a) to large wireless networks. In a large wireless network, we can always find system nodes in any small neighborhood. Hence, employ four system nodes that are in four different directions from the source: northwest, southwest, southeast, and northeast, as shown in Figure 14.8. Then, using secret sharing, generate three keys and the

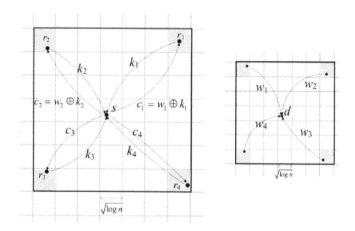

Figure 14.8: (Left) Around each source s, a "source base" is defined, which is a square region of size 7×7 cells. The four (shaded) corner cells are the relay cells, where nodes are selected to help initiate the transmission. The four relays do two-way exchanges with the source to receive four packets that form the secret message. The locations of the relays ensure that (compared to the source) no eavesdropper can be located closer to all relays at once, i.e., for any given eavesdropper e, $d(s, r_i) \leqslant d(e, r_i)$ for some $i \in \{1, 2, 3, 4\}$. (Right) The delivery of the four packets to the destination is shown. As is the case for the source phase, due to the location of the relays, no eavesdropper can be close enough to all relays at once to collect all four packets.

encoded information packet, as per the previous section. Finally, employ the two-way scheme of the previous paragraph to send the four parts of the secret message (the three keys and the encoded message) in four separate directions. Regardless of the location of the eavesdropper, he/she will miss one of these four messages and hence security is maintained—even in the presence of a near eavesdropper. All of the details of this quite complicated scheme are available from [30]. However, the key result is that, having solved the near eavesdropper problem, any number of noncollaborating eavesdroppers of any location can be tolerated while maintaining a secure per-session throughput of $O(1/\sqrt{n \, \log n})$.

14.5   Conclusion and Future Work

Information-theoretic security in wireless networks has attracted significant attention in recent years. Whereas most work has focused on small networks consisting of a transmitter (Alice), a receiver (Bob), an eavesdropper (Eve), and possibly one other system node (e.g., a relay), recent work has begun to focus on multihop wireless networks. There has been significant and important work on finite multihop networks, but this chapter has focused on the scaling of secrecy in asymptotically large wireless networks. In particular, we have reviewed significant work on the connectivity of large wireless networks, where we have largely considered work following Haenggi's secrecy graph formulation [13]. Next, we turned attention to the main topic of our interest: the design and analysis of schemes that achieve the maximum secure network capacity in the face of the largest number of eavesdroppers. Immediately, one realizes that the (insecure) network per-session throughputs of $O(1/\sqrt{n \log n})$ [16] and $O(1/n)$ [17] can be maintained in the presence of some number of eavesdroppers, and the goal is to maximize this number. Early results [20] worked under the assumption that eavesdropper locations were known, but our main interest has been on the case where eavesdropper locations are unknown. A sequence of results using first cooperative jamming, then a form of secret sharing, and then network coding, gradually improved the number of eavesdroppers that can be tolerated in the network. The conclusive result is that any number of noncollaborating eavesdroppers of arbitrary location can be tolerated while maintaining a secure per-session throughput of $O(1/\sqrt{n \log n})$.

Despite the conclusive nature of the results presented here to the research question originally posed in [25], there is still significant future work to be performed on both theoretical and practical issues. First, a critical assumption throughout the work is that the eavesdroppers do not collaborate. However, it is easy to envision application scenarios where a team of eavesdroppers spread out in the network, record signals, and then return to a common base to combine information and attempt to decode the secret message. Whereas some attention has been paid to this question in [23] and [27], the results are not nearly as extensive or conclusive as those for the noncollaborating case. Next, the results need to be transferred to the finite network case to be applied in practice. In particular, critical to the results is the availability of many system nodes: these provide the averaging of random network variations (e.g., fading) and the availability of system nodes with high enough probability in regions where they are needed to either jam or relay. Clearly, finite networks will not have such an availability of nodes, which will require a modification of the algorithms and likely the acceptance of more modest results.

Acknowledgment

This research was sponsored by the National Science Foundation under grants CNS-0905349 and CNS-1018464, and by the U.S. Army Research Laboratory and the U.K. Ministry of Defence under Agreement Number W911NF-06-3-0001. The views and conclusions contained in this document are those of the author(s) and should not be interpreted as representing the official policies, either expressed or implied, of the U.S. Army Research Laboratory, the U.S. Government, the U.K. Ministry of Defence, or the U.K. Government. The U.S. and U.K. Governments are authorized to reproduce and distribute reprints for Government purposes notwithstanding any copyright notation hereon.

References

[1] J. Talbot and D. Welsh, *Complexity and Cryptography: An Introduction*, Cambridge, 2006.

[2] J. Eisert and M. Wolf, "Quantum computing," in *Handbook of Nature-Inspired and Innovative Computing*, Springer, New York 2006.

[3] P. Shor, "Polynomial-time algorithms for prime factorization and discrete logarithms on a quantum computer," *SIAM Journal on Computing*, Vol. 26: pp. 1484–1509, October 1997.

[4] R. Benson, "The verona story," National Security Agency Central Security Service, Historical Publications (available via WWW).

[5] C. Shannon, "Communication theory of secrecy systems," *Bell Systems Technical Journal*, Vol. 28: pp. 656–715, 1949.

[6] A. Wyner, "The wire-tap channel," *Bell Systems Technical Journal*, Vol. 54: pp. 1355–1387, October 1975.

[7] Y. Liang, H. Poor, and S. Shamai (Shitz), *Information Theoretic Secrecy*, Now Publishers, Boston 2009.

[8] M. Bloch and J. Barros, *Physical Layer Security: From Information Theory to Security Engineering*, Cambridge University Press, 2011.

[9] U. Maurer, "Secret key agreement by public discussion from common information," *IEEE Transactions on Information Theory,* Vol. 39: pp. 733–742, May 1993.

[10] R. Ahlswede and I. Csiszár, "Common randomness in information theory and cryptography, I. Secret sharing," *IEEE Transactions on Information Theory,* Vol. 39: pp. 1121–1132, July 1993.

[11] S. Leung-Yan-Cheong and M. Hellman, "The Gaussian wire-tap channel," *IEEE Transactions on Information Theory,* Vol. 24: pp. 451–456, July 1978.

[12] P. Gopala, L. Lai, and H. El Gamal, "On the secrecy capacity of fading channels," *IEEE Transactions on Information Theory,* Vol. 54: pp. 4687–4698, October 2008.

[13] M. Haenggi, "The Secrecy Graph and Some of Its Properties," *Proceedings of IEEE International Symposium on Information Theory,* July, 2008.

[14] A. Sarkar and M. Haenggi, "Percolation in the Secrecy Graph," *Discrete Applied Mathematics* Vol. 161, Sept. 2013.

[15] R. Meester and R. Roy, *Continuum Percolation,* Cambridge University Press, 1996.

[16] P. Gupta and P. R. Kumar, "The capacity of wireless networks," *IEEE Transactions on Information Theory,* Vol. 46: pp. 388–404, February 2000.

[17] M. Franceschetti, O. Dousse, D. Tse, and P. Thiran, "Closing the gap in the capacity of wireless networks via percolation theory," *IEEE Transactions on Information Theory,* Vol. 53: pp. 1009–1018, March 2007.

[18] M. Grossglauser and D. Tse, "Mobility increases the capacity of ad hoc wireless networks," *IEEE/ACM Transactions on Networking,* Vol. 10: pp. 477–486, August 2002.

[19] A. Ozgur, O. Leveque, and D. Tse, "Hierarchical cooperation achieves optimal capacity scaling in ad hoc networks," *IEEE Transactions on Information Theory,* Vol. 53: pp. 3549–3572, October 2007.

[20] O. Koyluoglu, C. Koksal, and H. El Gamal, "On the secrecy capacity scaling in wireless networks," *IEEE Transactions on Information Theory,* Vol. 58: pp. 3000–3015, May 2012.

[21] R. Negi and S. Goel, "Secret communication using artificial noise," *Proceedings of the IEEE Vehicular Technology Conference (VTC),* 2005.

[22] S. Goel and R. Negi, "Guaranteeing secrecy using artificial noise," *IEEE Transactions on Wireless Communications,* Vol. 7, pp. 2180-2189, June 2008.

[23] S. Vasudevan, S. Adams, D. Goeckel, Z. Ding, D. Towsley, K. Leung, "Multi-User Diversity for Secrecy in Wireless Networks," Workshop on Information Theory and Applications, February 2010.

[24] D. Goeckel, S. Vasudevan, D. Towsley, S. Adams, Z. Ding, and K. Leung, "Artificial noise generation from cooperative relays for everlasting security in two-hop wireless networks," *IEEE Journal on Selected Areas in Communications: Special Issue on Advances in Military Communications and Networking,* Vol. 29: pp. 2067–2076, December 2011.

[25] S. Vasudevan, D. Goeckel, and D. Towsley, "Security versus capacity tradeoffs in large wireless networks using keyless secrecy," *ACM MobiHoc,* September 2010.

[26] C. Capar, D. Goeckel, B. Liu, and D. Towsley, "Cooperative Jamming to improve the connectivity of the 1-D secrecy graph," *Proceedings of the Conference on Information Sciences and Systems (CISS),* March 2011.

[27] C. Capar, D. Goeckel, B. Liu, and D. Towsley, "Secret communication in large wireless networks without eavesdropper location information," *IEEE InfoCom,* March 2012.

[28] A. Shamir, "How to share a secret," *Commun. ACM,* Vol. 22: pp. 612–613, November 1979.

[29] S. Srinivasan, S. Mathew, V. Erraguntla, and R. Krishnamurthy, "A 4gbps 0.57pj/bit process-voltage-temperature variation tolerant all-digital true random number generator in 45nm cmos," *Proceedings of the 22nd International Conference on VLSI Design,* January 2009.

[30] C. Capar and D. Goeckel, "Network coding for facilitating secrecy in large wireless networks," *Proceedings of the Conference on Information Sciences and Systems (CISS),* March 2012.

Index